P9-ARZ-639

# GREAT EVENTS

## 1900-2001

# GREAT EVENTS

## 1900-2001
### REVISED EDITION

## Volume 1
### 1900-1920

*From*

**The Editors of Salem Press**

SALEM PRESS, INC.

Pasadena, California          Hackensack, New Jersey

*Editor in Chief:* Dawn P. Dawson

| | |
|---|---|
| *Managing Editor:* R. Kent Rasmussen | *Research Supervisor:* Jeffry Jensen |
| *Manuscript Editor:* Rowena Wildin | *Acquisitions Editor:* Mark Rehn |
| *Production Editor:* Joyce I. Buchea | *Page Design and Graphics:* James Hutson |
| *Photograph Editor:* Philip Bader | *Layout:* William Zimmerman |
| *Assistant Editor:* Andrea E. Miller | Eddie Murillo |

*Cover Design:* Moritz Design, Los Angeles, Calif.

*Cover photos:* Center image—Corbis, Remaining images—AP/Wide World Photos
*Half title photos:* Library of Congress, Digital Stock, AP/Wide World Photos

© 2002 *Great Events: 1900-2001, Revised Edition*
© 1997 *The Twentieth Century: Great Scientific Achievements, Supplement* (3 volumes)
© 1996 *The Twentieth Century: Great Events, Supplement* (3 volumes)
© 1994 *The Twentieth Century: Great Scientific Achievements* (10 volumes)
© 1992 *The Twentieth Century: Great Events* (10 volumes)

∞ The paper used in these volumes conforms to the American National Standard for Permanence of Paper for Printed Library Materials, Z39.48-1992 (R1997).

**Library of Congress Cataloging-in-Publication Data**
Great events : 1900-2001 / editors of Salem Press.— Rev. ed.
    v. cm.
Includes index.
    ISBN 1-58765-053-3 (set : alk. paper) — ISBN 1-58765-054-1 (vol. 1 : alk. paper)
    1. History, Modern—20th century—Chronology. 2. Twentieth century. 3. Science—History—20th century—Chronology. 4. Technology—History—20th century—Chronology.
D421 .G627 2002
909.82—dc21

2002002008

First Printing

PRINTED IN THE UNITED STATES OF AMERICA

# CONTENTS

**v**

CONTENTS

page

# COMPLETE LIST OF CONTENTS

## VOLUME 1

## VOLUME 2

**xi**

# VOLUME 3

# VOLUME 4

# VOLUME 5

# VOLUME 6

# VOLUME 7

**xxvii**

## VOLUME 8

# PUBLISHER'S NOTE

The eight volumes of *Great Events: 1900-2001, Revised Edition* combine and update the contents of two Salem Press reference works and add completely new material that constitutes about 30 percent of the new whole. More than 1,050 articles derive from the ten volumes of *The Twentieth Century: Great Events* (1992) and its three-volume supplement (1996) and the ten volumes of *The Twentieth Century: Great Scientific Achievements* (1994) and its three-volume supplement (1997). About a quarter of these articles have been updated, some substantially, and 283 entirely new articles have been added.

The original reference sets were complementary works designed to provide comprehensive looks at the most significant events and scientific achievements of the twentieth century. They used a format combining clearly written and organized text and illustrations to present subject matter both inviting to the eye and engaging to readers from the middle-school level up. The current edition uses a modification of the original design that conserves space without losing the eye appeal of the original sets. The result is a compact, easy-to-use, comprehensive, and up-to-date reference set ideally suited to the needs of students and general library patrons.

In planning the *Great Events: 1900-2001, Revised Edition*, the first goal was to preserve as many of the original set's 1,072 articles as possible. However, ten articles from *Great Scientific Achievements* were dropped because they cover events that actually occurred in the late nineteenth century. Eight of these articles appeared in volume 1 of *Great Scientific Achievements*, two in volume 11 (the first volume of the supplement).

Another nine articles have been dropped for reasons of redundancy—because they are on subjects covered in both *Great Scientific Achievements* and *Great Events*. For example, both sets had articles on the construction of the Panama Canal, Ford's automobile assembly line, the development of the H-bomb, the nuclear accident at Chernobyl, the repair of the Hubble Space Telescope, and the opening of the English Chan-

nel tunnel. In each case, the Editors have selected one article for the revised edition, and nothing substantive has been lost.

Including their supplement volumes, the *Great Events* series covered events through mid-1995 and the *Great Scientific Achievements* series covered events up to the beginning of 1997. In selecting the 283 topics for the new essays appearing in this edition, the Editors have naturally emphasized events that have occurred since mid-1995. However, nearly sixty of the new topics cover events from before that date, with at least one new article for each decade. These new topics include the Galveston hurricane of 1900, the influenza pandemic of 1918, the 1921 murder conviction of Nicola Sacco and Bartolemeo Vanzetti, Amelia Earhart's disappearance in 1937, President Harry S. Truman's 1948 order to end segregation in the U.S. military, the conquest of Mount Everest in 1953, Thurgood Marshall's appointment to the U.S. Supreme Court in 1968, the Woodstock Festival of 1969, Patty Hearst's kidnapping in 1974, the *Exxon Valdez* oil spill in 1989, and Jean Chrétien's election as Canada's prime minister in 1993. The addition of the new essays on earlier years strengthens the set's already thorough coverage of the twentieth century's most important events.

The more than two hundred new articles covering events since the mid-1990's make *Great Events: 1900-2001* an exceptionally strong reference source on recent history. Every effort has been made to make the coverage as up to date into early 2002 as production restraints allow.

With its articles arranged in chronological order, *Great Events: 1900-2001* covers a wide variety of events, from national and world politics, wars, civil disturbances, disasters, crimes, social and economic developments to business events, scientific and medical breakthroughs, and even a few entertainment milestones. Coverage extends to the September 11, 2001, terrorist attacks on the United States and includes developments as recent as the accidental sinking of a Japanese fishing-school ship by a U.S. submarine, China's seizure of a U.S. reconnaissance plane, and the massacre of the Nepalese royal family by its own crown prince.

As in the original *Great Events* volumes, each article in the *Revised Edition* provides the ready-reference data for which Salem Press books are noted. Articles have headline-style titles that announce the events, such as "Jeffords Quits Republican Party, Giving Democrats Control of Senate." Top matter sections of each entry offer summary descriptions of the events and define their significance in a sentence or two. Following the descriptions are lists of the "What," "When," "Where," and "Who" of the events, providing readers with the vital data in shaded boxes on the first pages of the articles. The text of each one-thousand-word article is divided into three sections, the first and second of which recount the details of the event and its background, while the last—always subtitled "Consequences"—discusses both the event's short-term impact and its possible long-range ramifications.

More than 70 percent of the articles are illustrated—60 percent by eye-catching photographs or line drawings, the rest by other graphic elements, such as charts and tables. An entirely new feature in this edition is the addition of maps, which provide additional information and pinpoint the locations of more than 140 events, offering students precise visual geographical references.

In addition to adding new articles to the set, the Editors have updated more than a quarter of the original articles to make them current with later developments. Readers should note that many topics are the subject of more than a single article and that to get the full story on any given subject, they should consult the edition's multiple indexing tools. Help is provided in five different indexes.

The Category Index, which appears at the back of every volume, lists all article titles under more than sixty broad headings, such as Agriculture, Anthropology, Archaeology, Assassination, and Astronomy. Many articles are entered under more than one category. For example, the article on the September 11, 2001, terrorist attacks is listed under "Disasters" because of the scale of the devastation, under "International Relations" because of the event's many international implications, under "Terrorism" because of the nature of the event itself, and under "Transportation" because of the impact of the airline hijackings on the airline industry.

The other indexes, all of which are at the back of volume 8, include a Time Line, which lists all article topics in chronological order, with the full dates and titles of each article, as well as the places and categories assigned in the articles' top matter. The chronological arrangement of the articles not only serves as a time line but also casts light on the social, political, and economic evolution of the world from a collection of individual nations to a global community, in which developments in one region have repercussions in others.

The third indexing tool is the Geographical Index, which lists all the articles under the names of more than 225 states, nations, and continents, as well as extraplanetary terms, such as "Earth orbit," "Solar system," "Space," and the individual planet names. The purpose of this index is to bring together the articles most pertinent to each region. The listings are selective. For example, while many hundreds of articles concern events occurring in the United States, only those subjects of broad importance to the country as a whole are listed under "United States" in the Geographical Index. Likewise, only those articles pertinent to Europe as a whole are listed under "Europe."

The Personages Index lists the more than 3,000 individual names appearing under the "Who" heading in the top matter boxes, which list the persons who figure most prominently in each article. This index can serve as a convenient shortcut to finding events for which the reader may remember only the name of a person. For example, the name of Clarence E. Gideon will direct the reader to the article on the Supreme Court's *Gideon v. Wainwright* case. The name of Lee Harvey Oswald will lead the reader to the article on the assassination of President John F. Kennedy.

The indexes thus far described are all derived from information in the articles' boxed top matter. The general Subject Index includes all the names appearing in the personages index, as well as other names and key terms used in the texts of the articles.

When the original editions of *The Twentieth Century: Great Events* and *The Twentieth Century: Great Scientific Achievements* were published, contributors were listed in the first volumes but their bylines were not attached to individual articles. Articles in the supplement volumes and those newly commissioned for this edition all have contributor bylines. Although every effort has been made to insert correct bylines in all the original articles, this was not possible for some, which thus appear once again without bylines. However, the name of every scholar who contributed at least one article to any edition of *Great Events* is listed on the pages immediately following this note, whether the name appears in a byline or not. The contributor list contains a total of 573 names—itself an impressive tribute to the magnitude of this project. The Editors of Salem Press wish to thank all contributors—past and present—for making this reference work possible.

# CONTRIBUTORS

Amy Adelstein
*Independent Scholar*

Carl G. Adler
*East Carolina University*

Richard Adler
*University of Michigan, Dearborn*

Margaret I. Aguwa
*Michigan State University*

Craig W. Allin
*Cornell College*

Emily Alward
*Henderson District Libraries*

Stephen E. Ambrose
*Louisiana State University, New Orleans*

Michael S. Ameigh
*State University of New York College
at Oswego*

Eleanor B. Amico
*Independent Scholar*

David L. Ammerman
*Florida State University*

Frank Andritzky
*Concordia University*

David W. Appenbrink
*Chicago State University*

Stanley Archer
*Texas A&M University*

Gerald S. Argetsinger
*Rochester Institute of Technology*

Richard W. Arnseth
*Science Application International Corp.*

Paul Ashin
*Stanford University*

Mary W. Atwell
*Radford University*

Bryan Aubrey
*Independent Scholar*

James A. Baer
*Northern Virginia Community College*

Charles F. Bahmueller
*Center for Civic Education*

Iona C. Baldridge
*Lubbock Christian University*

JoAnn Balingit
*University of Delaware*

Grace A. Banks
*Chestnut Hill College*

Carl L. Bankston III
*Tulane University*

Russell J. Barber
*California State University,
San Bernardino*

David Barratt
*Independent Scholar*

Charles A. Bartocci
*Dabney S. Lancaster Community College*

Iraj Bashiri
*University of Minnesota*

Rose Ann Bast
*Mount Mary College*

Erving E. Beauregard
*University of Dayton*

Patricia A. Behlar
*Pittsburg State University*

Alvin K. Benson
*Brigham Young University*

Meredith William Berg
*Valparaiso University*

S. Carol Berg
*College of St. Benedict*

Christopher J. Biermann
*Oregon State University*

Cynthia A. Bily
*Adrian College*

Nicholas Birns
*New School University*

Michael S. Bisesi
*Medical College of Ohio*

Arthur Blaser
*Chapman University*

Arnold Blumberg
*Towson University*

Daniel P. Boehlke
*Gustavus Adolphus College*

Paul R. Boehlke
*Martin Luther College*

Nathaniel Boggs
*Alabama State University*

Steve D. Boilard
*Independent Scholar*

James J. Bolner
*Louisiana State University, Baton Rouge*

Jo-Ellen Lipman Boon
*Independent Scholar*

Lucy Jayne Botscharow
*Northeastern Illinois University*

Gordon L. Bowen
*Mary Baldwin College*

Wayne H. Bowen
*Ouachita Baptist University*

William D. Bowman
*Texas Christian University*

Suzanne Riffle Boyce
*Independent Scholar*

Michael R. Bradley
*Matlaw College*

Anthony D. Branch
*Golden Gate University*

John A. Britton
*Francis Marion College*

Kathleen S. Britton
*Florence-Darlington Technical College*

William S. Brockington, Jr.
*University of South Carolina, Aiken*

Alan Brown
*Livingston University*

Kendall W. Brown
*Brigham Young University*

Kenneth H. Brown
*Northwestern Oklahoma State
University*

M. Leann Brown
*University of Florida*

Michael L. Broyles
*Collin County Community College*

David S. Brumbaugh
*Northern Arizona University*

Anthony R. Brunello
*Eckerd College*

Maurice P. Brungardt
*Loyola University, New Orleans*

Faith Hickman Brynie
*Independent Scholar*

Thomas W. Buchanan
*Ancilla College*

David R. Buck
*West Virginia University*

Michael H. Burchett
*Limestone College*

Helen M. Burke
*Chestnut Hill College*

John T. Burns
*Bethany College*

Susan Butterworth
*Salem State College*

Joseph P. Byrne
*Belmont University*

Netiva Caftori
*Northeastern Illinois University*

Laura M. Calkins
*Oglethorpe University*

Edmund J. Campion
*University of Tennessee*

Mary Ellen Campion
*Independent Scholar*

Byron D. Cannon
*University of Utah*

Sheila Carapico
*University of Richmond*

Robert S. Carmichael
*University of Iowa*

José A. Carmona
*Daytona Beach Community College*

Jack Carter
*University of New Orleans*

Michael S. Casey
*Graceland University*

Gilbert T. Cave
*Lakeland Community College*

E. L. Cerroni-Long
*Eastern Michigan University*

Dennis Chamberland
*Independent Scholar*

Cheris Shun-ching Chan
*Northwestern University*

Frederick B. Chary
*Indiana University Northwest*

Victor W. Chen
*Chabot College*

David L. Chesemore
*California State University, Fresno*

Maxwell O. Chibundu
*University of Maryland School of Law*

S. M. Chiu
*Temple University*

Peng-Khuan Chong
*Plymouth State College*

Renny Christopher
*California State University, Stanislaus*

Ronald J. Cima
*Library of Congress*

Donald N. Clark
*Trinity University*

Douglas Clouatre
*Kennesaw State University*

S. Mary P. Coakley
*Georgian Court College*

Susan Coleman
*West Texas A&M University*

Richard H. Collin
*Louisiana State University, New Orleans*

Robert O. Collins
*University of California, Santa Barbara*

Jaime S. Colome
*California Polytechnic State University, San Luis Obispo*

Bernard A. Cook
*Loyola University, New Orleans*

Albert B. Costa
*Duquesne University*

Charles E. Cottle
*University of Wisconsin*

Arlene R. Courtney
*Western Oregon University*

John K. Cox
*Wheeling Jesuit University*

John A. Cramer
*Oglethorpe University*

Stephen Cresswell
*West Virginia Wesleyan College*

Richard A. Crooker
*Kutztown University*

Jennifrer L. Cruise
*College of St. Thomas*

David H. Culbert
*Louisiana State University, Baton Rouge*

Dennis Cumberland
*Independent Scholar*

Jeff Cupp
*Independent Scholar*

Donald E. Davis
*Illinois State University*

Mary Virginia Davis
*University of California, Davis*

Merle O. Davis
*Louisiana State University, Baton Rouge*

Nathaniel Davis
*Harvey Mudd College*

Roger P. Davis
*University of Nebraska, Kearney*

Ronald W. Davis
*Western Michigan University*

Scott A. Davis
*Mansfield University*

Dennis R. Dean
*University of Wisconsin, Parkside*

John H. DeBerry
*Memphis State University*

E. Gene DeFelice
*Purdue University, Calumet*

Tyler Deierhoi
*University of Tennessee*

Charles A. Desnoyers
*La Salle University*

CONTRIBUTORS

Thomas I. Dickson
*Auburn University*

Daniel D. DiPiazza
*University of Wisconsin*

Fredrick J. Dobney
*St. Louis University*

Stephen B. Dobrow
*Fairleigh Dickinson University*

Matthias Dörries
*University of California, Berkeley*

David Leonard Downie
*Columbia University*

John Duffy
*University of Maryland*

Steven I. Dutch
*University of Wisconsin, Green Bay*

Calvin Henry Easterling
*Oral Roberts University*

Jennifer Eastman
*Clark University*

Samuel K. Eddy
*Independent Scholar*

George R. Ehrhardt
*Duke University*

Harry J. Eisenman
*University of Missouri, Rolla*

Robert P. Ellis
*Independent Scholar*

David G. Engler
*Western Illinois University*

Suzanne Knudson Engler
*University of Southern California*

Victoria Erhart
*Catholic University of America*

Robert F. Erickson
*Independent Scholar*

Peter R. Faber
*United States Air Force Academy*

Daniel C. Falkowski
*Canisius College*

Laina Farhat
*Golden Gate University*

Elizabeth Fee
*Johns Hopkins University*

Alison E. Feeney
*Shippensburg University*

James E. Fickle
*Memphis State University*

John W. Fiero
*University of Southwestern Louisiana*

K. Thomas Finley
*State University of New York College at Brockport*

David G. Fisher
*Lycoming College*

Michael S. Fitzgerald
*Pikeville College*

George J. Fleming
*Calumet College*

Dale L. Flesher
*University of Mississippi*

George J. Flynn
*State University of New York College at Plattsburgh*

George Q. Flynn
*Independent Scholar*

William B. Folkestad
*Central Washington University*

Robert G. Font
*Font Geosciences Consulting*

Michael James Fontenot
*Southern University, Baton Rouge*

Robert J. Forman
*St. John's University*

Robert J. Frail
*Centenary College*

Donald R. Franceschetti
*University of Memphis*

Richard G. Frederick
*Independent Scholar*

Richard A. Fredland
*Indiana University, Indianapolis*

Jonathan M. Furdek
*Purdue University*

Anne Galantowicz
*El Camino College*

Michael J. Garcia
*Arapahoe Community College*

John C. Gardner
*Louisiana State University, Baton Rouge*

Keith Garebian
*Independent Scholar*

Thomas P. Gariepy
*Stonehill College*

Roberto Garza
*San Antonio College*

Larry N. George
*California State University, Long Beach*

Soraya Ghayourmanesh
*Queens Community College of the City University of New York*

Judith R. Gibber
*New York University*

Karl W. Giberson
*Eastern Nazarene College*

K. Fred Gillum
*Colby College*

Richard A. Glenn
*Millersville University*

Harold Goldwhite
*California State University, Los Angeles*

Douglas Gomery
*University of Maryland*

Nancy M. Gordon
*Independent Scholar*

Robert F. Gorman
*Southwest Texas State University*

Lewis L. Gould
*University of Texas, Austin*

Daniel G. Graetzer
*University of Washington Medical Center*

Hans G. Graetzer
*South Dakota State University*

John H. Greising
*Independent Scholar*

Noreen A. Grice
*Boston Museum of Science*

Robert J. Griffiths
*University of North Carolina, Greensboro*

Johnpeter Horst Grill
*Mississippi State University*

Charles F. Gritzner
*South Dakota State University*

Alan G. Gross
*Purdue University, Calumet*

Manfred Grote
*Purdue University, Calumet*

Gershon B. Grunfeld
*South Illinois University School of Medicine*

Ronald B. Guenther
*Oregon State University*

Sundaram Gunasekaran
*University of Wisconsin, Madison*

Lawrence W. Haapanen
*Lewis-Clark State College*

James M. Haas
*Independent Scholar*

Michael Haas
*California State University, Fullerton*

William J. Hagan, Jr.
*St. Anselm College*

Irwin Halfond
*McKendree College*

Darlene E. Hall
*Old Dominion University*

Sheldon Hanft
*Appalachian State University*

John R. Hanson II
*Independent Scholar*

Timothy Hanson
*University of Southwestern Louisiana*

Claude Hargrove
*Fayetteville State University*

William Harrigan
*Independent Scholar*

Robert Harrison
*University of Arkansas Community College, Batesville*

Fred R. van Hartesveldt
*Fort Valley State College*

Glenn Hastedt
*James Madison University*

Margaret Hawthorne
*Independent Scholar*

Robert M. Hawthorne, Jr.
*Independent Scholar*

Louis D. Hayes
*University of Montana*

James Hayes-Bohanan
*Bridgewater State College*

Judith E. Heady
*University of Michigan, Dearborn*

Noelle K. Heenan
*Bronx High School of Science*

Douglas Heffington
*Middle Tennessee State University*

Hans Heilbronner
*Independent Scholar*

John A. Heitmann
*University of Dayton*

Jean S. Helgeson
*Collin County Community College*

Peter B. Heller
*Manhattan College*

Arthur W. Helweg
*Western Michigan University*

Thomas E. Hemmerly
*Middle Tennessee State University*

Mary A. Hendrickson
*Wilson College*

Richard A. Hendry
*Westminster College*

Mark Henkels
*Western Oregon State College*

Diane Andrews Henningfeld
*Adrian College*

Howard M. Hensel
*U.S. Air Force, Air War College*

Charles E. Herdendorf
*Ohio State University*

James J. Herlan
*University of Maine*

Steve Hewitt
*Purdue University*

Keith Orlando Hilton
*University of the Pacific*

Kay Hively
*Independent Scholar*

Carl W. Hoagstrom
*Ohio Northern University*

Virginia L. Hodges
*Northeast State Technical Community College*

Samuel B. Hoff
*Delaware State University*

David Wason Hollar, Jr.
*Rockingham Community College*

Donald Holley
*University of Arkansas, Monticello*

Earl G. Hoover
*Independent Scholar*

Gerald Horne
*University of California, Santa Barbara*

Howard L. Hosick
*Washington State University*

Ruth H. Howes
*Ball State University*

John L. Howland
*Bowdoin College*

Mary Hrovat
*Indiana University, Bloomington*

Ronald K. Huch
*University of Papua New Guinea*

Theodore C. Humphrey
*California State Polytechnic University, Pomona*

Diane White Husic
*East Stroudsburg University*

Mahmood Ibrahim
*California State Polytechnic University, Pomona*

John Quinn Imholte
*University of Minnesota, Morris*

Thomas H. Irwin
*Independent Scholar*

Tseggai Isaac
*University of Missouri, Rolla*

Robert Jacobs
*Central Washington University*

Andrew Jamison
*California State University, Northridge*

George P. Jan
*University of Toledo*

Robert J. Janosik
*Occidental College*

Allan Jenkins
*University of Nebraska, Kearney*

Jeffrey A. Joens
*Florida International University*

Charles W. Johnson
*University of Tennessee*

Lloyd Johnson
*Campbell University*

Peter D. Johnson, Jr.
*Auburn University*

Richard C. Jones
*Rollins College*

Joseph S. Joseph
*University of Cyprus*

Thomas W. Judd
*State University of New York College
at Oswego*

Pamela R. Justice
*Collin County Community College*

Richard C. Kagan
*Hamline University*

Rajiv Kalra
*Minnesota State University,
Moorhead*

Karen E. Kalumuck
*Independent Scholar*

Edward Kannyo
*Wells College*

Christopher J. Kauffmann
*Independent Scholar*

Burton Kaufman
*Kansas State University*

Edward P. Keleher
*Purdue University, Calumet*

Rodney D. Keller
*Ricks College*

John P. Kenny
*Bradley University*

Kimberley H. Kidd
*East Tennessee State University*

Leigh Husband Kimmel
*Independent Scholar*

Cassandra Kircher
*Elon College*

Glenn J. Kist
*Rochester Institute of Technology*

Richard S. Knapp
*Belhaven College*

Barry L. Knight
*Independent Scholar*

Jeffrey A. Knight
*Mount Holyoke College*

Paul W. Knoll
*University of Southern California*

James Knotwell
*Wayne State College*

Grove Koger
*Boise Public Library*

Kevin B. Korb
*Indiana University, Bloomington*

Theodore P. Kovaleff
*Independent Scholar*

Ludwik Kowalski
*Montclair State College*

Gregory C. Kozlowski
*DePaul University*

Felicia Krishna-Hensel
*Auburn University*

P. Krishnan
*University of Delaware*

Marty Kuhlman
*West Texas A&M University*

Craig B. Lagrone
*Birmingham Southern College*

Shlomo Lambroza
*Independent Scholar*

Philip E. Lampe
*University of the Incarnate Word*

Ralph L. Langenheim, Jr.
*University of Illinois, Urbana-
Champaign*

Victor J. LaPorte, Jr.
*University of Central Florida*

Eugene S. Larson
*Pierce College*

Michael M. Laskier
*Spertus College of Judaica*

Edward T. Lee
*Florida International University*

Harvey S. Leff
*California State Polytechnic University,
Pomona*

Jean-Robert Leguey-Feilleux
*St. Louis University*

Denyse Lemaire
*Rowan University*

Saul Lerner
*Purdue University, Calumet*

Leon Lewis
*Appalachian State University*

Thomas Tandy Lewis
*Mount Senario College*

Xiaobing Li
*University of Central Oklahoma*

Peter D. Lindquist
*Independent Scholar*

Huping Ling
*Truman State University*

Guoli Liu
*College of Charleston*

Roger D. Long
*Eastern Michigan University*

Ronald W. Long
*West Virginia Institute of Technology*

Robert Lovely
*University of Wisconsin, Madison*

William C. Lowe
*Mount St. Clare College*

Arthur L. Lowrie
*University of South Florida*

David C. Lukowitz
*Hamline University*

Arthur F. McClure
*Central Missouri State University*

Matthew G. McCoy
*Arizona State University*

Dana P. McDermott
*Independent Scholar*

Scott McElwain
*University of San Francisco*

Mary Beth McGranaghan
*Chestnut Hill College*

David MacInnes, Jr.
*Guilford College*

William J. McKinney
*Indiana University, Bloomington*

Paul Madden
*Hardin-Simmons University*

Paul D. Mageli
*Independent Scholar*

Frank N. Magill
*Independent Scholar*

Russell M. Magnaghi
*Northern Michigan University*

David W. Maguire
*Mott Community College*

Cynthia Keppley Mahmood
*University of Maine*

Joseph T. Malloy
*Hamilton College*

Theo Edwin Maloy
*West Texas A&M University*

Nancy Farm Mannikko
*Independent Scholar*

Carl Henry Marcoux
*University of California, Riverside*

Renee Marlin-Bennett
*American University*

Lyndon C. Marshall
*College of Great Falls*

Sherri Ward Massey
*University of Central Oklahoma*

Thomas D. Matijasic
*Prestonsburg Community College*

Grace Dominic Matzen
*Malloy College*

Maureen S. May
*Rochester Institute of Technology*

Daniel J. Meissner
*University of Wisconsin, Madison*

Jonathan Mendilow
*Rider University*

Beth A. Messner
*Ball State University*

Joan E. Meznar
*University of South Carolina*

Andre Millard
*University of Alabama, Birmingham*

Gordon L. Miller
*Independent Scholar*

Randall L. Milstein
*Oregon State University*

George R. Mitchell
*Independent Scholar*

Ellen F. Mitchum
*Space Center*

Susan J. Mole
*Siena Heights College*

Rex O. Mooney
*Louisiana State University, Baton Rouge*

William V. Moore
*College of Charleston*

John H. Morrow, Jr.
*University of Georgia*

Rodney C. Mowbray
*University of Wisconsin, La Crosse*

Otto H. Muller
*Alfred University*

Turhon A. Murad
*California State University, Chico*

Charles Murphy
*Independent Scholar*

Alice Myers
*Simon's Rock of Bard College*

J. Paul Myers, Jr.
*Trinity University*

Taha Mzoughi
*Governor's School of Sciences and Math*

Vidya Nadkarni
*University of San Diego*

Indira Nair
*Carnegie Mellon University*

John Panos Najarian
*William Paterson College*

Peter Neushul
*University of California, Santa Barbara*

Anthony J. Nicastro
*West Chester University*

Brian J. Nichelson
*Independent Scholar*

Veronica Nmoma
*University of North Carolina, Charlotte*

Joseph L. Nogee
*University of Houston*

Burl L. Noggle
*Louisiana State University*

Cathal J. Nolan
*Boston University*

Norma C. Noonan
*Augsburg College*

Cynthia Clark Northrup
*University of Texas, Arlington*

Edward B. Nuhfer
*University of Wisconsin, Platteville*

Paul G. Nyce
*Eastern Nazarene College*

Donnacha Ó Beacháin
*University College, Dublin*

Robert C. Oberst
*Nebraska Wesleyan University*

Charles H. O'Brien
*Western Illinois University*

Marilyn Bailey Ogilvie
*Oklahoma Baptist University*

Dele A. Ogunseitan
*University of California, Irvine*

Deepa Mary Ollapally
*Swarthmore College*

Kathleen K. O'Mara
*State University of New York College at Oneonta*

Kenneth O'Reilly
*University of Alaska*

Maria Pacheco
*Buffalo State College*

Maria A. Pacino
*Azusa Pacific University*

William A. Paquette
*Tidewater Community College*

Robert J. Paradowski
*Rochester Institute of Technology*

Gordon A. Parker
*University of Toledo*

Keith Krom Parker
*Western Montana College*

Anita B. Pasmantier
*Seton Hall University*

Jerry A. Pattengale
*Azusa Pacific University*

D. G. Paz
*University of North Texas*

Thomas R. Peake
*King College*

Richard R. Pearce
*Illinois State University*

Joseph G. Pelliccia
*Bates College*

William A. Pelz
*Institute of Working Class History*

Robert T. Pennock
*University of Pittsburgh*

Louis G. Perez
*Illinois State University*

María Elizabeth Pérez y
   González
*Brooklyn College*

Andrés I. Pérez y Mena
*Long Island University*

Nis Petersen
*Jersey City State College*

Alan P. Peterson
*Gordon College*

Mark Anthony Phelps
*Drury University*

John R. Phillips
*Purdue University, Calumet*

Doris F. Pierce
*Purdue University*

Julio Cèsar Pino
*Kent State University*

Harry Piotrowski
*Towson University*

Robert T. Plant
*University of Miami*

George Plitnik
*Frostburg State University*

Betty L. Plummer
*Independent Scholar*

Bernard John Poole
*Independent Scholar*

Francis Poole
*University of Delaware*

Philip R. Popple
*Independent Scholar*

David L. Porter
*William Penn University*

Bernard Possidente, Jr.
*Skidmore College*

Ellen Powell
*Independent Scholar*

John Powell
*Cumberland College*

Luke A. Powers
*Tennessee State University*

Victoria Price
*Lamar University*

James W. Pringle
*Independent Scholar*

Carolyn V. Prorok
*Slippery Rock University*

Maureen J. Puffer-Rothenberg
*Valdosta State University*

George F. Putnam
*Independent Scholar*

Wen-yuan Qian
*Blackburn College*

Srinivasan Ragothaman
*University of South Dakota*

Steven J. Ramold
*Doane College*

R. Kent Rasmussen
*Independent Scholar*

Eugene L. Rasor
*Emory & Henry College*

John David Rausch, Jr.
*West Texas A&M University*

Marc D. Rayman
*Jet Propulsion Laboratory*

John D. Raymer
*Purdue University*

Robert Reeves
*Independent Scholar*

Dennis Reinhartz
*University of Texas, Arlington*

Rosemary M. Canfield Reisman
*Charleston Southern University*

N. A. Renzetti
*Jet Propulsion Laboratory
   California Institute of Technology*

Richard Rice
*James Madison University*

Betty Richardson
*Southern Illinois University,
   Edwardsville*

Douglas W. Richmond
*University of Texas, Arlington*

Brian L. Roberts
*Northeast Louisiana University*

Charles W. Rogers
*Southwestern Oklahoma State University*

Daniel W. Rogers
*Somerset Community College*

Deborah L. Rogers
*Somerset Community College*

Karl A. Roider
*Louisiana State University, Baton Rouge*

Jennifer E. Rosenberg
*Independent Scholar*

Courtney B. Ross
*Louisiana State University, Baton Rouge*

René R. Roth
*University of Western Ontario*

Kelly Rothenberg
*Independent Scholar*

Joseph R. Rudolph, Jr.
*Towson University*

Frank Louis Rusciano
*Rider College*

Irene Struthers Rush
*Independent Scholar*

Frank A. Salamone
*Iona College*

Helen Salmon
*University of Guelph*

José M. Sánchez
*Independent Scholar*

Eve N. Sandberg
*Oberlin College*

Richard H. Sander
*Northwestern University School of Law*

Amandeep Sandhu
*University of Victoria*

Robert M. Sanford
*University of Southern Maine*

Kwasi Sarfo
*York College*

Richard Sax
*Madonna University*

Elizabeth D. Schafer
*Independent Scholar*

William J. Scheick
*University of Texas, Austin*

Rosemary Scheirer
*Chestnut Hill College*

Nancy Schiller
*University of Buffalo*

J. Christopher Schnell
*Southeast Missouri State University*

Harold A. Schofield
*Independent Scholar*

Margaret S. Schoon
*Indiana University Northwest*

John Richard Schrock
*Emporia State University*

Joseph Albert Schufle
*New Mexico Highlands University*

Thomas C. Schunk
*Independent Scholar*

Catherine Scott
*Agnes Scott College*

Ralph C. Scott
*Towson University*

Rose Secrest
*Independent Scholar*

Robert W. Seidel
*Los Alamos National Laboratory*

Asit Kumar Sen
*Texas Southern University*

Roger Sensenbaugh
*Indiana University, Bloomington*

Elizabeth Algren Shaw
*Independent Scholar*

John M. Shaw
*Education Systems Incorporated*

Martha Sherwood-Pike
*University of Oregon*

L. B. Shriver
*Independent Scholar*

Stephen J. Shulik
*Clarion University of Pennsylvania*

R. Baird Shuman
*University of Illinois, Urbana-
Champaign*

Gary W. Siebein
*University of Florida*

Michael W. Simpson
*Wayland Baptist University*

Sanford S. Singer
*University of Dayton*

Paul P. Sipiera
*William Rainey Harper College*

Peter D. Skiff
*Bard College*

Jane A. Slezak
*Fulton-Montgomery Community College*

Genevieve Slomski
*Independent Scholar*

Clyde Smith
*South Carolina Governor's School for
Science and Mathematics*

Michael S. Smith
*Independent Scholar*

Roger Smith
*Linfield College*

Ira Smolensky
*Monmouth College*

Larry Smolucha
*Benedictine University*

Jean M. Snook
*Memorial University of Newfoundland*

Alvin Y. So
*Hong Kong University of Science &
Technology*

Katherine R. Sopka
*Four Corners Analytic Sciences*

Beatrice Spade
*Independent Scholar*

Kenneth S. Spector
*St. Louis University School of
Medicine*

Ronald N. Spector
*Office of the Chief of Military History*

Joseph L. Spradley
*Wheaton College*

Grace Marmor Spruch
*Rutgers University*

Alene Staley
*St. Joseph's College*

Michael A. Steele
*Wilkes University*

Leon Stein
*Roosevelt University*

William F. Steirer, Jr.
*Clemson University*

Joan C. Stevenson
*Western Washington University*

Robert J. Stewart
*California Maritime Academy*

Ruth Goring Stewart
*Independent Scholar*

Glenn Ellen Starr Stilling
*Appalachian State University*

Anthony N. Stranges
*Texas A&M University*

Leslie A. Stricker
*Park University*

Taylor Stults
*Independent Scholar*

Susan A. Stussy
*Madonna University*

Charles E. Sutphen
*Blackburn College*

Larry N. Sypolt
*West Virginia University*

Renée Taft
*Council for the International Exchange of Scholars*

Robert D. Talbott
*University of Northern Iowa*

Joyce Tang
*University of California, Berkeley*

Gerardo G. Tango
*Independent Scholar*

Eric R. Taylor
*University of Southwestern Louisiana*

Jeremiah R. Taylor
*Independent Scholar*

Christel N. Temple
*University of Maryland*

David R. Teske
*Independent Scholar*

Wilfred Theisen
*St. John's University*

Emory M. Thomas
*University of Georgia*

Jack Ray Thomas
*Bowling Green State University*

Virginia Thompson
*Towson University*

Larry Thornton
*Hanover College*

H. Christian Thorup
*Cuesta College*

Leslie V. Tischauser
*Prairie State College*

Russell R. Tobias
*Independent Scholar*

Evelyn Toft
*Fort Hays State University*

Robin S. Treichel
*Oberlin College*

Paul B. Trescott
*Southern Illinois University*

Mfanya Donald Tryman
*Mississippi State University*

William M. Tuttle
*University of Kansas*

Jiu-Hwa Lo Upshur
*Eastern Michigan University*

Charles F. Urbanowicz
*California State University, Chico*

Jonathan G. Utley
*University of Tennessee*

Garrett L. Van Wicklen
*University of Georgia*

Kevin B. Vichcales
*Western Michigan University*

Charles L. Vigue
*University of New Haven*

Indu Vohra
*DePauw University*

Thomas Vollman
*St. Mary's College of Maryland*

Harry E. Wade
*Independent Scholar*

Shirley Ann Wagner
*Fitchburg State College*

Thomas J. Edward Walker
*Pennsylvania College of Technology*

Bennett H. Wall
*Tulane University*

Peter J. Walsh
*Fairleigh Dickinson University*

Richard L. Warms
*Southwest Texas State University*

William L. Waugh, Jr.
*Georgia State University*

Martha Ellen Webb
*University of Nebraska, Lincoln*

Amy K. Weiss
*Appalachian State University*

Marcia J. Weiss
*Point Park College*

Henry G. Weisser
*Colorado State University*

Eric B. Welch
*Mississippi State University*

T. K. Welliver
*Clarion University of Pennsylvania*

Allen Wells
*Bowdoin College*

Winifred Whelan
*St. Bonaventure University*

Thomas Whigham
*University of Georgia*

Nan White
*Maharishi International University*

George Martin Whitson III
*University of Texas, Tyler*

Edwin G. Wiggins
*Webb Institute*

Thomas A. Wikle
*Oklahoma State University*

Donald H. Williams
*Hope College*

Bradley R. A. Wilson
*University of Cincinnati*

John Wilson
*Independent Scholar*

Raymond Wilson
*Fort Hays State University*

Shawn Vincent Wilson
*Independent Scholar*

Theodore A. Wilson
*United States Naval Academy*

John D. Windhausen
*St. Anselm College*

Michael Witkoski
*University of South Carolina*

Thomas P. Wolf
*Indiana University Southeast*

Frank Wu
*University of Wisconsin, Madison*

Clifton K. Yearley
*State University of New York at Buffalo*

Jay R. Yett
*Orange Coast College*

Ivan L. Zabilka
*Independent Scholar*

Edward A. Zivich
*Calumet College*

# Ehrlich and Metchnikoff Pioneer Field of Immunology

> *Élie Metchnikoff and Paul Ehrlich advanced new theories concerning the body's cellular defense mechanisms, providing a springboard for future advances in immunology.*

**What:** Medicine
**When:** Early 1900's
**Where:** Russia, France, and Germany
**Who:**
ÉLIE METCHNIKOFF (1845-1916), a Russian biologist, microbiologist, and pathologist
PAUL EHRLICH (1854-1915), a German research physician and chemist

## The Cellular Response

In the late nineteenth century, the field of immunology, or the study of how the body fights disease, was a young science that was fraught with seemingly incongruous results. The field thus benefited greatly from the work of Élie Metchnikoff and Paul Ehrlich, who shared the 1908 Nobel Prize in Physiology or Medicine. Their work paved the way for the effective development and use of vaccinations, antibiotics, and even organ transplants.

Metchnikoff correctly interpreted and advanced the concept of phagocytosis as a major mechanism by which a particular kind of cell, the phagocyte, combats (by engulfing) foreign particles and disease organisms invading the body. He first used the term "phagocyte," derived from the Greek, in 1893. Metchnikoff was educated as a zoologist, but his studies led him increasingly into the field of pathology. In 1865, he made the first observation that would lead to his concept of phagocytosis as a disease-fighting mechanism. Metchnikoff examined the process of intracellular digestion in the roundworm *Fabricia*. Metchnikoff correctly interpreted the relationship between phagocytic digestion and phagocytic defense mechanisms.

In 1882, while studying starfish larvae, Metchnikoff observed mobile cells engulf foreign bodies introduced into a larva. He noted that these cells arose from the mesoderm layer (middle layer of the embryo) rather than the endoderm layer, which is associated with the digestive system.

Metchnikoff demonstrated that human white blood cells also develop from the mesodermal layer and serve the role of attacking foreign bodies, particularly bacteria. These ideas were revolutionary because, at the time, one school of thought held that leukocytes (white blood cells) were responsible for nurturing and spreading bacterial infection throughout the body. Indeed, the observation of many white blood cells in the blood of patients (the process of inflammation) who died of infection seemed to contradict Metchnikoff's phagocytosis theory. The inflammatory response was believed to be operative only in higher forms of animal life that possessed cardiovascular systems. Metchnikoff had demonstrated the principle of inflammatory response in lower life-forms devoid of such a system.

It was in the study of the higher animal systems that Metchnikoff faced his most significant challenge in understanding phagocytosis and disease. He studied the process of infection by the anthrax bacillus. His observations appeared to conflict because phagocytosis seemed to be limited—very virulent bacilli were not attacked, while weaker bacilli were. Complicating the study was the observation that resistant animals exhibited active phagocytosis, while susceptible animals displayed no phagocytosis.

## The Humoral Response

While Metchnikoff wrestled with the problem of establishing phagocytosis as a mechanism of

Hulton Archive

*Élie Metchnikoff.*

defense to disease, Ehrlich studiously examined antitoxins. His first major accomplishment was to improve the diphtheria antitoxin discovered and developed by the German bacteriologist Emil Adolf von Behring. Ehrlich's research on antitoxin and immunity led to the development of the concept of specific cellular responses toward toxins and antitoxins. This theory led to the concept of cell receptors—the basis of the cell's chemical affinity for certain chemical substances.

Whereas Metchnikoff studied the factors associated with phagocytosis and thus the cellular aspect of immunity, Ehrlich studied the factors associated with the humoral aspects of immunity—immunity based in the body fluids. Ehrlich's research on toxins and antitoxins aided

later studies and the development of an understanding of antigens and antibodies. Ehrlich's work helped to explain the relationships between toxins and antitoxins.

Ehrlich's work established that animals can build an immunity to toxic substances by administration of initially minute doses that are gradually increased over time. Furthermore, he demonstrated in mice the transference of this immunity to offspring through maternal milk. From his work on diphtheria toxin, he developed a standardized method for determining the proper dosage of an antitoxin.

Ehrlich's work in immunity arose from his study of blood, which he would analyze by staining with various dyes. These studies convinced him that cells will bind with certain dyes with which they share distinct chemical affinities. The blood components he discovered and the staining methods he developed not only prepared him for his immunity research but also marked Ehrlich as the founder of modern hematology. Additionally, the Wassermann test for syphilis, developed by the German bacteriologist August von Wassermann in 1906, was a direct outgrowth of Ehrlich's immunological research and views.

## Consequences

Prior to the phagocytosis doctrine advanced by Metchnikoff, the commonly held view was that resistance to bacterial infection resided in chemical properties of the blood. This view enjoyed reinforcement because the presence of certain antibodies in blood had been demonstrated. In 1903, the English scientists Sir Almroth Edward Wright and Stewart Douglas demonstrated the presence of substances in blood that, by binding to the bacterial surface, seem to prepare bac-

teria for phagocytosis by white blood cells. Thus, the phagocytosis doctrine appeared to require a precondition: a precoated bacterium. In the ensuing years, the role of phagocytosis and antibody formation became better defined.

Following these studies—all of which rested on his fundamental belief in chemical affinities between a cell and chemical substances—Ehrlich went on to study ways of curing disease by chemical means. His early work and successes led to his most celebrated application of salvarsan to cure syphilis.

Together, the achievements of these two scientists were critical to the understanding of the mechanics of immunity and chemotherapy. As a consequence of their work and its refinement by others, modern medicine can fight microorganisms on their own ground—the molecular level—and win.

*Eric R. Taylor*

# Einthoven Develops Early Electrocardiogram

> *Willem Einthoven led the study of the electrical currents of the heart by inventing the string galvanometer, which evolved into the modern-day electrocardiogram.*

**What:** Medicine; Photography
**When:** Early 1900's
**Where:** Leiden, The Netherlands
**Who:**
WILLEM EINTHOVEN (1860-1927), a Dutch physiologist and winner of the 1924 Nobel Prize in Physiology or Medicine
AUGUSTUS D. WALLER (1856-1922), a German physician and researcher
SIR THOMAS LEWIS (1881-1945), an English physiologist

## Horse Vibrations

In the late 1800's, there was substantial research interest in the electrical activity that took place in the human body. Researchers studied many organs and systems in the body, including the nerves, eyes, lungs, muscles, and heart. Because of a lack of available technology, this research was tedious and frequently inaccurate. Therefore, the development of the appropriate instrumentation was as important as the research itself.

The initial work on the electrical activity of the heart (detected from the surface of the body) was conducted by Augustus D. Waller and published in 1887. Many credit him with the development of the first electrocardiogram. Waller used a Lippmann's capillary electrometer (named for its inventor, the French physicist Gabriel-Jonas Lippmann) to determine the electrical charges in the heart and called his recording a "cardiograph." The recording was made by placing a series of small tubes on the surface of the body. The tubes contained mercury and sulfuric acid. As an electrical current passed through the tubes, the mercury would expand and contract. The resulting images were projected onto photographic paper to produce the first cardiograph. Yet Waller had only limited success with the device and eventually abandoned it.

In the early 1890's, Willem Einthoven, who became a good friend of Waller, began using the same type of capillary tube to study the electrical currents of the heart. Einthoven also had a difficult time working with the instrument. His laboratory was located in an old wooden building near a cobblestone street. Teams of horses pulling heavy wagons would pass by and cause his laboratory to vibrate. This vibration affected the capillary tube, causing the cardiograph to be unclear. In his frustration, Einthoven began to modify his laboratory. He removed the floorboards and dug a hole some ten to fifteen feet deep. He lined the walls with large rocks to stabilize his instrument. When this failed to solve the problem, Einthoven, too, abandoned the Lippmann's capillary tube. Yet Einthoven did not abandon the idea, and he began to experiment with other instruments.

## Electrocardiographs over the Phone

In order to continue his research on the electrical currents of the heart, Einthoven began to work with a new device, the d'Arsonval galvanometer (named for its inventor, the French biophysicist Arsène d'Arsonval). This instrument had a heavy coil of wire suspended between the poles of a horseshoe magnet. Changes in electrical activity would cause the coil to move; however, Einthoven found that the coil was too heavy to record the small electrical changes found in the heart. Therefore, he modified the instrument by replacing the coil with a silver-coated quartz thread (string). The movements could be recorded by transmitting the deflections through a microscope and projecting them on photographic film. Einthoven called the new instrument the "string galvanometer."

In developing his string galvanometer, Einthoven was influenced by the work of one of his teachers, Johannes Bosscha. In the 1850's, Bosscha had published a study describing the technical complexities of measuring very small amounts of electricity. He proposed the idea that a galvanometer modified with a needle hanging from a silk thread would be more sensitive in measuring the tiny electric currents of the heart.

By 1905, Einthoven had improved the string galvanometer to the point that he could begin using it for clinical studies. In 1906, he had his laboratory connected to the hospital in Leiden by a telephone wire. With this arrangement, Einthoven was able to study in his laboratory electrocardiograms derived from patients in the hospital, which was located a mile away. With this source of subjects, Einthoven was able to use his galvanometer to study many heart problems. As a result of these studies, Einthoven identified the following heart problems: blocks in the electrical conduction system of the heart; premature beats of the heart, including two premature beats in a row; and enlargements of the various chambers of the heart. He was also able to study how the heart behaved during the administration of cardiac drugs.

A major researcher who communicated with Einthoven about the electrocardiogram was Sir Thomas Lewis, who is credited with developing the electrocardiogram into a useful clinical tool. One of Lewis's important accomplishments was his identification of atrial fibrillation, the overactive state of the upper chambers of the heart. During World War I, Lewis was involved with studying soldiers' hearts. He designed a series of graded exercises, which he used to test the soldiers' ability to perform work. From this study, Lewis was able to use similar tests to diagnose heart disease and to screen recruits who had heart problems.

## Consequences

As Einthoven published additional studies on the string galvanometer in 1903, 1906, and 1908, greater interest in his instrument was generated around the world. In 1910, the instrument, now called the "electrocardiograph," was installed in the United States. It was the foundation of a new laboratory for the study of heart disease at The Johns Hopkins University.

As time passed, the use of the electrocardiogram—or "EKG," as it is familiarly known—increased substantially. The major advantage of the EKG is that it can be used to diagnose problems in the heart without incisions or the use of needles. It is relatively painless for the patient; in comparison with other diagnostic techniques, moreover, it is relatively inexpensive.

Recent developments in the use of the EKG have been in the area of stress testing. Since many heart problems are more evident during exercise, when the heart is working harder, EKGs are often given to patients as they exercise, generally on a treadmill. The clinician gradually increases the intensity of work the patient is doing while monitoring the patient's heart. The use of stress testing has helped to make the EKG an even more valuable diagnostic tool.

*Bradley R. A. Wilson*

The Nobel Foundation

*Willem Einthoven.*

# Hilbert Presents His Twenty-three Problems

*David Hilbert proposed twenty-three problems that motivated and directed mathematical research.*

**What:** Mathematics
**When:** 1900
**Where:** Paris, France
**Who:**
DAVID HILBERT (1862-1943), a German
mathematician

## The Foundational Period

By the end of the nineteenth century, most of the foundations of classical mathematics had been formulated. The fundamental groundwork of mathematical analysis had been established. The axiomatic approach, which involves using definitions, assumptions, and deductions with proofs, had become standard. (Axioms are mathematical statements that are assumed to be true.) Calculus and the theory of functions were formalized in this manner. "Abstract" algebra had developed throughout the nineteenth century. The basis of set theory had been laid out by Georg Cantor (1845-1918). Logic was in a rudimentary stage, in spite of its long history. New geometries and theories of curves had been conceived. With the foundations of mathematics solidified and new theories continually being proposed, the time had come to explore the deeper aspects of these theories.

David Hilbert, one of the major mathematicians of this period, established the groundwork for many of these theories. Among his published papers and monographs were complete treatises on number theory, logic, and geometry. Hilbert was known for presenting lectures that stimulated research, especially among his students at Göttingen, Germany. Many of his students would become great researchers and founders of schools of mathematics. The concise description of complex, original problems introduced by means of an interesting and motivating development of ideas was Hilbert's style. His address to

the international mathematical congress was a fine example of this style.

## Hilbert's Challenge to Mathematicians

Hilbert's address to the international mathematical congress of 1900 presented ten of a total of twenty-three problems in topics at the forefront of mathematical research. Hilbert challenged the audience to solve these problems. Later, all twenty-three problems were published, translated into several other languages, and distributed widely. Generations of mathematicians have worked on solving those problems. In mathematics, such work involves formally proving a theorem, not merely discovering patterns or confirming hypotheses with further evidence.

The first problem asked whether infinities exist that are larger than the set of all whole numbers (for example, $\{1, 2, 3, 4, \ldots\}$) and yet smaller than the set of all real numbers (the real numbers are all positive numbers, all negative numbers, and zero). It went even further, asking about the nature of the structure of the set of real numbers. A partial solution was found by Kurt Gödel (1906-1978) in 1940, and Paul Cohen (1934-    ) solved the problem completely in 1963.

The second problem was to determine whether the axioms of arithmetic were consistent. Could false conclusions be reached by using the assumptions of arithmetic? This is the basic issue for all mathematics. Gödel proved in 1931 that the answer to this question is yes.

Hilbert's third mathematical problem involved three-dimensional geometric figures. It was promptly solved. Problem four asked about the nature of geometries that are similar to the standard model of the universe. By changing the basic assumptions of space, what new properties and conclusions could be proved about that system? Some problems were very broad. The sixth problem asked for the formulation of a collection of axioms from which physical laws in general could be derived. Hilbert may have hoped that such an elegant "core" of assumptions could be used to derive mathematical physics.

The seventh problem dealt with numbers raised to the power of other numbers whose digits have no simple pattern. Even when this problem was solved, it led to many other problems that are still unsolved. In mathematics, even simple operations can generate difficult questions.

Some problems were not original but resulted in a resurgence of interest and new attempts to solve them. The eighth problem is a historic problem dating back to Bernhard Riemann (1826-1866). It deals with the frequency and distribution of prime numbers (prime numbers are numbers that can be divided only by themselves and 1).

Some problems originated in antiquity but have modern applications. Hilbert's tenth problem, "Determination of the Solvability of a Diophantine Equation," asks whether an important category of equations has a solution as a finite sequence of arithmetic operations resulting in integer (the integers are zero and the negative and positive whole numbers) answers. Although it is quite old, this problem has applications in such areas as factory production scheduling.

Not all of Hilbert's problems have been solved in the conventional sense. In the last third of the twentieth century, Hilbert's tenth problem was shown to be unsolvable. Even if all future human and computer effort were applied to this one problem, no general answer would ever be achieved. This is particularly ironic, because Hilbert seemed to hope that all mathematics could be reduced to a concise set of assumptions and methodically derived by deductive proof.

## Consequences

After Hilbert's twenty-three problems were published, the wheels of progress turned with a focused purpose. Hilbert's questions ushered in the modern period of mathematics. The progress in twentieth century mathematics that was made in attempts to solve these problems is significant in both quality and quantity. These questions required new methods and systems of reasoning, methods of greater abstraction, and more careful construction. These features characterize modern mathematics, in which small sets of assumptions prove very general results.

*David Hilbert.*

Several of Hilbert's problems had the effect of reducing and clarifying key issues in mathematics.

Whole new fields of mathematics were created to build the powerful tools and deductions needed to find answers to these problems. In May, 1974, a symposium reviewed the new subdisciplines created in the process of resolving Hilbert's problems. Progress in many fields can be measured in terms of the degree of success that has been achieved in solving Hilbert's problems in those fields.

The most profound results of Hilbert's problems (and perhaps of all modern mathematics) were "negative solutions." The answer to Hilbert's second problem was Kurt Gödel's famous incompleteness theorem, which states that arithmetic cannot be proved to be consistent. Basically, this means that some questions are unanswerable. These problems cannot be solved by humans and computers, not for any mystical reason, but because they are infinitely complex. Hilbert's legacy is rich indeed.

*John Panos Najarian*

# Landsteiner Discovers Human Blood Groups

*Karl Landsteiner investigated the chemistry of the immune system and discovered the A-B-O blood groups, the most significant advance toward safe blood transfusions.*

**What:** Medicine
**When:** 1900-1901
**Where:** Vienna, Austria
**Who:**
KARL LANDSTEINER (1868-1943), an Austrian pathologist, immunologist, and winner of the 1930 Nobel Prize in Physiology or Medicine
PHILIP LEVINE (1900-1987), a physician, immunoserologist, and Landsteiner's student
ALEXANDER S. WEINER (1907-1976), an immunoserologist who, with Landsteiner, discovered the rhesus blood system

## Repelling Foreign Invaders

In the late 1800's, immunology was developing rapidly, as scientists examined the various physiological changes associated with bacterial infection. Some pathologists studied the ways in which cells helped the body to fight disease; others studied the roles of noncellular factors. In 1886, for example, the British biologist George Nuttall showed how substances in blood serum fight bacteria and other microorganisms that enter the human body. (Serum is the part of the blood that becomes separated when a clot forms.)

It was during this time that terms such as "antibody" and and "antigen" were introduced. Once a disease-producing organism invades the body, the body reacts by producing helpful substances, or antibodies. These are produced in the blood or tissues and weaken or destroy bacteria and other organic poisons. Antigens are any substances that, after entering the body, stimulate the production of antibodies. The latter then go about their work of fighting these potentially harmful invaders.

Another researcher, Max von Gruber, a bacteriologist at the Hygiene Institute in Vienna, was particularly interested in how the serum of one individual initiates the clumping, or agglutination, of foreign cells that it encounters. He and a student, Herbert Edward Durham, discovered that antibodies are what cause the agglutination of disease organisms in the blood.

## Discovering Blood Types

All this research was applied by scientists trying to figure out a way of making blood transfusions safer. It is now known that before transfusing blood from one person to another, the blood types of each person must be determined. For a transfusion to work, the donor and the recipient must have compatible blood types; otherwise, the recipient's immune system will reject the new blood. If it does, agglutination will occur, and clots will form in the recipient's blood; the result can be fatal for the recipient.

The discovery of the existence of blood groups was a necessary first step in understanding the mechanics of transfusion. Samuel Shattock, an English pathologist, first came close to discovering human blood groups. In 1899 and 1900, he described the clumping of red cells in serum taken from patients with acute pneumonia and certain other diseases. Since he could not find the clumping in the serums of healthy persons, he concluded that his results reflected a disease process.

In 1900 and 1901, Karl Landsteiner synthesized the results of such earlier experiments and provided a simple but correct explanation. He took blood samples from his colleagues, separated the red blood cells from the serum, and suspended the samples in a saline solution. He then mixed each person's serum with a sample from every cell suspension. Agglutination (clumping) occurred in some cases; there was no reaction in others. From the pattern he observed, he hypothesized that there were two types of red blood cell, A and B, whose serum would agglutinate the other type of red cell. There was another group, C (in later papers, group O), whose serum agglutinated red blood cells of both types A and B, but whose red blood cells were not agglutinated by serum from individuals with either A or B. He concluded that there were two types of antibodies, now called "anti-A" and "anti-B," found in persons of blood types B and A, respectively, and together in persons with blood type C. Successful transfusing was thus understood as dependent upon accurate blood-type matching.

In 1902, two students of Landsteiner, Alfred von Decastello and Adriano Sturli, working at Medical Clinic II in Vienna with more subjects, tested blood with the three kinds of cells. Four out of 155 individuals had no antibodies in their serums (2.5 percent), but their cells were clumped by the other types of serums. This fourth rare kind of blood was type AB, because both A and B substances are present on red cells. Decastello and Sturli proved also that the red cell substances were not part of a disease process when they found the markers equally distributed in 121 patients and 34 healthy subjects.

## Using Blood Groups

Landsteiner anticipated forensic uses of the blood by observing that serum extracted from fourteen-day-old blood that had dried on cloth would still cause agglutination. He suggested that the reaction could be used to identify blood. He noted also that his results could explain the devastating reactions that occurred after some blood transfusions. Human-to-human transfusions had replaced animal-to-human transfusions, but cell agglutination and hemolysis (dissolution of red blood cells) still resulted after some transfusions using human donors. In a brief paper, Landsteiner interpreted agglutina-

tion as a normal process rather than the result of disease. He thus laid the basis for safe transfusions and the science of forensic serology; he became known as the "father of blood groups."

Landsteiner's experiments were performed at room temperature in dilute saline suspensions. These made possible the agglutination reaction of anti-A and anti-B antibodies to antigens on red blood cells, but hid the reaction of warm "incomplete antibodies" (small antibodies that coat the antigen but require a third substance before agglutination occurs) to other, yet undetected antigens such as the rhesus antigens, which are important for understanding hemolytic disease of the newborn.

*Karl Landsteiner.*

## Consequences

The most important practical outcome of the discovery of blood groups was the increased safety of blood transfusions. In 1907, Ottenberg was the first to apply Landsteiner's discovery by matching blood types for a transfusion. A New York pathologist, Richard Weil, argued for testing blood to ensure compatibility.

Subgroups of blood type A were discovered in 1911, but it was not until 1927 that Landsteiner, now working at the Rockefeller Institute in New York, and his student, Philip Levine, discovered additional blood group systems. They injected different red blood cells into rabbits and eventually obtained antibodies that could distinguish human blood independently from A-B-O differences. The new M, N, and P factors were not important for blood transfusion but were used for resolving cases of disputed parentage. More scientists eventually became aware of the multiple applications of Landsteiner's blood group research, and in 1930, he was awarded the Nobel Prize in Physiology or Medicine.

*Joan C. Stevenson*

# Russell and Whitehead's *Principia Mathematica* Launches Logistic Movement

*Bertrand Russell and Alfred North Whitehead attempted to formulate all mathematics within formal logic and set theory.*

**What:** Mathematics
**When:** 1900-1910
**Where:** Cambridge, England
**Who:**
BERTRAND RUSSELL (1872-1971), an English mathematician and philosopher
ALFRED NORTH WHITEHEAD (1861-1947), an English mathematician and philosopher
GIUSEPPE PEANO (1858-1932), an Italian mathematician and logician
GOTTLOB FREGE (1848-1925), a German mathematician and logician

## Mathematical Logic Slumbers

For about 2,100 years, from the time of Euclid to the nineteenth century, the relationship between logic and mathematics was simple: Logic was something the mathematician used to discover and to present mathematics. In the "axiomatic method," the mathematician states a few "axioms," or statements about the subject of the theory with which reasonable people would be likely to agree, and then deduces further statements, called "theorems," from these axioms. The deductions are governed by the "rules of inference" allowed by the logic.

By the early nineteenth century, it began to dawn on geometers that the objects and relations referred to in geometrical axioms did not necessarily correspond to "real" objects and relations in only one "reality," or model of the axioms. For example, in projective geometry, which has axioms that refer to points and lines, one can find models of those axioms in which points look like lines and lines look like points. In other models

of geometrical axioms, sometimes the points or lines do not look like anything at all; they may be functions or formulas that cannot be pictured in ordinary ways.

The axiomatic method thus gained an unexpected charge of power. Axioms formed by abstraction from one reality might have meaning and be valid in many different realities. Therefore, if the rules of inference were valid and were correctly applied, the theorems proceeding from those axioms would be truths that were applicable in many different settings.

The price of this new power was that informal proofs based on diagrams depicting scenes from only one of the axioms' models were no longer acceptable, because such proofs might innocently use properties of that particular model that were not shared by other models, thus giving rise to theorems that were not valid in every model of the axioms. Proofs had to be more precise.

## Mathematical Logic Awakens

The first great success in improving the quality of proofs was *The Mathematical Analysis of Logic*, by George Boole (for whom Boolean algebra is named), which was published in 1847. The primary idea contained in the volume had been examined by Gottfried Wilhelm Leibniz and a succession of others, but it was Boole who finally crystallized it. The idea was that logical inference should be algebraic. In other words, it should consist of a series of transformations that would be executed on strings of symbols according to certain rules; the meanings that were assigned to those symbols should not affect the rules that govern the transformations.

This idea was exactly what was needed to breathe new life into the axiomatic method.

Boole's work became much honored, although it was not much applied, because most mathematicians continued to use informal proofs based on informal postulates. The idea that it might be possible to create a formal axiomatization of all mathematics, however, began to generate excitement. Mathematicians wondered what axioms would underlie such an endeavor and what previously unimagined realities would be laid open to the human mind by such an undertaking.

In the 1870's and 1880's, Georg Cantor proved some very surprising results that related to infinite sets, and this led some mathematicians to seize upon set theory as the foundation from which all mathematics might arise by formal deduction. Sets seemed to be fundamental both mathematically and philosophically. The difficulty was that, before Cantor, no one had known how to begin. For example, how could the positive integers (the positive whole numbers 1, 2, 3, and so forth) be defined in an axiomatic theory of sets? Cantor did not demonstrate precisely how to do this, but he paved the way for others to follow.

After Cantor, Gottlob Frege, a German logician, set out to formulate arithmetic in an axiomatic theory of sets. At the same time, the Italian Giuseppe Peano was developing an axiomatic theory of arithmetic of a less fundamental kind.

In the summer of 1900, Bertrand Russell, then a young lecturer at Cambridge University, attended a philosophical symposium in Paris in the company of Alfred North Whitehead, his colleague and former teacher. At that time, Russell was acquainted with the work of Boole and Cantor but not with that of Frege. Russell and Whitehead heard Peano speak in Paris, and they were so impressed with the sophistication of his work that they resolved to follow and to surpass him, to create a more formal axiomatic theory of arithmetic. The following year, Russell discovered

*Bertrand Russell.*

Library of Congress

that Frege had already done this, but he also discovered a shattering contradiction in Frege's system. Now known as Russell's paradox, the contradiction arises because in Frege's system it is permissible to speak of the set R of sets that are not members of themselves. By the defining property of R, R is a member of itself *only* if it is not a member of itself.

Russell and Whitehead were both stimulated and chastened by this discovery. After ten years of hard work, they produced *Principia Mathematica*, three large volumes of formal logical deduction that purport to develop arithmetic from a very bare foundation. They took great pains to avoid contradictions such as Russell's paradox. It may never be known whether their system is free of contradictions. The effort involved in creating *Principia Mathematica* was enormous. For example, it required more than two hundred pages of dense logic to prove that one plus one equals two.

## Consequences

*Principia Mathematica* is to mathematics what Egypt's pyramids are to architecture—a monument to an obsession that still exists, but in an evolved form. The lasting mathematical contribution of *Principia* is to set theory, although modern set theory is not based on the foundation created by *Principia*. After *Principia*, formal axiomatics became an object, rather than a tool, of mathematical study. The fact that the work had been done, rather than the substance of the work itself (which is somewhat stupefying), stimulated the efforts that led to the development of Gödel's theorems in the 1930's and the brave new era of mathematical logic.

*Peter D. Johnson, Jr.*

# United States Reforms Extend the Vote

*Various laws and amendments adopted by individual states and by Congress gave common people a greater share in government decision making.*

**What:** Civil rights and liberties; Political reform
**When:** 1900-1913
**Where:** United States
**Who:**
GROVER CLEVELAND (1837-1908), president of the United States from 1885 to 1889 and 1893 to 1897
ROBERT MARION LAFOLLETTE (1855-1925), governor of Wisconsin from 1901 to 1906
WILLIAM U'REN (1859-1949), a newspaper editor from Oregon who has been called "the father of direct democracy"

## The Need for Change

The U.S. Declaration of Independence affirmed that all men are created equal, and in the middle of the nineteenth century Abraham Lincoln said that government must be "of the people, by the people, and for the people." Still, even at the beginning of the twentieth century the American voting system fell short of these ideals.

By 1890, the right to vote had been extended to all adult male citizens. Only in a few Western states, however, were women allowed the same right. United States senators were elected by state legislatures rather than by the people, and there were no legal means for people to impeach public officials—to seek to remove elected persons from office because of misconduct. Moreover, voters had no way of forcing state legislatures to take action on specific issues.

Under the common system of casting votes, ballots of different colors were printed by the political parties or independent candidates and were distributed to the voters. As voters deposited their ballots in the appropriate boxes, they were watched by their employers or people who held local political power. Clearly, this system encouraged fraud and bribery: Some officials bought people's votes, and employers could threaten employees with the loss of their jobs if they did not vote as the employers wished.

Candidates were generally chosen in party caucuses—gatherings dominated by a wealthy few. The common citizen had no say in the choice of candidates whose names appeared on the ballot. Because there were no controls on campaign spending, the rich had great influence over the course of political campaigns. In 1907, when Simon Guggenheim revealed publicly what he had spent to get elected U.S. senator from Colorado, the nation was shocked. Many agreed that the system needed to be changed.

## Reforms and Reformers

The secret ballot (also called the Australian ballot) had been first introduced in the state of Massachusetts in 1888. Through the efforts of President Grover Cleveland and others, this system had come into nationwide use by the election of 1910. Now voters could make choices according to their convictions, without being pressured by their employers or others.

Under the leadership of Governor Robert LaFollette, the state of Wisconsin pioneered the use of primaries to select those who would run in the general election. In this way voters gained a voice in the nomination of candidates. The primary system soon spread across the country.

In the first decade of the twentieth century, more and more states gave women the vote and enacted other important reforms. Laws designed to curb corrupt election practices and to limit campaign spending came onto the books in many states, and similar federal laws followed.

Oregon was a leader in the movement to make voting a true exercise of democracy. A key figure in the movement was William U'Ren, who never held a major political office yet was responsible for significant reforms in Oregon and ultimately across the United States. A blacksmith who later became a newspaper editor and a lawyer, U'Ren had read how "direct democracy" was practiced in Switzerland and committed himself to introducing it in his state. Mostly as a result of his efforts, Oregon adopted the initiative and referendum in 1902, the direct primary in 1904, and the recall in 1910.

Initiative and referendum made it possible for a majority vote among the people—the electorate—to pass laws when the legislature was unable or unwilling to do so. They also gave voters the power to veto unpopular laws. With the power of recall, voters could quickly remove an elected official if they were displeased with his conduct.

Recall was adopted more readily by states west of the Mississippi. Eastern states, usually more conservative, did not generally favor recall but did come to adopt initiative and referendum.

During this same period, there was a drive to amend the U.S. Constitution to allow a direct vote for U.S. senators. Across the country many had become aware that because the Senate was not answerable to the people, it had turned into a "rich men's club" that had opposed many pro-gressive laws, such as those abolishing child labor and revising the tariff system.

Western states were again in the forefront of the movement toward change. Several of them began holding primary elections to select Senate nominees, and in some of these states there were laws forcing the state legislature to abide by the voters' choices. By 1912, twenty-nine states had adopted this way of giving the voters greater power. Finally, the Senate submitted a constitutional amendment to the states for ratification. The Seventeenth Amendment, providing for direct election of senators, was ratified on May 31, 1913.

## Consequences

With the ratification of the Seventeenth Amendment, a number of major political reforms were in place across the United States. Though the guarantee of all female citizens' right to vote would have to await the passage of the Nineteenth Amendment in 1920, many states, particularly those in the West, were establishing that right within their borders. Through the secret ballot, the primary, the initiative, the referendum, the recall, and laws limiting campaign spending, political power was placed in the hands of American voters as it never had been before.

# Evans Unearths Minoan Civilization

Sir Arthur Evans changed concepts of Greek history through his excavations at Knossos.

**What:** Archaeology
**When:** March 23, 1900
**Where:** Kephála (Knossos), Crete
**Who:**
SIR ARTHUR EVANS (1851-1941), an
    English archaeologist
DUNCAN MACKENZIE (1859-1935), a
    Scottish archaeologist
MICHAEL VENTRIS (1922-1956), an English
    cryptographer

## The Hard Road to Crete

Near the age of fifty, Sir Arthur Evans discovered a civilization that was older than that of the Greeks. He had been interested in archaeology for most of his life. When he was only twenty, he had wanted to write an archaeological history of the Balkans (the countries on the Balkan peninsula in Southeastern Europe that later came to be known as Yugoslavia and Greece). When he was twenty-seven, he moved to Ragusa (later Dubrovnik) with his wife, Margaret Freeman; they lived there for four years, his money coming from his father and from writing newspaper articles for *The Manchester Guardian.* Because these articles were in favor of Balkan independence from Austrian rule, Evans was eventually arrested, sent to prison, and expelled from the country in April, 1882.

Back in England, Evans's family got him a job as Keeper of the Ashmolean Museum in Oxford. Evans was able to organize the museum's many archaeological, geological, and historical objects and make the institution financially secure. In 1894, he supervised the construction of its handsome building.

Evans often took trips to the Mediterranean looking for possible archaeological sites. He had met Heinrich Schliemann there in 1873, only three years after Schliemann had excavated the ancient city of Troy. Evans was excited about finding even older cities, and after the sudden death of his wife in 1893, his trips to the Mediterranean began to last even longer.

What led Evans to Crete, though, was a set of three- and four-sided sealstones from Crete that he found in an antiques shop in Athens, Greece. They were covered with a form of picture writing that Evans would eventually call "Linear A." He had seen writing like it on two vases found by a Greek archaeologist named Chrestos Tsountas at Mycenae on the Greek Peloponnese. Evans was sure that the writing was from an independent Cretan culture that existed thousands of years before the Greeks became powerful. He wanted to dig at Kephála (the modern name for the area around Knossos), where the palace of Minos was thought to have stood. Evans called the culture he was looking for "Minoan," not only because of the palace and the legend of the mythical bull-headed beast known as the Minotaur, but also to set it apart from cultures that existed at the time of the Trojan War.

Evans's friend Federico Halbherr, an Italian archaeologist, convinced him that important discoveries would be found at Knossos. In 1878, Halbherr had found several large storage jars, called pithoi, near there. Schliemann had wanted to look in this spot as early as 1878, but the owner of the land would not allow it. The Turkish authorities who governed Crete also stood in the way, but Evans cut through every problem by buying the land for himself, using money from his family and from the Cretan Exploration Fund he had established. Not many archaeologists before or since have owned the land they were excavating.

## Uncanny Success

Evans's ability to manage money and his uncanny sense of where to dig led to immediate success. On the first day of digging (March 23,

1900), he found walls and pottery fragments that were only 33 centimeters underground. On the second day, he located an ancient house and fragments of plaster that had once belonged to a kind of painting called a fresco. On the third day, he uncovered smoke-blackened walls and broken pottery, including the rims of large storage jars similar to those found by Halbherr. Exactly one week after he had begun digging, he found clay jars with the same writing on them that he had seen in the shop in Athens. Because of his amazing success, he was able to continue raising enough money to keep excavating through both world wars.

Evans discovered two kinds of writing at Knossos. No one has been able to read Linear A, which is hieroglyphic, but Linear B is different. Evans had hoped to decipher it before he died, and he had kept several thousand tablets at his home. Twelve years after Evans's death, Michael Ventris, who had been a cryptographer during World War II, was able to prove that each character in Linear B stands for a certain syllable, rather than a whole word, as in Linear A. Alice Kober, an American who had tried to decipher it before, had thought that the upright lines in the writing were used to divide words. By using this idea and trying to make connections between characters that repeated, Ventris broke the code and discovered that the tablets were mostly inventory lists. The lists were written on mud brick, which would have been appropriate for inventory lists, since the words could easily be changed. Some of the lists were of food, while others had to do with trade with other countries. The trade lists show that the Minoan civilization was not as independent as Evans had believed.

Evans had thought that the Minoans were independent of the Greeks until a se-

ries of earthquakes forced them to move the government onto the mainland. He thought that this move may have been the start of Greek civilization. Later archaeologists thought that earthquakes had nothing to do with it. They argued that the Greeks had invaded Knossos, probably as early as 1500 B.C.E. Evans rejected this idea as long as he lived, but most archaeologists accept it.

## Consequences

Evans tried to restore the Minoan palace and the frescoes he found on the site. Where there was not enough left of the Minoan palace Evans was trying to restore, he had his architect, Theodore Fyfe, rebuild it, using much less evidence than modern archaeologists would have accepted. In the same way, he had Émile Gilliéron restore the frescoes. Gilliéron chose unusually bright colors, including a particular shade of red that came to be called "Minoan red." These restorations became famous among wealthy patrons of the arts.

Although this popular attention brought in substantial sums of money, both through dona-

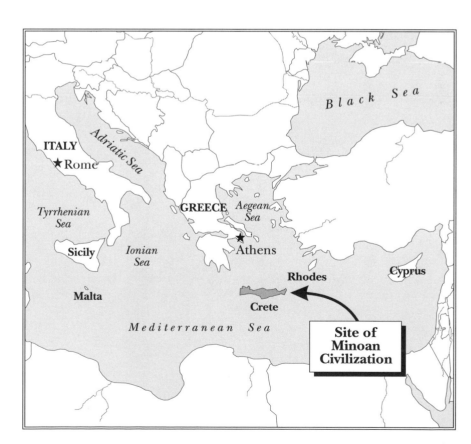

EVANS UNEARTHS MINOAN CIVILIZATION

tions and from sales of reproductions of Minoan art, many professionals who were connected with the project were embarrassed, especially Duncan Mackenzie, the Scottish archaeologist who was in charge of the daily operations. Mackenzie was often dismissed for arguing with Evans about the latter's sometimes outrageous methods. Nevertheless, Mackenzie's day books, which described in detail every day of the excavation, are still used as a model for how to draw archaeological finds.

The most important effect that Evans's work had on archaeology was to change the understanding of the time and importance of Cretan civilization. Evans believed until he died that this civilization began earlier than 3500 B.C.E. and was more powerful than the Greek civilization at the time. Most archaeologists disagree on the date and on how powerful the Minoans were, but without Evans's work, the debate would never have begun.

*Robert J. Forman*

# Boxer Rebellion Tries to Drive Foreigners from China

*With the Boxer Rebellion, the Chinese people tried to free themselves from the influence of foreigners.*

**What:** Civil strife

**When:** June-September, 1900

**Where:** North China, especially Shandong (Shantung) and Hebei (Chihli) provinces

**Who:**

CIXI (TZ'U-HSI; 1835-1908), empress dowager of China

GUANGXU (KUANG-HSÜ; 1871-1908), emperor of China from 1875 to 1908

COUNT ALFRED VON WALDERSEE (1832-1904), commander of the international force sent to China

## Staking Claims in China

After the First Opium War with Great Britain (1839-1842), China was pressured more and more by foreign powers. Foreigners were given the rights to control trade, collect customs money, and run the courts in dozens of Chinese cities, called "Treaty Ports."

After Hong Kong was given to the British in 1842, wars and threats forced the Qing (Ch'ing; Manchu) Dynasty to give up other territories. Russia had special influence in Manchuria and Central Asia; France took control of Indochina in 1884; and Japan humiliated China in a war in 1894 and took Taiwan as a prize. Soon Germany was also competing for influence in China.

Sometimes the Chinese tried to rise up against this foreign domination, but the empire's armies and ocean fleets were no match for the military forces of foreign powers. In the 1890's, many secret societies and militia worked to oppose foreigners—especially in the northern provinces of Hebei, Shandong, and Shaanxi (Shensi), where there were many Christian missionaries and other foreigners.

Some of the most active secret societies were part of the Yihe Quan (I-ho Ch'üan; Association of Righteousness and Harmony, usually called the "Righteous and Harmonious Fists"). In its ceremonies, this group practiced the ancient Chinese art of *taiji chuan* (*t'ai-chi ch'uan*), which included shadowboxing. Thus the foreigners nicknamed this group the "Boxers."

The Boxers had always been opposed to foreign control. In fact, at first they worked against the Qing Dynasty, because the Manchus (who had founded the dynasty and still held most of the important court positions) did not come from the Han ethnic group. More and more, however, the Boxers began to oppose missionaries—especially after the Germans started dominating Shandong in 1898.

## Reform and Rebellion

The Qing government was in a difficult position. The Boxers and other groups wanted foreigners out, while the foreign powers insisted that the government stand firm against the Boxers.

For a short time in the summer of 1898, it seemed as if some of the problems would be resolved. Emperor Guangxu, who had recently come of age, began to reform the Chinese government. In September, however, this "One Hundred Days of Reform" came to a sudden end. Guangxu's aunt, Empress Dowager Cixi, and her chief adviser, Ronglu (Jung-lu), staged a coup. Guangxu was arrested, and Cixi became ruler of China. She was committed to getting rid of foreign control.

With the support of many officials in North China, the Boxers began to attack foreign railroads and settlements. Their motto became "Fuqing, mieyang" ("Fu-Ch'ing, mieh-yang"; support the Qing, exterminate the foreigners). During the winter of 1899-1900, the foreign powers in China protested the Boxers' attacks and began threatening to send troops.

The empress dowager believed the Boxers when they claimed that the foreigners' bullets could not harm them, and she called upon the army and people to defend the country from a foreign invasion that was sure to come. Now the Boxers in Beijing, Hebei, and Shaanxi joined the fight. Hundreds of missionaries and thousands of Chinese Christians were tortured and killed.

In early June, a force of foreign troops sent from Tianjin (Tientsin) was turned back by Boxers and Chinese army units. The German minister to China, Count Clemens von Kettler, was shot down in the streets of the capital. On June 21, 1900, the Qing government declared war on all the treaty powers in China. The Boxer militia was commanded to attack the section of Beijing where the foreign diplomats lived.

Chinese officials throughout the empire were told to cooperate with the Boxers in attacking foreigners. Those in North China obeyed, but government and army leaders in many other parts of the country did their best to avoid open attacks. Some of them felt sympathy for the captive emperor or doubted the Boxers' fighting abilities.

By late July, a powerful international force of twenty thousand men—including Germans, Japanese, Americans, British, Russians, French, Austrians, and Italians—had been brought together in Tianjin under the command of Count Alfred von Waldersee. After two weeks of fighting, they made their way to Beijing, entered the city through an unguarded sewer gate, and ended the siege of the diplomatic section.

The empress dowager and her court fled to Xi'an (Sian), and most government forces surrendered quickly. The Boxers, who had not proved reliable in battle, escaped to the countryside of North China.

Angry at the Boxers' cruel treatment of foreigners and Chinese Christians, the allied troops began looting and burning the suburbs of Beijing and Tianjin. International forces occupied the capital until September, 1901, and the empress dowager did not return until the beginning of 1902.

The final peace treaty, the Boxer Protocols, was accepted by the Chinese on January 16, 1901. Under the treaty, officials who had cooperated with the Boxers were to be dismissed, exiled, or even executed. Foreigners were to occupy the area between Beijing and Tianjin, and China was to pay $333 million in war reparations over a period of thirty-nine years.

## Consequences

Because of the Boxer Rebellion, the Qing Dynasty lost the respect of the Chinese people as well as the respect of foreign powers. Many people with skills and education avoided serving a government that seemed so incompetent. Nationalism became a stronger force among the Chinese people, especially in Chinese communities overseas.

The empire had severe money problems. Its income was not enough to pay the war debt, so taxes had to be increased, and loans were taken from banks in Western countries.

The empress dowager now reluctantly opened the way for many of the "Hundred Days" reforms. Education was modernized, along with army training. Chinese officials toured the West, studying the different systems of government. They made a plan for a constitutional monarchy in China, and there were elections in 1909 and 1910 for regional and national parliaments.

For a number of young Chinese, however, these reforms came too late. A number of republican, nationalist, reform, and secret society organizations joined in the Revolutionary Alliance, led by Sun Yat-sen. This coalition managed to topple the empire on October 10, 1911.

*Charles A. Desnoyers*

# Reed Discovers Cause of Yellow Fever

*Walter Reed established that yellow fever was caused by an unknown infectious agent being transmitted by the* Aëdes *mosquito.*

**What:** Medicine
**When:** June, 1900-February, 1901
**Where:** Quemados, near Havana, Cuba
**Who:**
CARLOS JUAN FINLAY (1833-1915), a
    Cuban physician
WALTER REED (1851-1902), a physician,
    bacteriologist, and major in the U.S.
    Army Medical Corps
JESSE WILLIAM LAZEAR (1866-1900), a
    physician, bacteriologist, and
    entomologist
JAMES CARROLL (1854-1907), a physician
    and bacteriologist
ARISTIDES AGRAMONTE (1869-1931), a
    physician and pathologist
WILLIAM CRAWFORD GORGAS (1854-1920),
    a physician and colonel in the U.S.
    Army Medical Corps

## One Invasion Halted by Mosquitoes

In the 1800's, yellow fever had been known for centuries along the west coast of Africa. There it showed little effect in the native population, probably because generations of exposure to the disease had resulted in a high degree of inherent resistance. Europeans, however, had virtually no resistance to yellow fever, and thousands died of the disease during efforts to develop the natural resources of Africa. So deadly was yellow fever that the continent of Africa became known to Europeans as the "white man's grave."

Yellow fever was a particularly frightening disease. It attacked suddenly, causing a high fever, headache, and nausea. The eyes became inflamed, and the skin took on a yellow pallor. Many victims bled internally, producing a black vomit. Anywhere from 30 to 70 percent of these victims died, with most deaths occurring on the sixth or seventh day following infection. In 1802, an epidemic in Santo Domingo on the island of Hispaniola in the Caribbean, killed twenty-nine thousand of the thirty-three thousand troops sent there by Emperor Napoleon I for a planned invasion up the Mississippi River in North America. As a result of these devastating losses, Napoleon changed his plans, and the next year he negotiated the Louisiana Purchase with U.S. president Thomas Jefferson.

Two theories developed as to the cause and spread of yellow fever. Some considered it to be a contagious disease that could be spread directly from one person to another. Others believed it to be caused by an odor (the miasma theory), filth, or rotting vegetables. The Philadelphia epidemic of 1793, for example, was attributed to a shipment of spoiled coffee beans.

In 1881, Carlos Juan Finlay, a Cuban physician, became a strong advocate of the mosquito transmission theory that had been proposed in 1848 by Josiah Nott. Finlay thought that it was probably the *Culex* mosquito (the name was later changed to *Aëdes*) that was spreading an unknown infectious agent. He performed experiments in which some individuals were bitten by mosquitoes and came down with yellow fever. His unsophisticated experiments, however, were easily discounted by his detractors.

In 1898, at the end of the Spanish-American War, William Crawford Gorgas was appointed chief sanitary officer for Havana. Gorgas set out to rid Havana of the filth and disease left by the war. As a result of his campaign, typhoid and dysentery were significantly reduced. Yellow fever temporarily declined but struck again in 1900, this time bypassing the filthiest areas, where there was a high level of immunity from previous epidemics, and causing devastation in the cleaner sections of town, including the United States Army headquarters.

## Two Hundred Dollars for Yellow Fever

It was at this time that Walter Reed was appointed to head a commission to study yellow fever in Cuba. The United States Army Yellow Fever Board was made up of Reed, James Carroll, Jesse William Lazear, and Aristides Agramonte, a native Cuban. Their initial work was to attempt to identify the cause of the disease.

Despite initial skepticism concerning the mosquito hypothesis, at the request of the Yellow Fever Board, on August 1, 1900, Finlay supplied mosquito eggs for Reed to study. Since no animals were known to acquire yellow fever, the board knew they must rely upon human experi-

*Walter Reed.*

mentation. It was agreed that the first humans involved in the experiments would include the members of the board. Lazear was to be in charge of the mosquitoes.

On September 13, Lazear was feeding one of the experimental mosquitoes on a patient when another mosquito came through the window and landed upon the doctor's hand. Rather than chance disrupting the experiment, he allowed this wild mosquito to feed upon his blood. Five days later, Lazear developed symptoms of yellow fever. His condition deteriorated rapidly, and on September 25, 1900, he died. When Reed requested approval to build an experiment station in an open field away from the town of Quemados, Cuba, it was decided that the site would be named "Camp Lazear."

Camp Lazear was established on November 20, 1900, and consisted of seven floored hospital tents and support facilities. Each of the tents was made mosquito-proof and was inhabited by one to three nonimmune volunteers who were paid one hundred dollars for participation and another one hundred dollars if they developed yellow fever as a result of the experiments. From November 20 through December 30, six volunteers were bitten by infected mosquitoes. Five developed yellow fever, and all survived. Even though these carefully controlled experimental cases provided strong evidence that mosquitoes could carry yellow fever, one more experiment was required.

Reed constructed a mosquito-proof frame building separated into two rooms by a center wire screen mesh partition. Each room was occupied by a nonimmune volunteer. In one room only were mosquitoes that carried the disease. As expected, the volunteer bitten by these mosquitoes became ill, while the volunteer in the room free of mosquitos remained healthy. The Yellow Fever Board was now satisfied with its experiments. The *Journal of the American Medical Association* published Reed's results in February, 1901.

## Consequences

It was clear that the way to control yellow fever was to eliminate exposure to mosquitoes. Gorgas quickly set out to do just that. The approach was to protect infected patients from mosquito bites and to eliminate mosquito breeding places. In March, 1901, Gorgas set out with great enthusiasm by launching a house-to-house attack on the mosquito. All water containers were to be emptied, covered, or layered with kerosene. Failure to comply meant a ten-dollar fine.

Gorgas's tactics were first met with derision by the populace. During the summer of 1901, however, the number of yellow fever cases in Havana was drastically reduced. By October, for the first time in decades, no cases were being reported. Within a few years, the entire Western Hemisphere was for the most part rid of yellow fever. One of the most significant results was that it became possible to build the Panama Canal.

*Daniel W. Rogers*

# Zeppelin Builds Rigid Airship

Ferdinand von Zeppelin made the first full-sized rigid airship, a vehicle that played a major role in World War I and in international air traffic until 1938.

**What:** Space and aviation
**When:** July 2, 1900
**Where:** Manzell, Lake Constance, Germany
**Who:**
FERDINAND VON ZEPPELIN (1838-1917), a retired German general
THEODOR KOBER (1865-1930), Zeppelin's private engineer

## Early Competition

When the Montgolfier brothers launched the first hot-air balloon in 1783, engineers began working on ways to use machines to control the speed and direction of balloons. They thought of everything: rowing through the air with silk-covered oars; building movable wings; using a rotating fan, an airscrew, or a propeller powered by a steam engine (1852) or an electric motor (1882). At the end of the nineteenth century, the internal combustion engine was invented. It promised higher speeds and more power. Up to this point, however, the balloons were not rigid.

A rigid airship could be much larger than a balloon and could fly farther. In 1890, a rigid airship designed by David Schwarz of Dalmatia was tested in St. Petersburg, Russia. The test failed because there were problems with inflating the dirigible. A second test, in Berlin in 1897, was only slightly more successful, since the hull leaked and the flight ended in a crash.

Schwarz's airship was made of an entirely rigid aluminum cylinder. Ferdinand von Zeppelin had a different idea: His design was based on a rigid frame. Zeppelin knew about balloons from having fought in two wars in which they were used: the American Civil War of 1861-1865 and the Franco-Prussian War of 1870-1871. He wrote

down his first "thoughts about an airship" in his diary on March 25, 1874, inspired by an article about flying and international mail. Zeppelin soon lost interest in this idea of civilian uses for an airship and concentrated instead on the idea that dirigible balloons might become an important part of modern warfare. He asked the German government to fund his research, pointing out that France had a better military air force than Germany did.

In 1893, in order to get more money, Zeppelin tried to persuade the German military and engineering experts that his invention was practical. Even though a government committee decided that his work was worth a small amount of funding, the army was not sure that Zeppelin's dirigible was worth the cost. Finally, the committee chose Schwarz's design. In 1896, however, Zeppelin won the support of the powerful Union of German Engineers, which in May, 1898, gave him 800,000 marks to form a stock company called the Association for the Promotion of Airship Flights. In 1899, Zeppelin began building his dirigible in Manzell at Lake Constance. In July, 1900, the airship was finished and ready for its first test flight.

## Several Attempts

Zeppelin, together with his engineer, Theodor Kober, had worked on the design since May, 1892, shortly after Zeppelin's retirement from the army. They had finished the rough draft by 1894, and though they made some changes later, this was the basic design of the Zeppelin. An improved version was patented in December, 1897.

In the final prototype, called the LZ 1, the engineers tried to make the airship as light as possible. They used a light internal combustion engine and designed a frame made of the light metal aluminum. The airship was 128 meters

*Ferdinand von Zeppelin's creation, a rigid airship, flew for the first time on July 2, 1900.*

Hulton Archive

long and had a diameter of 11.7 meters when inflated. Twenty-four zinc-aluminum girders ran the length of the ship, being drawn together at each end. Sixteen rings held the body together. The engineers stretched an envelope of smooth cotton over the framework to reduce wind resistance and to protect the gas bags from the sun's rays. Seventeen gas bags made of rubberized cloth were placed inside the framework. Together they held more than 120,000 cubic meters of hydrogen gas, which would lift 11,090 kilograms. Two motor gondolas were attached to the sides, each with a 16-horsepower gasoline engine, spinning four propellers.

The test flight did not go well. The two main questions—whether the craft was strong enough and fast enough—could not be answered because little things kept going wrong; for example, a crankshaft broke and a rudder jammed. The first flight lasted no more than eighteen minutes, with a maximum speed of 13.7 kilome-

ters per hour. During all three test flights, the airship was in the air for a total of only two hours, going no faster than 28.2 kilometers per hour.

Zeppelin had to drop the project for some years because he ran out of money, and his company was dissolved. The LZ 1 was wrecked in the spring of 1901. A second airship was tested in November, 1905, and January, 1906. Both tests were unsuccessful, and in the end the ship was destroyed during a storm.

By 1906, however, the German government was convinced of the military usefulness of the airship, though it would not give money to Zeppelin unless he agreed to design one that could stay in the air for at least twenty-four hours. The third Zeppelin failed this test in the autumn of 1907. Finally, in the summer of 1908, the LZ 4 not only proved itself to the military but also attracted great publicity. It flew for more than twenty-four hours and reached a speed of more than 60 kilometers per hour. Caught in a storm at the end of this flight, the airship was forced to land and exploded, but money came from all over Germany to build another.

## Consequences

Most rigid airships were designed and flown in Germany. Of the 161 that were built between 1900 and 1938, 139 were made in Germany, and 119 were based on the Zeppelin design.

More than 80 percent of the airships were built for the military. The Germans used more than one hundred for gathering information and for bombing during World War I (1914-1918). Starting in May, 1915, airships bombed Warsaw, Poland; Bucharest, Romania; Salonika, Greece; and London, England. This was mostly a fear tactic, since the attacks did not cause great damage, and the English antiaircraft defense improved quickly. By 1916, the German army had lost so many airships that it stopped using them, though the navy continued.

Airships were first used for passenger flights in 1910. By 1914, the Delag (German Aeronautic Stock Company) used seven passenger airships for sightseeing trips around German cities. There were still problems with engine power and weather forecasting, and it was difficult to move the airships on the ground. After World War I, the Zeppelins that were left were given to the Allies as payment, and the Germans were not al-lowed to build airships for their own use until 1925.

In the 1920's and 1930's, it became cheaper to use airplanes for short flights, so airships were useful mostly for long-distance flight. A British airship made the first transatlantic flight in 1919. The British hoped to connect their empire by means of airships starting in 1924, but the 1930 crash of the R-101, in which most of the leading English aeronauts were killed, brought that hope to an end.

The United States Navy built the *Akron* (1931) and the *Macon* (1933) for long-range naval reconnaissance, but both airships crashed. Only the Germans continued to use airships on a regular basis. In 1929, the world tour of the *Graf Zeppelin* was a success. Regular flights between Germany and South America started in 1932, and in 1936, German airships bearing Nazi swastikas flew to Lakehurst, New Jersey. The tragic explosion of the hydrogen-filled *Hindenburg* in 1937, however, brought the era of the rigid airship to a close. The U.S. secretary of the interior vetoed the sale of nonflammable helium, fearing that the Nazis would use it for military purposes, and the German government had to stop transatlantic flights for safety reasons. In 1940, the last two remaining rigid airships were destroyed.

*Matthias Dörries*

# Hurricane Levels Galveston, Texas

*The deadliest hurricane in Texas history hit the port city of Galveston, Texas, leveling homes and destroying thousands of lives.*

**What:** Disasters
**When:** September 8, 1900
**Where:** Galveston, Texas
**Who:**
ISAAC MONROE CLINE (1861-1955), Weather Bureau observer in Galveston
RICHARD SPILLANE (1864-1936), Galveston newspaperman
GEORGE BANNERMAN DEALEY (1863-1946), general manager of *The Dallas News*

## Queen City of the Southwest

In 1900, Galveston was popularly regarded as the Queen City of the Southwest. With a population of 37,000 people and all major railroads converging on its small island off the coast south of Houston, Galveston operated as the marketplace for the agricultural hinterland. The Cotton Exchange, the Custom House, and the docks provided employment for immigrants who continued to arrive daily. With approximately 4,000 foreigners disembarking every year, the city developed an atmosphere of diversity. *The Galveston News,* which owned *The Dallas News,* provided political, agricultural, and business news for the entire state.

## The Storm

On September 4, 1900, the winds off the coast of Galveston gained speed as a strong equinoctial hurricane from the West Indies moved past Cuba and out of the Caribbean Sea. As the storm traveled through the Florida Keys, the wind speeds increased to 130 miles per hour (209 kilometers per hour), leaving dozens of wrecked boats behind as the storm headed for Texas. During the afternoon and evening of September 7, Isaac Monroe Cline, director of the Galveston Weather Bureau, watched as his barometer continued to fall. Realizing that the storm was approaching, he raced along the beach yelling to people to get off the island as quickly as they could. Many tourists and islanders heeded his warnings, while others stayed to watch the ocean swells rising above the streets.

At 3:00 A.M., the news of the storm's impending destruction was broadcast over the telegraph wire. Within hours, the tide had risen over half the island, where the land elevation peaked at only 8.7 feet (2.6 meters). At the height of the storm, the barometric pressure registered 28.55 inches (72.5 centimeters), the lowest pressure ever recorded. The ocean swept across the narrow strip of land, cresting at 15.7 feet (4.8 meters) before the waters began to subside. The force of the waves destroyed buildings, snapping them like twigs.

As people struggled to reach safety, some were rescued by the Sisters of Charity, who pulled them out of the water from the second-story balcony of St. Mary's Orphan Asylum. Others, including Cline's pregnant wife, disappeared in the floodwaters. The nuns in the orphanage vowed to hold on to the children, tying a string around their waists in case the water continued to rise. Taking shelter in the upstairs girls' dormitory, the nuns and the children heard the adjacent boys' dorm collapse under the pressure of the sea. Within a short time, the structure they were in also collapsed. Ten nuns and ninety children died; only three boys managed to survive.

Most of the other survivors in the town weathered the storm in sturdier buildings along the Strand, where the Tremont Hotel and *The Galveston News* offices were located. After the weather improved, Richard Spillane, a Galveston newspaperman, reached the mainland on the first boat off the island and relayed to the rest of

**27**

the state information concerning initial deaths and estimates of property damage. His first account stated that approximately one thousand had died, but during the next few weeks, that number climbed to more than six thousand.

## Consequences

As the water receded, the survivors began the process of locating loved ones and burying the dead. Between six thousand and twelve thousand people died in the storm, either washed out to sea or drowned. For two months, authorities forced a "dead gang" to burn the dead on funeral pyres located on the beach, having to resort to bribes of whiskey and threats of being shot if the workers refused to continue until the job was complete. Additional corpses continued to wash ashore as the tides came in. The hurricane destroyed more than thirty-six hundred buildings with damages reaching $20 million.

George Dealey of *The Dallas News* organized a relief effort to provide the city with food and supplies, collecting more than $28,000 in three weeks. The city of Dallas contributed another $100,000, and the state of Texas also provided relief.

After the cleanup, the residents of Galveston decided to rebuild their city. Precautions against

Galveston County Historical Museum

*Part of the Sacred Heart Church remained standing after a hurricane passed through Galveston, Texas.*

future storms included the construction of a 16-foot (4.9-meter) seawall and the raising of the island to 25 feet (7.6 meters) above sea level at a cost of $15 million. For years, the residents endured the catwalks and stench as construction workers dredged 16 million cubic yards (12.2 million cubic meters) of sand from the Houston ship channel to raise the island. At the same time, all buildings, water, sewer, and gas lines were also raised.

For the next sixty years, the city continued to extend the seawall from the original section, located between Sixth and 39th Streets, to cover an area of almost 10 miles (16 kilometers). In 1915,

the wall prevented a second major disaster from befalling Galveston when another hurricane, of almost the same force, hit the island. In September of 1961, the seawall once again reduced the amount of flood damage after another devastating storm.

Even though residents rebuilt the city, it never regained its former prominence. Businesses, seeking a safer environment, relocated to Houston after the construction of the ship channel. Galveston remains a remnant of its past, enjoyed today mainly by tourists.

*Cynthia Clark Northrup*

# Planck Articulates Quantum Theory

*Max Planck's study of light prepared the way for the development of quantum mechanics.*

**What:** Physics
**When:** December 14, 1900
**Where:** Berlin, Germany
**Who:**
MAX PLANCK (1858-1947), a German physicist who won the 1918 Nobel Prize in Physics
ALBERT EINSTEIN (1879-1955), a German-born American physicist
NIELS BOHR (1885-1962), a Danish physicist

### The Ultraviolet Catastrophe

By the end of the nineteenth century, many physicists believed that the study of physics was complete. Much progress had recently been made in such areas as electricity, thermodynamics, optics, and electromagnetic radiation. It seemed that there was nothing left to do except to find more accurate ways of calculating known values. Several experimental oddities, however, could not be explained. One of these was "black body radiation."

When a piece of metal becomes very hot, it begins to turn a dull red that gets redder as it gets hotter. If the temperature is increased even more, the metal becomes white and eventually blue. There is a continual shift of the color of a heated object from red through white to blue as it is heated to higher and higher temperatures. In terms of frequency (the number of waves of radiation that pass a given point per unit of time), the radiation emitted from the heated object goes from a lower to a higher frequency as the temperature increases, because red is in a lower frequency range of the spectrum than is blue.

The observed colors are the frequencies that are being emitted in the greatest proportion. Moreover, any heated body will exhibit a frequency spectrum (a range of different intensities for each frequency). A theoretical body that can completely absorb all the radiation falling upon it is called a *black body*; the radiation it gives off when it is heated is called "black body radiation."

In theory, if the black body were heated past the blue stage, it would give off an invisible light, called ultraviolet light, which has an even higher frequency than that of blue light. Strangely, this did not happen in experiments. Physicists called this failure the "ultraviolet catastrophe."

### Planck's Solution

On December 14, 1900, a soft-spoken professor named Max Planck stood before the German Physics Society and offered a solution: What if the metal did not absorb energy, such as heat, in a continuous way, but only in certain fixed amounts or quantities? This "quantum" theory contradicted every theory of light that existed at the time.

It had been assumed that light waves behaved like mechanical waves. Mechanical waves, much like waves in a pond, are a collection of all possible waves at all frequencies, with higher-frequency waves being the most visible. In the mechanical wave theory, all waves appear when energy is introduced, with high-frequency waves naturally being present in greater amounts. The ultraviolet catastrophe observed in black body radiation showed this theory to be incorrect.

Planck stated that light waves do not behave like mechanical waves. He postulated that the reason for the discrepancy lay in the relationship between energy and wave frequency. The energy that is either absorbed or emitted as light depends in some fashion on the frequency of the light that is emitted. Somehow, the heat energy supplied to the glowing material failed to excite higher-frequency light waves unless the temperature of that body was very high. The high-frequency waves simply cost too much energy to be produced.

To prove his point, Planck created a formula that reflected this dependence of the energy upon the frequency of the waves. His formula said that energy and frequency are directly proportional, related by a proportionality constant that is now called "Planck's constant." Higher frequency meant higher energy. Consequently, unless the energy of the heated body was high enough, the higher-frequency light was not seen. In other words, only a fixed amount of energy was available at a given temperature.

Moreover, the release of that energy could be made only in exact amounts, by dividing up the energy exactly. Small divisions of the energy, resulting in large numbers of units, were favored over large divisions of small numbers of units. Lower frequencies (small units of energy) were favored over higher frequencies. In this way, black body radiation was explained.

## Consequences

Planck's new idea was so revolutionary that even he had trouble believing it. In 1905, however, Albert Einstein would use the quantum theory to explain the "photoelectric effect." In the photoelectric effect, when light strikes the surface of a piece of metal, electrons are ejected. Einstein observed that the number of ejected electrons depends not on the intensity of the light but on the frequency of the light. As a result, raising the frequency of the light striking the metal surface causes more electrons to be ejected. Blue light will knock off more electrons than will red light. This explanation won for Einstein the 1921 Nobel Prize in Physics.

In 1913, Niels Bohr used the ideas of the quantum theory to explain the structure of atoms. When atoms are excited, they release radiation.

*Max Planck.*

Bohr reasoned that such radiation can be caused only by an energy change in the electrons of those atoms. The energy level of an atom, according to Bohr, can be changed only if that atom absorbs or releases a certain exact quantity of energy. For his effort, Bohr received the 1922 Nobel Prize in Physics.

*Scott A. Davis*

# Commonwealth of Australia and Dominion of New Zealand Are Born

*Australia and New Zealand, which had been British colonies, became self-governing states.*

**What:** Political independence
**When:** 1901, 1907
**Where:** Australia, New Zealand, and Great Britain
**Who:**

SIR EDMUND BARTON (1849-1920), first prime minister of the Australian Federation

SIR GEORGE HOUSTON REID (1845-1918), premier of New South Wales from 1894 to 1899

SIR JOSEPH GEORGE WARD (1856-1930), premier of New Zealand from 1906 to 1912

JOSEPH CHAMBERLAIN (1836-1914), British secretary of state for the colonies from 1895 to 1903

VICTOR ALEXANDER BRUCE, NINTH EARL OF ELGIN (1849-1917), British secretary of state for the colonies from 1905 to 1908

## Australia Becomes a Federation

Though both the United States and Australia were once colonies of Great Britain, their paths to self-government were quite different. The British government had not wanted the American colonies to become independent, but the American people rose up and fought to win their freedom. After the humiliation of being forced to yield control of the thirteen North American colonies, British policy toward its colonies changed. Throughout the nineteenth century, when other British colonies populated by white settlers sought independence, the government of Great Britain was generally cooperative.

The Canadian colonies, which became the Dominion of Canada in 1867, benefited first from this new approach. Australia was next. Neither these countries nor New Zealand had to fight a revolutionary war to gain freedom from Great Britain's control.

What is now Australia had become home to several diverse groups of people. Its original inhabitants were wandering bands of aborigines. The first settlers from abroad were English and Irish convicts whose punishment was forced exile to this large island. When gold was discovered there in 1851, a great wave of British immigrants came of their own accord to seek a share of the new wealth. By 1889, the island had been divided into six colonies: Victoria, New South Wales, Queensland, South Australia, Tasmania, and Western Australia.

In February, 1890, premiers of the six colonies gathered in Melbourne to discuss the possibility of federation—uniting under a central government that would permit each colony (or state) a degree of independence. The colonial legislatures sent delegates to the first Australasian Convention a year later, in March, 1891, in Sydney. In April, the convention approved a draft constitution for a federal system of government.

Yet the colonial legislatures were not enthusiastic about the proposed constitution and did not take any action on it. Those who were involved in the growing trade-union movement thought the constitution would not give enough power to the common people; others said it took too many powers away from the individual colonies.

Within each colony, however, there were "Federal Leagues" keeping alive the vision of an Australian federation. In August, 1894, George Houston Reid, premier of New South Wales, called for another conference of premiers to plan a new constitutional convention. The conference of premiers, which met in Hobart, Tasmania, in January, 1895, decided that this time the delegates who would draft a constitution would be elected by the people in each of the six colonies. The resulting constitution would be put to a popular vote in each colony, and once approved by the people, it would be submitted to the British government for final approval.

Elections for constitutional delegates were held in every colony except Queensland. The federal constitutional convention began on March 22, 1897, in Adelaide, and concluded on March 17, 1898, in Melbourne. The rivalry between smaller, poorer colonies and those that were larger and wealthier caused most of the convention's controversy. Finally, however, a draft constitution was put together and brought before the voters of each colony.

Amendments to safeguard the interests of New South Wales, the largest colony, were needed before the new constitution could receive enough votes in all colonies. Finally, however, the document was ready to be presented to the British colonial secretary, Joseph Chamberlain, and other officials in London.

Chamberlain introduced the necessary legislation in the British parliament, and on July 9, 1900, Queen Victoria signed the new act creating the Commonwealth of Australia. On January 1, 1901, Edmund Barton was appointed as its first prime minister.

## New Zealand Follows Suit

The two islands of New Zealand were originally inhabited by the Maoris, a people related to the Polynesians. The first British settlers were mostly traders, adventurers, and shipwrecked sailors. Though Great Britain annexed the islands in 1840, the Maoris launched wars against the white colonists, and it was not until 1870 that the British army finally crushed the resistance.

Beginning in 1856, New Zealand's government was appointed by the British governor, but it was responsible to the popularly elected branch of the legislature. Twenty years later, fur-

*Crowds watch a parade in celebration of the creation of the Australian commonwealth in Centennial Park, Sydney.*

**33**

ther political reforms took power away from the provincial governments and gave it to a central administration. With regard to internal affairs, New Zealand had become fully self-governing.

For the most part, the New Zealanders were quite loyal to Great Britain; still, a desire for independence grew among them. They were beginning to see themselves as a country separate from England. Their government began to claim larger rights, such as the right to negotiate trade agreements with foreign states. Another step toward full independence came when Premier Richard Seddon proclaimed New Zealand's annexation of the Cook Islands, in 1901.

Premiers of all the self-governing British colonies gathered in London in 1907, at what was called the Imperial Conference. There New Zealand's premier, Sir Joseph Ward, urged the British government to change the official status of New Zealand from that of a "colony" to that of a "dominion."

British colonial secretary Victor Alexander Bruce, Lord Elgin, realized that his government could continue to count on New Zealand's support of its interests. After all, New Zealand had voluntarily sent troops to fight on Great Britain's side in the Boer War of 1899-1902 in southern Africa. Ward obtained ready agreement from Lord Elgin and other British officials, and when he returned home, the New Zealand parliament gave its formal approval to the change in name. It would be known from that time onward as the Dominion of New Zealand.

**Consequences**

Australia and New Zealand achieved independence from Great Britain through a process of negotiation rather than war. Ties of loyalty to the British Empire remained. Both countries fought on the side of Great Britain in World War I and World War II. Yet their military and economic ties gradually weakened throughout the twentieth century, as the empire disintegrated and new alliances were formed. In the twenty-first century, what special ties remain between Australia and New Zealand and their "mother country" are ties of language and culture, not governmental or economic control.

# Booth Invents Vacuum Cleaner

*The invention of the home vacuum cleaner enables people to remove dust and dirt from carpeting and other surfaces by means of suction.*

**What:** Engineering
**When:** 1901
**Where:** United States
**Who:**
H. CECIL BOOTH (1871-1955), the holder of the earliest patent on a power-driven vacuum cleaner in 1901
JAMES MURRAY SPANGLER (1848-1915), the developer of the first practical household vacuum cleaner in 1908
WILLIAM H. HOOVER (1849-1932), the founder, with Spangler, of the Hoover Company, a well-known manufacturer of vacuum cleaners
HIRAM H. HERRICK, the patentee of the first mechanical carpet sweeper in 1858
MELVILLE R. BISSELL (1843-1889), the inventor and marketer of the Bissell carpet sweeper in 1876

## Putting Electricity to Work

Electricity was becoming widely available in the United States by the end of the nineteenth century, and with it came many new gadgets that used this clean and relatively cheap source of power to run motors, to create heat or cold air, to provide light, and to perform many other tasks. Some were used in industry, others in the home. Among the many inventions that found application in both places was the vacuum cleaner.

As air is forced from one place to another through the blades of a fan, a partial vacuum is created. The vacuum causes more air to rush in, filling the void, as demonstrated when an electric fan is turned on, causing air to move through it. The same principle is applied in a vacuum cleaner. A fan inside the machine forces air through one side, at the same time drawing in more air through the other side, the nozzle of the machine. Loose dust and other particles are caught up in the rush of air through the system. Beater bars, brushes, and other devices loosen dirt that is then drawn into the cleaner, where it is trapped in a disposable bag. Once the bag is filled, it is replaced.

Vacuum cleaners are of two types: stand-up models that are pushed along, sucking up debris through a wide but thin suction opening at the front, and canister models with long, flexible hoses attached to small suction brushes or other accessories. Both are used in the home, and each is fairly lightweight.

The first home vacuum cleaner appeared in 1901. It was patented by H. Cecil Booth, who described it as a suction dust-removal machine. Booth's vacuum cleaner looked more like a lawn mower than the modern electric vacuum cleaner. It was large and awkward to use, with a 5-horsepower piston-driven motor mounted atop the front. Also, it was extremely heavy, an upright model without attachments.

It should be noted that large, mobile industrial vacuum cleaners were in use as early as the 1890's. These large suction cleaners were mounted on wagons pulled by horses and connected to the vacuum nozzle by long, flexible hoses that were strung through doorways and windows to reach the surfaces to be cleaned. These machines provided vacuum-cleaning services for schools, hotels, theaters, and other buildings subjected to high levels of foot traffic. There were also smaller, nonelectric piston-driven consumer models available before 1900 that created more work than they saved for the homemaker. As a result, they were never accepted. Others were hand-operated, with bellows for pumping air. There were also vacuum-based airblowers and some cleaners that ran on compressed air.

## New and Improved

Booth's vacuum cleaner, although clumsy and difficult to manage, provided the basis for the improved models that were to come. Ten years after the Booth patent was filed, newer, lighter, and more powerful electric motors began to appear. This meant that vacuum cleaners could be much more portable. In 1908, James Murray Spangler patented a much smaller and more popular version of the vacuum cleaner, later forming a partnership with his cousin, William H. Hoover. The new enterprise was called the Hoover Company.

In 1924, the first portable vacuum cleaner appeared, originating in Sweden. This early portable came with a flexible hose that could be attached for cleaning hard-to-reach places, furniture, draperies, and wall hangings. Other attachments followed over the years, some using powerful jets of air to loosen dust and dirt and others with special rake brushes to clean deep-pile shag rugs.

As electricity reached more and more homes during the early twentieth century, the vacuum cleaner became the carpet sweeper of choice. Modern home construction often included built-in vacuum systems that allowed the operator to plug a hose into receptacles placed strategically throughout the house. Because the electric motor and fan design can be made to operate at a wide range of suction levels, applications were found quickly for the vacuum cleaner in industry.

## Consequences

Unlike many home appliances that have come and gone or evolved into contraptions that scarcely resemble their earliest ancestors, modern vacuum cleaners are very similar in design to the Booth vacuum of 1901. They operate on the same principle of suction and are used for the same purpose: to eliminate dust and small particles from living spaces. Over the years, vacuum cleaners have increased in popularity as more and more floors have been covered with wall-to-wall carpeting, including those in many commercial and institutional settings.

As with most home appliances, continuous evolution of the technology has brought new uses for the vacuum cleaner. Today, it often takes the form of rechargeable handheld appliances that mount on walls and are used to clean drawers, kitchen cabinets, automobile interiors, and any number of other out-of-the-way places. Tiny vacuum cleaners are used in high-tech industrial applications for cleaning dust and particles from microcircuitry, and very large vacuum fans remove particles from the air in and around industrial manufacturing areas to protect the health of workers. Booth's vacuum cleaner was the forerunner of a long line of vacuum innovations that have made homes and workplaces cleaner and safer.

*Michael S. Ameigh*

# Dutch Researchers Discover Cause of Beriberi

Dutch physicians discovered that beriberi was caused by a nutritional deficiency in a diet of polished rice, an understanding that led to the development of vitamins.

**What:** Medicine; Health; Food science
**When:** 1901
**Where:** Javanese Medical School, Batavia, Java
**Who:**
CHRISTIAAN EIJKMAN (1858-1930), a Dutch physician and Nobel laureate who showed that a diet of polished rice contributed to beriberi
GERRIT GRIJNS (1865-1944), a Dutch physician and assistant to Eijkman who proposed that the deficiency of a protective substance in polished rice caused beriberi
ROBERT KOCH (1843-1910), a German physician, bacteriologist, and Nobel laureate who taught Eijkman about bacteriology

## "I Cannot"

Beriberi is a disease caused by a deficiency of vitamin $B_1$, or thiamine, in the diet. The name of the disease is Sinhalese for "I cannot": People afflicted with severe beriberi are too sick to do even the simplest things. Its symptoms include stiffness of the lower limbs, paralysis, severe pain, gradual breakdown of the muscles, anemia, mental confusion, enlargement of the heart, and death resulting from heart failure.

Beriberi has been known for thousands of years; reference to a disease with its symptoms is found in Chinese medical literature of 2700 B.C.E. Asia is the main beriberi zone because of the dietary combination of high tea consumption, use of rice as the main cereal grain, and the common practice of eating raw fish. Raw fish contains an enzyme that will destroy thiamine; tea also contains a chemical that thwarts the vitamin's actions.

In 1887, an important early observation on the dietary origins of beriberi was reported by Takagi Kanehiro, the director-general of the Japanese navy. Kanehiro became interested in beriberi because as many as a third of all Japanese sailors suffered symptoms of the disease. After careful study, Kanehiro became convinced that the cause of such cases was the standard naval diet of polished rice and fish, foods that were low in protein. He ordered the addition of red meat, vegetables, and whole-grain wheat to the naval diet, thus effecting a substantial reduction in the incidence of naval beriberi.

## Chickens with Beriberi

The work of Christiaan Eijkman and his assistant, Gerrit Grijns, began in the late nineteenth century. The Dutch East Indies (now Indonesia) was becoming a dangerous beriberi zone. The disease was spreading in epidemic proportions in the armed forces, prisons, and general population. For example, conditions were so bad in Javanese prisons that a jail sentence was considered to be almost a death sentence. In some afflicted people, cardiac insufficiency and massive edema (swelling) of the legs predominated. In many other patients, the progressive paralysis of the legs was the main problem.

Fortuitously, Eijkman observed a disease similar to beriberi in chickens; he named the disease *polyneuritis gallinarium*. In the course of studying the birds, Eijkman had them moved—and the disease disappeared. Eijkman soon discovered that, as a consequence of the move, the food given to the birds had changed in one important respect. Originally, the birds had been fed on leftover boiled white rice from the officers' ward

PhotoDisc

*Asian women plant rice in a field. The consumption of polished white rice rather than brown rice contributed to making beriberi a common affliction in Asia.*

in the military hospital. After their relocation, the chickens were given unpolished rice because the cook at the new site "refused to give any military rice to civilian fowl."

On the strength of this observation, Eijkman set out to determine whether the polished rice was the cause of the disease. He found this to be the case, because the feedings of unpolished rice cured the disease. Eijkman postulated that a chemical—perhaps a toxin from intestinal bacteria—was the actual causative agent. Although this concept was not correct, Eijkman's efforts began the scientific research that later showed that thiamine taken from the outer layer (the pericarp) of unpolished rice protected against beriberi.

In 1896, Eijkman left Java and returned to the Netherlands because of ill health. At that time, Grijns took over the study of beriberi. In 1901, Grijns proposed that the disease was a result of

the lack of some natural nutrient substance that was found in unpolished rice and in other foods. The work that Eijkman began and Grijns continued is thus often regarded as the basis for the modern theory of vitamins.

## Consequences

Eijkman's observations that beriberi was caused by excessive dietary use of polished rice and that it could be cured by a diet of unpolished rice were critically important to the understanding of nutrition. A sequence of events was set into motion that led to the development of many aspects of modern nutrition theory, to the evolution of biochemical explanations for several serious nutritional diseases, and to the virtual elimination of beriberi.

In 1912, the American biochemist Casimir Funk proposed that beriberi and other nutritional diseases such as scurvy and rickets were deficiency diseases caused by a lack of substances that were each a "vitamine." Funk admittedly coined the term because it "would sound well and serve as a catch word." Soon the term was both accepted and shortened to its current spelling. Subsequently, it was found that there were several types of isolated vitamins, such as A, B, C, and D, named in order of their discovery. A key event was the preparation of a pure antiberiberi factor, thiamine (or vitamin $B_1$), and determination of its structure by Robert R. Williams, starting in 1935. This effort led to the commercial synthesis of the vitamin by pharmaceutical companies and to its wide distribution.

The availability of the pure vitamin allowed examination of its metabolic effects. Researchers soon discovered that thiamine was an essential coenzyme that was required for the biological action of many important enzymes, or biological catalysts. The lack of function of these catalysts was shown eventually to cause beriberi. Similar results with other vitamins led to the understanding of the roles of vitamins and to the establishment of the basic precepts of nutrition.

Despite the fact that the need for thiamine has been known for many years, a significant percentage of the public is believed to eat barely enough thiamine to meet dietary needs. As a consequence, many foods low in thiamine (such as white bread, breakfast cereals, pasta, and white flour) have been enriched with the vitamin as a health precaution.

*Sanford S. Singer*

# Elster and Geitel Demonstrate Natural Radioactivity

*Julius Elster and Hans Friedrich Geitel pioneered research in radioactivity.*

**What:** Earth science
**When:** 1901
**Where:** Wolfenbüttel, Germany
**Who:**
JULIUS ELSTER (1854-1920), a German physicist
HANS FRIEDRICH GEITEL (1855-1923), a German physicist
ANTOINE-HENRI BECQUEREL (1852-1908), a French physicist
ERNEST RUTHERFORD (1871-1937), an English physicist
MARIE CURIE (1867-1934), a French physicist
PIERRE CURIE (1859-1906), a French physicist
HANS GEIGER (1882-1945), a German physicist

## Discovering Radioactivity

Radioactivity was first observed in 1896 by the French physicist Antoine-Henri Becquerel. He found that uranium ore emitted radiation strong enough to blacken covered photographic plates and to discharge a charged electroscope, a device for detecting the presence of electricity. In 1898, the French physicists Pierre and Marie Curie announced the discovery of two new radiation-emitting elements, polonium and radium. In 1903, Ernest Rutherford, an English physicist, made the discovery that radiation from such elements was composed of three different kinds of energetic rays: alpha rays with a positive charge, which were ionized helium atoms; beta particles with a negative charge, which were high-energy electrons; and gamma rays with no charge, which were high-energy photons.

Natural radioactivity is the property possessed by roughly fifty elements—such as radium, thorium, and uranium—of spontaneously emitting alpha or beta rays, and sometimes gamma rays, by the disintegration of the nuclei of atoms. During the disintegration process, alpha or beta particles are emitted, and atoms of a new element are formed. This new element is lighter than its predecessor and possesses different chemical and physical properties. Each radioactive element also has its own predictable rate of disintegration.

## Measuring Radioactivity

When a charged particle passes through matter, it causes excitation (the raising of electrons to higher energy states) and ionization (loss of electrons) of the molecules of the material. This property is the basis of nearly all the instruments used for the detection of such particles and the measurement of their energies. For example, alpha particles produce tiny flashes of light when passing through zinc sulfide.

Julius Elster and Hans Friedrich Geitel discovered a method of counting alpha particles based on this fact. They prepared a screen by dusting small crystals of zinc sulfide containing a very small amount of copper impurity on a slip of glass. Each alpha particle produces one tiny flash, so the number of alpha particles that fell on a detecting screen per unit of time was given directly by the number of flashes counted per unit of time; the counting was done by a microscope. This was a precise but difficult method that worked well for counting alpha particles in the presence of other radiations because the zinc sulfide screen was comparatively insensitive to beta and gamma rays.

During this period of major discovery, Elster and Geitel were actively conducting research on radioactivity in rocks, springs, and air. Elster was on the faculty of the Gymnasium in Wolfenbüttel,

Germany, from 1881 to 1919. Geitel was a teacher at Strosse Schule in Wolfenbüttel. One of Elster and Geitel's first collaborations was in 1880, when they carried out a systematic study of the electrification of hot bodies. From this early work and their perfection of instruments for detection and measurement, Elster and Geitel moved on to a study of other sources of radiation. A basic fact is that the earth is heated from within by the energy released when uranium, thorium, and other radioactive elements undergo natural nuclear disintegration. As such disintegration occurs, these elements find their way into rocks, soil, water, and air. The total energy released through nuclear disintegration over the earth's history is more than one hundred calories per gram of earth material.

In 1901, Elster and Geitel discovered that it is possible to produce excited radioactivity from the atmosphere simply by exposing a highly charged wire to a negative potential in the atmosphere for many hours. They found that the radioactivity could dissolve with exposure to acids and that the wire would be left unchanged, a discovery that had an important bearing on the theory of atmospheric electricity. In 1903, the two scientists discovered a property of alpha rays that proved of great importance in radioactive measurement. If a screen coated with small crystals of phosphorescent zinc sulfide is exposed to alpha rays, a brilliant luminosity is observed. Further, the study of penetrating radiation had its origin in their observations that there was a definite transport of electrical charge to an insulated system in the presence of radioactivity, even after precautions had been taken to reduce electrical leakage over the insulators.

Elster and Geitel's discoveries—though less heralded than those of some of their contemporaries—were a major contribution to nuclear physics and particularly to the understanding and detection of the omnipresence of radioactive elements in the environment.

## Consequences

The discoveries of Elster and Geitel significantly affected the field of geology. In natural radioactivity, the unstable nucleus emits several types of high-energy particles and also releases energy in the form of electromagnetic waves similar to light energy. This process results in radioactive decay, whereby a naturally radioactive atom is changed to an atom of another element. This means that the number of atoms of the radioactive element decreases with the passage of time. If the rate of disintegration is known, then measuring the amounts of the parent and daughter elements will reveal the age of the mineral containing the parent. This is the principle of radioactive dating.

Radiation detection also made it possible to locate uranium ores. The early work by Elster, Geitel, Becquerel, the Curies, and Rutherford resulted in the development of the most widely used radiation-detection instrument, the Geiger counter. The Geiger counter consists of a metal cylinder enclosed in a glass tube filled with a gas at low pressure. The entrance of a charged particle ionizes the gas enough to cause a current flow, a momentary pulse of voltage. Geiger counters are equipped with special cylinders for counting alpha, beta, and gamma particles. The counter, invented by Johannes Geiger in 1913 and perfected in 1926 by Walther Müller, is still widely used.

The early work of Elster and Geitel in the second decade of the twentieth century led to the discovery that the atomic nucleus is a source of large quantities of energy. Humans have learned to use nuclear energy in many different ways: for medical therapy, for power in industry, for energy to propel submarines and ships, for research in biological sciences, and for weapons of great destruction.

*Earl G. Hoover*

# First Synthetic Vat Dye Is Synthesized

*The synthesis of vat dyes ended centuries of efforts to obtain the brilliant colors displayed in nature.*

**What:** Chemistry
**When:** 1901
**Where:** Ludwigshafen, Germany
**Who:**
RENÉ BOHN (1862-1922), a synthetic organic chemist
KARL HEUMANN (1850-1894), a German chemist who taught Bohn
ROLAND SCHOLL (1865-1945), a Swiss chemist who established the correct structure of Bohn's dye
AUGUST WILHELM VON HOFMANN (1818-1892), an organic chemist
SIR WILLIAM HENRY PERKIN (1838-1907), an English student in Hofmann's laboratory

## Synthesizing the Compounds of Life

From prehistoric times until the mid-nineteenth century, all dyes were derived from natural sources, primarily plants. Among the most lasting of these dyes were the red and blue dyes derived from alizarin and indigo.

The process of making dyes took a great leap forward with the advent of modern organic chemistry in the early years of the nineteenth century. At the outset, this branch of chemistry, dealing with the compounds of the element carbon and associated with living matter, hardly existed, and synthesis of carbon compounds was not attempted. Considerable data had accumulated showing that organic, or living, matter was basically different from the compounds of the nonliving mineral world. It was widely believed that although one could work with various types of organic matter in physical ways and even analyze their composition, they could be produced only in a living organism.

Yet, in 1828, the German chemist Friedrich Wöhler found that it was possible to synthesize the organic compound urea from mineral compounds. As more chemists reported the successful preparation of compounds previously isolated only from plants or animals, the theory that organic compounds could be produced only in a living organism faded.

One field ripe for exploration was that committed to exploiting the uses of coal tar. Here, August Wilhelm von Hofmann was an active worker. He and his students made careful studies of this complex mixture. The high-quality stills they designed allowed for the isolation of pure samples of important compounds for further study.

Of greater importance was the collection of able students Hofmann attracted. Among them was Sir William Henry Perkin, who is regarded as the founder of the dyestuffs industry. In 1856, Perkin undertook the task of synthesizing quinine (a bitter crystalline alkaloid used in medicine) from a nitrogen-containing coal tar material called toluidine. Luck played a decisive role in the outcome of his experiment. The sticky compound Perkin obtained contained no quinine, so he decided to investigate the simpler related compound aniline. A small amount of the impurity toluidine in his aniline gave Perkin the first synthetic dye, Mauveine.

## Searching for Structure

From this beginning, the great dye industries of Europe, particularly Germany, grew. The trial-and-error methods gave way to more systematic searches as the structural theory of organic chemistry was formulated.

As the twentieth century began, great progress had been made, and German firms dominated the industry. Badische Anilin-und Soda-Fabrik (BASF) was incorporated at Ludwigshafen in 1865 and undertook extensive explorations of both alizarin and indigo. A chemist, René Bohn, had made important discoveries in 1888, which helped the company recover lost ground in the alizarin field. In 1901, he undertook the synthesis of a dye he hoped would combine the desirable attributes of both alizarin and indigo.

As so often happens in science, nothing like the expected occurred. Bohn realized that the beautiful blue crystals that resulted from his synthesis represented a far more important product. Not only was this the first synthetic vat dye, Indanthrene, ever prepared, but also, by studying the reaction at higher temperature, a useful yellow dye, Flavanthrone, could be produced.

The term "vat dye" is used to describe a method of applying the dye, but it also serves to characterize the structure of the dye, because all currently useful vat dyes share a common unit. One fundamental problem in dyeing relates to the extent to which the dye is water-soluble. A beautifully colored molecule that is easily soluble in water might seem attractive given the ease with which it binds with the fiber; however, this same solubility will lead to the dye's rapid loss in daily use.

Vat dyes are designed to solve this problem by producing molecules that can be made water-soluble, but only during the dyeing or vatting process. This involves altering the chemical structure of the dye so that it retains its color throughout the life of the cloth.

By 1907, Roland Scholl had showed unambiguously that the chemical structure proposed by Bohn for Indanthrene was correct, and a major new area of theoretical and practical importance was opened for organic chemists.

## Consequences

Bohn's discovery led to the development of many new and useful dyes. The list of patents issued in his name fills several pages in *Chemical Abstracts* indexes.

The true importance of this work is to be found in a consideration of all synthetic chemistry, which may perhaps be represented by this particular event. More than two hundred

dyes related to Indanthrene are in commercial use. The colors represented by these substances are a rainbow making nature's finest hues available to all. The dozen or so natural dyes have been synthesized into more than seven thousand superior products through the creativity of the chemist.

Despite these desirable outcomes, there is doubt whether there is any real benefit to society from the development of new dyes. This doubt is the result of having to deal with limited natural resources. With so many urgent problems to be solved, scientists are not sure whether to search for greater luxury. If the field of dye synthesis reveals a single theme, however, it must be to expect the unexpected. Time after time, the search for one goal has led to something quite different—and useful.

*K. Thomas Finley*

*A worker tests a strip of fabric to make sure that the dyes do not run or fade.*

# Ivanov Develops Artificial Insemination

> *Ilya Ivanovich Ivanov developed practical techniques for the artificial insemination of farm animals, revolutionizing livestock breeding practices throughout the world.*

**What:** Agriculture
**When:** 1901
**Where:** Russia
**Who:**
Lazzaro Spallanzani (1729-1799), an Italian physiologist
Ilya Ivanovich Ivanov (1870-1932), a Soviet biologist
R. W. Kunitsky, a Soviet veterinarian

## Reproduction Without Sex

The tale is told of a fourteenth-century Arabian chieftain who sought to improve his mediocre breed of horses. Sneaking into the territory of a neighboring hostile tribe, he stimulated a prize stallion to ejaculate into a piece of cotton. Quickly returning home, he inserted this cotton into the vagina of his own mare, who subsequently gave birth to a high-quality horse. This may have been the first case of "artificial insemination," the technique by which semen is introduced into the female reproductive tract without sexual contact.

The first scientific record of artificial insemination comes from Italy in the 1770's. Lazzaro Spallanzani was one of the foremost physiologists of his time, well known for having disproved the theory of spontaneous generation, which states that living organisms can spring "spontaneously" from lifeless matter. There was some disagreement at that time about the basic requirements for reproduction in animals. It was unclear if the sex act was necessary for an embryo to develop, or if it was sufficient that the sperm and eggs come into contact. Spallanzani began by studying animals in which union of the sperm and egg normally takes place outside the body of the female. He stimulated males and females to release their sperm and eggs, then mixed these sex cells in a glass dish. In this way, he produced young frogs, toads, salamanders, and silkworms.

Next, Spallanzani asked whether the sex act was also unnecessary for reproduction in those species in which fertilization normally takes place inside the body of the female. He collected semen that had been ejaculated by a male spaniel and, using a syringe, injected the semen into the vagina of a female spaniel in heat. Two months later, she delivered a litter of three pups, which bore some resemblance to both the mother and the male that had provided the sperm.

It was in animal breeding that Spallanzani's techniques were to have their most dramatic application. In the 1880's, an English dog breeder, Sir Everett Millais, conducted several experiments on artificial insemination. He was interested mainly in obtaining offspring from dogs that would not normally mate with one another because of difference in size. He followed Spallanzani's methods to produce a cross between a short, low, basset hound and the much larger bloodhound.

## Long-Distance Reproduction

Ilya Ivanovich Ivanov was a Soviet biologist who was commissioned by his government to investigate the use of artificial insemination on horses. Unlike previous workers who had used artificial insemination to get around certain anatomical barriers to fertilization, Ivanov began the use of artificial insemination to reproduce thoroughbred horses more effectively. His assistant in this work was the veterinarian R. W. Kunitsky.

In 1901, Ivanov founded the Experimental Station for the Artificial Insemination of Horses. As its director, he embarked on a series of experiments to devise the most efficient techniques for breeding these animals. Not content with the demonstration that the technique was scientifi-

44

cally feasible, he wished to ensure further that it could be practiced by Soviet farmers.

If sperm from a male were to be used to impregnate females in another location, potency would have to be maintained for a long time. Ivanov first showed that the secretions from the sex glands were not required for successful insemination; only the sperm itself was necessary. He demonstrated further that if a testicle were removed from a bull and kept cold, the sperm would remain alive. More useful than preservation of testicles would be preservation of the ejaculated sperm. By adding certain salts to the sperm-containing fluids, and by keeping these at cold temperatures, Ivanov was able to preserve sperm for long periods.

Ivanov also developed instruments to inject the sperm, to hold the vagina open during insemination, and to hold the horse in place during the procedure. In 1910, Ivanov wrote a practical textbook with technical instructions for the artificial insemination of horses. He also trained some three hundred veterinary technicians in the use of artificial insemination, and the knowledge he developed quickly spread throughout the Soviet Union. Artificial insemination became the major means of breeding horses.

Until his death in 1932, Ivanov was active in researching many aspects of the reproductive biology of animals. He developed methods to treat reproductive diseases of farm animals and refined methods of obtaining, evaluating, diluting, preserving, and disinfecting sperm. He also began to produce hybrids between wild and domestic animals in the hope of producing new breeds that would be able to withstand extreme weather conditions better and that would be more resistant to disease. His crosses included hybrids of ordinary cows with aurochs, bison, and yaks, as well as some more exotic crosses of zebras with horses.

Ivanov also hoped to use artificial insemination to help preserve species that were in danger of becoming extinct. In 1926, he led an expedition to West Africa to experiment with the hybridization of different species of anthropoid apes.

## Consequences

The greatest beneficiaries of artificial insemination have been dairy farmers. Some bulls are able to sire genetically superior cows that produce exceptionally large volumes of milk. Under natural conditions, such a bull could father at most a few hundred offspring in its lifetime. Using artificial insemination, a prize bull can inseminate ten to fifteen thousand cows each year. Since frozen sperm may be purchased through the mail, this also means that dairy farmers no longer need to keep dangerous bulls on the farm. Artificial insemination has become the main method of reproduction of dairy cows.

In the 1980's, artificial insemination gained added importance as a method of breeding rare animals. Animals kept in zoo cages, animals that are unable to take part in normal mating, may still produce sperm that can be used to inseminate a female artificially. Some species require specific conditions of housing or diet for normal breeding to occur, conditions not available in all zoos. Such animals can still reproduce using artificial insemination.

*Judith R. Gibber*

# Kipping Discovers Silicones

> *Frederic Stanley Kipping discovered silicones, which nearly fifty years later became widely used because of their unique properties.*

**What:** Chemistry
**When:** 1901-1904
**Where:** Nottingham, England
**Who:**
FREDERIC STANLEY KIPPING (1863-1949), a Scottish chemist and professor
EUGENE G. ROCHOW (1909- ), an American research chemist
JAMES FRANKLIN HYDE (1903-1999), an American organic chemist

## Synthesizing Silicones

Frederic Stanley Kipping, in the first four decades of the twentieth century, made an extensive study of the organic (carbon-based) chemistry of the element silicon. He had a distinguished academic career and summarized his silicon work in a lecture in 1937 ("Organic Derivatives of Silicon"). Since Kipping did not have available any naturally occurring compounds with chemical bonds between carbon and silicon atoms (organosilicon compounds), it was necessary for him to find methods of establishing such bonds. The basic method involved replacing atoms in naturally occurring silicon compounds with carbon atoms from organic compounds.

While Kipping was probably the first to prepare a silicone and was certainly the first to use the term "silicone," he did not pursue their commercial possibilities. Yet his careful experimental work was a valuable starting point for all subsequent workers in organosilicon chemistry, including those who later developed the silicone industry.

On May 10, 1940, chemist Eugene G. Rochow of the General Electric (GE) Company's corporate research laboratory in Schenectady, New York, discovered that methyl chloride gas, passed over a heated mixture of elemental silicon and copper, reacted to form compounds with silicon-carbon bonds. Kipping had shown that these silicon compounds react with water to form silicones.

The importance of Rochow's discovery was that it opened the way to a continuous process that did not consume expensive metals, such as magnesium, or flammable ether solvents, such as those used by Kipping and other researchers. The copper acts as a catalyst, and the desired silicon compounds are formed with only minor quantities of by-products. This "direct synthesis," as it came to be called, is now done commercially on a large scale.

## Silicone Structure

Silicones are examples of what chemists call "polymers." Basically, a polymer is a large molecule made up of many smaller molecules that are linked together. At the molecular level, silicones consist of long, repeating chains of atoms. In this molecular characteristic, silicones resemble plastics and rubber.

Silicone molecules have a chain composed of alternate silicon and oxygen atoms. Each silicon atom bears two organic groups as substituents, while the oxygen atoms serve to link the silicon atoms into a chain. The silicon-oxygen backbone of the silicones is responsible for their unique and useful properties, such as the ability of a silicone oil to remain liquid over an extremely broad temperature range and to resist oxidative and thermal breakdown at high temperatures.

A fundamental scientific consideration with silicone, as with any polymer, is to obtain the desired physical and chemical properties in a product by closely controlling its chemical structure and molecular weight. Oily silicones with thousands of alternating silicon and oxygen atoms have been prepared. The average length of the molecular chain determines the flow characteristics (viscosity) of the oil. In samples with very long chains, rubberlike elasticity can be achieved by cross-linking the silicone chains in a controlled manner and adding a filler such as silica. High degrees of cross-linking could produce a hard, intractable material instead of rubber.

The action of water on the compounds produced from Rochow's direct synthesis is a rapid method of obtaining silicones but does not provide much control of the molecular weight. Further development work at GE and at the Dow-Corning Company showed that the best procedure for controlled formation of silicone polymers involved treating the crude silicones with acid to produce a mixture from which high yields of an intermediate called "D4" could be obtained by distillation. The intermediate D4 could be polymerized in a controlled manner by use of acidic or basic catalysts. Wilton I. Patnode of GE and James F. Hyde of Dow-Corning made important advances in this area. Hyde's discovery of the use of traces of potassium hydroxide as a polymerization catalyst for D4 made possible the manufacture of silicone rubber, which is the most commercially valuable of all the silicones today.

*Frederic Stanley Kipping.*

## Consequences

Although Kipping's discovery and naming of the silicones occurred from 1901 to 1904, the practical use and impact of silicones started in 1940, with Rochow's discovery of direct synthesis.

Production of silicones in the United States came rapidly enough to permit them to have some influence on military supplies for World War II (1939-1945). In aircraft communication equipment, extensive waterproofing of parts by silicones resulted in greater reliability of the radios under tropical conditions of humidity, where condensing water could be destructive. Silicone rubber, because of its ability to withstand heat, was used in gaskets under high-temperature conditions, in searchlights, and in the engines on B-29 bombers. Silicone grease ap-

plied to aircraft engines also helped to protect spark plugs from moisture and promote easier starting.

After World War II, the uses for silicones multiplied. Silicone rubber appeared in many products from caulking compounds to wire insulation to breast implants for cosmetic surgery. Silicone rubber boots were used on the moon walks where ordinary rubber would have failed.

Silicones in their present form owe much to years of patient developmental work in industrial laboratories. Basic research, such as that conducted by Kipping and others, served to point the way and catalyzed the process of commercialization.

*John R. Phillips*

# Banach Sets Stage for Abstract Spaces

*On the basis of his observation that many mathematical structures had features in common with ordinary three-dimensional structures, Stefan Banach constructed abstract spaces of tremendous beauty and utility.*

**What:** Mathematics
**When:** 1901-1932
**Where:** Poland
**Who:**
BERNHARD RIEMANN (1826-1866), a German mathematician
DAVID HILBERT (1862-1943), a German mathematician
STEFAN BANACH (1892-1945), a Polish mathematician

## Classical Geometries

A one-dimensional space can be visualized as a straight line. To locate points on that line, one selects a point on the line and calls it the "origin." Other points can be located relative to the origin. One direction from the origin is labeled positive; the other, negative. A point on the line is specified by means of its distance from the origin and by determining whether it is positive or negative. The resulting value, $x$, is called the "coordinate" of the point.

A two-dimensional space is a plane. Points on a plane are located by choosing an origin and two lines that intersect in the origin and are at right angles. On a plane, it takes two coordinates ($x$ and $y$) to locate a point relative to the origin. The point is found by moving $x$ units on the first axis and then $y$ units parallel to the second axis.

A point in three-dimensional space is located in a similar fashion, by adding another axis and using three coordinates ($x$, $y$, and $z$). Each time another dimension is added, another coordinate is required.

Early in the twentieth century, Albert Einstein and Hermann Minkowski decided to add time as a fourth dimension, and they developed the space-time system, which uses four coordinates ($x$, $y$, $z$, and $t$). A space-time-pressure-temperature system would be a six-dimensional space with the coordinates $x$, $y$, $z$, $t$, $p$, and $T$. There are also different ways to define the distance between points. In a plane, for example, one normally takes the length of the straight line connecting two points to be the distance between those two points. A space with this distance is called "Euclidean," named for the Greek mathematician Euclid. A driver traveling through a city, however, will follow a "non-Euclidean" route to travel a certain distance. Any particular type of space is characterized by its system of coordinates and its method of measuring the distance between points.

## Abstract Spaces

An important step in extending the concept of a space was taken in 1901 by David Hilbert when he realized that the classical approach to geometry was based on assumptions about the relationships that existed among fundamental elements such as points, lines, and planes. He even constructed an infinite dimensional analogue of the Euclidean space described above, which is an example of what is now known as a Hilbert space.

In the succeeding years, mathematicians experimented with spaces of more than three dimensions, which were characterized abstractly in terms of coordinates. Many different ways of measuring distances were also introduced. These spaces of four or more dimensions cannot be visualized in the same way that ordinary space can, but they can be worked with in the same way and can be discussed in similar language. Thus, a "point" is simply a finite or infinite set of numbers. One can speak of the distance between points or the neighborhood of a point. The picture in one's mind is actually two- or three-dimensional, but the structure and formalism are more general than such a conception would indicate.

This process of moving from a concrete situation to one of increasing generality is an example of mathematical abstraction. The range of application is increased enormously when such a process is used. In elementary arithmetic, one starts by adding concrete objects (for example, adding two apples and two apples to get four apples). This process is not, however, restricted to apples (or any other type of object); instead, the general step is $2 + 2 = 4$. This same procedure is used with abstract spaces.

In 1932, all these developments were brought together by Stefan Banach. The seeds of many of his ideas are contained in the work of Bernhard Riemann, which profoundly influenced Albert Einstein. Banach began with a set and then assumed that there was an important class of spaces, now called Banach spaces, about which many questions could be posed and answered. Formulating the theory in an abstract manner made it possible to find connections between seemingly unrelated problems and to use the language of geometry in many areas of mathematics, physics, and engineering.

## Consequences

One of the primary characteristics of modern mathematics has been its abstract nature. This has led to its ability to treat ever broader classes of problems. Nowhere is this fact more clear than in its treatment of abstract spaces. The general approach to such spaces has led to a unification of many areas of science. Engineering problems are routinely solved in this framework. Vast areas of modern technology are based on it. Physicists use these ideas in their treatment of problems in classical mechanics, electromagnetic theory, and atomic physics. Mathematicians obtain solutions to problems in a Hilbert- or Banach-space setting on a daily basis.

Although the abstract-spaces approach is very powerful, its abstraction has made it increasingly difficult—if not impossible—for most people to understand and appreciate its results. It will be a challenge for succeeding generations to make abstract results, which have so profoundly affected the lives of modern humans, accessible to the general public.

*Ronald B. Guenther*

# Insular Cases Determine Status of U.S. Possessions Overseas

*Supreme Court decisions about whether tariff laws were applicable to overseas possessions of the United States helped clarify the status of such territories as Puerto Rico, the Philippines, and the Virgin Islands.*

**What:** Law
**When:** May 27, 1901
**Where:** Washington, D.C.
**Who:**
HENRY BILLINGS BROWN (1836-1913), associate justice of the United States from 1890 to 1906
EDWARD DOUGLASS WHITE (1845-1921), associate justice of the United States from 1894 to 1910
MELVILLE WESTON FULLER (1833-1910), chief justice of the United States from 1888 to 1910

### New Colonies, New Problems

From its founding through the nineteenth century, the United States expanded across the North American continent. Through the annexation and settlement of new territories, the nation gradually added states. The problems and questions that arose often had to do with states' rights versus the rights of the federal government.

In 1898, however, the United States suddenly became a colonial power. Through a joint resolution of Congress, Hawaii was annexed; at the end of the Spanish-American War, Spain ceded Puerto Rico, Guam, and the Philippine Islands to the United States as well.

With these new overseas possessions came new questions. Was it right for the United States to acquire overseas territories at all? If it was constitutionally and ethically allowed, how should such territories be governed? Did the rights guaranteed to Americans under the Constitution apply to these lands and their people? The Constitution required, for example, that all taxes, duties, and imposts be uniform throughout the United States. Did that apply to overseas colonies as well?

Those who opposed American imperialism adopted a slogan, "The Constitution follows the flag." Wherever the United States governed, they argued, the Constitution had to be in force. If a territory had no prospect of becoming a full-fledged state, it should not be annexed. Opponents of imperialism—mainly Democrats—believed that under the Constitution, the federal government did not have the power to take over land that would be held permanently as a colony. Colonialism, they said, was in contradiction to the Declaration of Independence itself.

Supporters of colonial expansion—mostly Republicans—said that the United States, by winning the Spanish-American War, had already become a world power. Overseas possessions would strengthen the nation's military security and give it more control of the seas. Furthermore, the colonies could help the United States gain economic strength.

### The Court's Decisions

The Supreme Court became part of the debate when it heard the "Insular Tariff Cases" for six days in January, 1901, and announced its opinions on May 27, 1901.

*DeLima v. Bidwell* had to do with whether general tariff laws applied to imports from Puerto Rico once the treaty with Spain had been signed and Puerto Rico had become a U.S. territory. Should the normal tariffs on overseas imports be levied on goods coming from Puerto Rico? In stating the majority opinion, Justice Henry Billings Brown said, "We are therefore of the opinion that at the time these duties were levied, Puerto Rico was not a foreign country within the meaning of the tariff laws but a territory of the United States, that the duties were illegally exacted and that the plaintiffs are entitled to recover them back." Four of the nine justices, however, dissented from this opinion, insisting that Puerto Rico was not "a part" of the United States, and so the tariff laws still applied.

The companion case, *Downes v. Bidwell*, had to do with whether a special tariff law applicable only to Puerto Rican goods (the Foraker Act) was valid. Though the decision in this case seems to contradict the *DeLima v. Bidwell* decision, Justice Brown again delivered the majority opinion. He took up not only the question of the tariff but also the larger issue of the constitutionality of colonization. The courts, he said, should be careful not to put obstacles in the way of the "American Empire." It might be desirable to annex distant possessions "inhabited by alien races, differing from us in religion, customs, laws, methods of taxation and modes of thought." Treating such possessions differently from states, he said, was not forbidden in the Constitution. He went on to rule that Puerto Rico belonged to the United States but was not "a part" of the United States.

*Henry Billings Brown.*

Library of Congress

The Foraker Act, then, was constitutional.

Justice Edward Douglass White wrote an opinion in support of Brown's. In this opinion he explained his theory of "incorporation," which later came to be generally accepted by the Court. Though Puerto Rico did belong to the United States, he said, it had not been "incorporated" into the nation, so the Constitution did not apply to it.

Chief Justice Melville Weston Fuller wrote a dissenting opinion, supported by three other justices. The Constitution did follow the flag, he claimed, and Puerto Rico should not be subject to special taxes or tariffs.

Later cases took up questions of the rights of criminal defendants in the territories. Was the Bill of Rights in force for the inhabitants of overseas colonies? In *Hawaii v. Mankichi* (1903), Justice Brown said that it was not in force for Hawaii at that time. In Hawaii, then, a defendant could be convicted by only nine "guilty" votes out of a jury of twelve. On the basis of his "incorporation" theory, Justice White agreed.

In 1905, the Court decided that Alaska had been "incorporated" into the United States to the extent that its people did enjoy constitutional guarantees. In the *Rasmussen v. United States* case, then, the Court reversed the conviction of an Alaskan defendant found guilty on the basis of a six-member jury. The Court continued to make these kinds of distinctions. Though Congress had granted U.S. citizenship to Puerto Ricans, the Court ruled in 1918 (*Puerto Rico v. Tapia*) that these citizens were not necessarily protected by the Constitution.

## Consequences

The Insular Cases, in effect, gave Supreme Court approval to the U.S. pursuit of overseas empire. Until they were "incorporated," these colonies and their people were not guaranteed constitutional rights.

Alaska and Hawaii, the two territories that were found by the Supreme Court to have been incorporated, became states of the Union in 1959 and 1960, respectively. The Philippines, Puerto Rico, and the Virgin Islands never were incorporated, though their inhabitants were declared U.S. citizens.

*James J. Bolner*

**51**

# Theodore Roosevelt Becomes U.S. President

> *Taking up the United States presidency after the fatal shooting of William McKinley, Theodore Roosevelt brought new energy and prestige to the office.*

**What:** National politics
**When:** September 14, 1901
**Where:** Mainly Washington, D.C.
**Who:**
THEODORE ROOSEVELT (1858-1919),
    president of the United States from
    1901 to 1909

## Inheriting the Office

In Buffalo, New York, on September 6, 1901, an anarchist named Leon Czolgosz shot and seriously wounded William McKinley, president of the United States. Rushing to the side of the stricken president, Vice President Theodore Roosevelt was reassured that though the wound was severe, McKinley would probably recover.

Thinking to communicate this reassurance to the American people through his actions, Roosevelt left for a mountain-climbing expedition in New York's Adirondacks. While still in the mountains, he received a report by special messenger that the president was dying. By buckboard and train, Roosevelt hurried back to Buffalo but was unable to reach the city before McKinley's death. On the afternoon of September 14, 1901, Theodore Roosevelt took the oath of office as president of the United States.

It was a dramatic beginning for a dramatic career. As he faced his new responsibilities, Roosevelt's situation was not easy. He had risen to the presidency through a tragedy, not by the vote of the people. He was a Republican, but the Republican Party did not by any means promise its full support. In fact, Senator Marcus A. Hanna of Ohio, who had been President McKinley's chief supporter, seemed to be planning to challenge Roosevelt for the Republican presidential nomination in 1904. Within the Senate, four Republicans—Nelson W. Aldrich of Rhode Island, William B. Allison of Kansas, Orville H. Platt of Connecticut, and John C. Spooner of Wisconsin—had come to exercise a veto power over proposals brought before Congress. These men, known as the "oligarchs," expected to keep their special privileges during the Roosevelt presidency.

There were larger problems within the country. Industry had grown rapidly during the nineteenth century, and so had cities. Unjust labor practices and the problems of city life were placing new burdens on all levels of government. At the state and local levels, many reform-minded political leaders had come to power. They were demanding help from the federal government.

## Meeting the Challenges

The president attacked his problems with skill and energy. Within the Republican Party, he was able to gain allies and undercut Senator Hanna's power. He made friends with the four Senate "oligarchs," so that they came to be his supporters rather than his enemies.

Roosevelt also identified himself with the reformers who were calling for changes in the practices of big industry. In March, 1902, he ordered the U.S. attorney general to file suit against the Northern Securities Company, one of the corporate giants of the railroad industry. Later that year, when an anthracite coal strike threatened to leave city dwellers without fuel during the winter months, Roosevelt intervened personally to help resolve the dispute.

Actions such as these helped establish Roosevelt's political authority within the nation. Now he was able to give more attention to problems overseas. Because of his vision for the building of a canal that would link the Atlantic and Pacific oceans, he backed the actions of a revolutionary junta (group of leaders) that led Panama in seceding from Colombia. The way was cleared for the construction of the Panama Canal.

The Monroe Doctrine, formed by President James Monroe in the early 1800's, specified that the United States would resist European governments that tried to interfere in the affairs of independent nations of North and South America. Roosevelt now added what became known as the Roosevelt Corollary to the Monroe Doctrine: He claimed for the United States the right to exercise an international police power in Latin America in order to maintain the status quo and punish any government that did not conduct its affairs with decency. The Roosevelt Corollary was officially rejected by President Herbert Hoover in 1930, but to a large extent U.S. presidents have continued to follow its principles in dealing with Latin American states.

Roosevelt acted with similar decisiveness elsewhere in the world. In 1905, he took on the role of an arbitrator to help bring an end to the Russo-Japanese War. When his effort proved successful, he was awarded the Nobel Peace Prize.

### Consequences

Early in Theodore Roosevelt's presidency, he showed great skill in dealing with powerful senators and with reform-minded politicians across the nation. He combined stubborn confidence with an ability to negotiate—qualities that served him well as he began to address international problems. Roosevelt was one of the first U.S. presidents to claim a prominent place for the United States—and for himself—in world affairs.

*Theodore Roosevelt.*

Library of Congress

Roosevelt had a dramatic flair that captured the imagination of the American people. By focusing attention on his office and himself, he brought new power and prestige to the presidency and made the White House the focal point of the U.S. government.

*Rex O. Mooney*

# Marconi Receives First Transatlantic Radio Transmission

> *Guglielmo Marconi received the first wireless telegraph signal sent across the Atlantic, showing that long-distance communication through open space was possible.*

**What:** Communications
**When:** December 12, 1901
**Where:** Poldhu, England, and St. John's, Newfoundland, Canada
**Who:**
GUGLIELMO MARCONI (1874-1937), an Italian scientist and inventor
JAMES CLERK MAXWELL (1831-1879), a Scottish scientist
SAMUEL F. B. MORSE (1791-1872), an American portrait painter and scientist
JOSEPH HENRY (1797-1878), an American physicist
HEINRICH HERTZ (1857-1894), a German physicist

## Telegraphic History

On December 12, 1901, Guglielmo Marconi was in St. John's, Newfoundland, Canada, to receive a Morse code signal to be transmitted to him from Poldhu, Cornwall, England, a distance of 3,440 kilometers across the Atlantic Ocean. Humankind was about to enter the age of worldwide electronic communication.

The idea of electric telegraphy had been around since the nineteenth century, and for years people had worked to send electric signals farther and farther through wire cables. Marconi's achievement was based on the work of many other people.

Samuel F. B. Morse was an American artist and inventor who had sailed to Europe in 1832 to study art. On the way home, a fellow passenger and American, the chemist Charles T. Jackson, introduced Morse to electromagnetism, which is the basis of the telegraph. In 1835, Morse invented a code, now known as the "Morse code," and a device that would send the code over wire cables. On May 25, 1844, Morse sent a message over an electric telegraph line that was 64 kilometers long from Washington, D.C., to Baltimore, Maryland. The message he sent was "What hath God wrought?"

Morse was not the only person working on telegraphy; he borrowed ideas from another American, the physicist and inventor Joseph Henry, who, in 1835, had invented an electrical relay that was very similar to the device Morse had made to send his code. Henry unfortunately did not patent his invention, so Morse is usually the one who gets the credit.

The United States was not the only country using telegraphy. In 1837, the Great Western Railway in England used an early telegraphic signal (with wires strung along next to railroad lines) to find out how fast the trains were going. In 1852, Germany and France agreed to allow telegraph wires to cross their borders for messages.

When telegraphy first became popular, people imagined connecting wires all around the globe. In 1850, a well-insulated copper wire was laid in the English Channel for telegraphic signals between France and England, and in 1854 people began to think about laying a wire cable across the Atlantic Ocean. It took three tries between 1857 and 1858 to lay more than 590,500 kilometers of iron wire across the ocean. On August 13, 1858, President James Buchanan of the United States and Queen Victoria of England sent messages to each other across this cable. It took about sixteen and one-half hours for Queen Victoria's ninety-word message to cross the Atlantic. Unfortunately, the cable worked for only one month. Two more cables were laid, and by the end of the nineteenth century, more than

90 million telegrams a year were being sent across the Atlantic Ocean.

## Marconi Goes to Work

No researcher works alone, and everyone builds on and borrows from the work of other people. In addition to the work of Morse and Henry, Marconi based his ideas on the work of the Scottish mathematician and physicist James Clerk Maxwell, as well as on that of the German physicist Heinrich Hertz. Maxwell pointed out that electricity and magnetism were simply different forms of electromagnetic radiation. He also showed that what is called "light" is one form of electromagnetic radiation. This was the beginning of the idea that electromagnetic waves other than light could be sent and received through space.

Electric current running through a wire does not travel at the speed of light. It can travel only as fast as the wire and the sending device will let it. If signals could be sent through space, without wires, the signals would travel at the speed of light. This is what Marconi achieved across the Atlantic Ocean on December 12, 1901.

In 1894, Marconi became aware of Hertz's work of using a transmitter to send electromagnetic waves through space. Marconi came up with the idea of using this transmitter to send signals—telegraphy without the wires. There was little support for this project in Italy, so Marconi went to England. He thought that his transmitter would be very useful between ships and shore, and since England had more ships at the time than any other country, he thought that England was most likely to support his idea. On June 2, 1896, Marconi applied for, and received, the world's first patent for wireless telegraphy.

Marconi experimented with his invention from 1896 to 1901, using the code that Morse had created.

In 1894, Marconi had sent a signal across 2.4 kilometers. By January, 1901, he was able to send a signal 299 kilometers.

Many people were excited about Marconi's invention and wanted to see it work over even longer distances. By the end of 1901, he was ready to send a signal across the Atlantic. Many people thought that this was impossible, but Marconi built an antenna in England that was 48 meters tall to send the signal. Then he traveled to Canada. There, during a winter gale, Marconi used a kite to lift a trailing antenna that was 152 meters long. On December 12, 1901, in St. John's, Marconi received the letter *S* from Poldhu.

*Guglielmo Marconi operates a wireless radio.*

## Consequences

When Marconi received the letter *S*, the world was forever changed, since the sending of signals was no longer limited to wires and signals could travel at the speed of light. Not everyone believed him at first, and many did not like the fact that he had not really invented something new, but had simply put together the inventions of other people. Still, Marconi kept experimenting, and in February, 1902, he did it again, using the same antenna in Poldhu. This time the signal was received by the SS *Philadelphia*, which was more than 3,230 kilometers away in the Atlantic Ocean. By 1903, newspaper stories from New York City were being sent to *The Times* of London.

Although there is no question that Marconi's invention was based on the work of Henry, Morse, Maxwell, Hertz, and others, Nobel Prizes are not awarded to large groups of researchers. In 1909, Marconi shared the Nobel Prize in Physics with Karl Ferdinand Braun, who had invented the cathode-ray tube in 1897.

As soon as Marconi introduced the wireless telegraph, other inventions followed, which were combined with Marconi's idea to create other inventions. In 1906, Reginald Aubrey Fessenden, an American radio engineer, invented a way to control the wavelengths being sent and received and thus invented radio. In 1908, A. Swinton published a brief letter in *Nature* called "Distant Electric Vision," and television as a form of "wireless telegraphy" came about. The first televisions were demonstrated in England in 1926 and in the United States in 1927.

Marconi's invention grew into a global network that now allows nearly anyone in the world to contact nearly anyone else almost instantly, so long as both have the right equipment. Fiber-optic cables have replaced copper wire cables in many applications, and signals are sent through space to and from satellites orbiting the earth. Although the world is the same size it has always been, radio, telephones, and satellites have made it seem to shrink considerably since that first transatlantic telegraphic transmission in 1901.

*Charles F. Urbanowicz*

# Kennelly and Heaviside Theorize Existence of Ionosphere

Arthur Edwin Kennelly and Oliver Heaviside independently proposed the existence of an electrified layer in the upper atmosphere that would reflect radio waves around the curved surface of the earth.

**What:** Earth science
**When:** 1902
**Where:** Cambridge, Massachusetts, and Newton Abbot, England
**Who:**
ARTHUR EDWIN KENNELLY (1861-1939), a British American electrical engineer and Harvard professor
OLIVER HEAVISIDE (1850-1925), an English physicist and electrical engineer
BALFOUR STEWART (1828-1887), a Scottish physicist
GUGLIELMO MARCONI (1874-1937), an Italian electrical engineer and cowinner of the 1909 Nobel Prize in Physics for demonstrating radio transmission across the Atlantic Ocean
SIR EDWARD VICTOR APPLETON (1892-1965), an English physicist who won the 1947 Nobel Prize in Physics for his discovery of the ionosphere in 1924
HEINRICH HERTZ (1857-1894), a German physicist

## Guiding Radio Waves over the Horizon

On December 12, 1901, only thirteen years after Heinrich Hertz had discovered radio waves, Guglielmo Marconi succeeded in transmitting radio signals from Cornwall, England, to Newfoundland, Canada. This historic event was difficult to explain, since it was known that radio signals consisted of electromagnetic waves like light, which travel in nearly straight lines. Thus, there was considerable discussion among scientists as to how Marconi's signals could travel around the curved surface of the Atlantic Ocean. Several scientists tried to show that electromagnetic waves of sufficiently long wavelength could bend around the earth's curvature by diffraction (the small tendency of waves to spread around obstacles). Calculations showed, however, that diffraction theories were not enough to explain Marconi's results.

The correct explanation was suggested almost simultaneously in 1902 by Arthur Edwin Kennelly in the United States and Oliver Heaviside in England. They independently proposed the existence of a layer in the upper atmosphere that could conduct electricity and reflect radio waves back to Earth. Successive reflections between this conducting layer and the surface of the earth could guide the waves around the earth's curvature. Heaviside also suggested that the conductivity of this region might result from the presence of electrically charged particles, called ions, in the upper atmosphere caused by solar radiation. This layer would later be labeled the "ionosphere."

The idea of such an electrically charged conducting shell in the upper atmosphere had already been suggested by Balfour Stewart twenty years earlier in 1882. In his study of terrestrial magnetism, Stewart had proposed that electrical currents flowing high in the atmosphere could explain the small daily changes in the earth's magnetic field. Such variations would be caused by the tidal movements in the surrounding "sea of air," which were caused by solar and gravitational influences. This explanation of fluctuations in the earth's magnetism, combined with many peculiar features of shortwave radio, seemed to confirm the existence of the Kennelly-Heaviside layer, but these phenomena provided only indirect evidence.

## Measuring the Heights of the Ionosphere

The first direct evidence for the existence of the ionosphere was obtained by Sir Edward Victor Appleton, with the assistance of Miles A. F. Barnett, in 1924. At the University of Cambridge, Appleton had studied radio signals from the new British Broadcasting Company (BBC) station in London and had noticed the typical variations in their strength. When he took up a new position at the University of London in 1924, Appleton arranged to use the new transmitter at Bournemouth after midnight, with receiving apparatus located at Oxford University. By varying the transmitter frequency, he hoped to detect any changes in signal strength that resulted when ground waves and waves reflected from the ionosphere (assuming they existed) interfered with one another.

On December 11, 1924, Appleton and Barnett observed the regular fading in and out of the signal as the frequency of the transmitter was slowly increased. From the data they collected, they calculated that the reflection that they had discovered was from a height of about 100 kilometers. This confirmation of the Kennelly-Heaviside prediction was published in 1925 in *Nature* under the title "Local Reflection of Wireless Waves from the Upper Atmosphere."

In 1926, Appleton found that the ionization of the Heaviside layer (E layer) was sufficiently reduced before dawn by recombination of electrons with positive ions to allow penetration by radio waves. Reflection, however, was still observed as originating from a higher layer where the air was too thin for efficient recombination. The height of that layer, now called the "Appleton layer" (F layer), was measured at about 230 kilometers above the earth. This result was published in *Nature* in 1927 under the title "The Existence of More than One Ionized Layer in the Upper Atmosphere." Moreover, observations during a solar eclipse in 1927 showed that the height of the Heaviside layer changed, revealing that ionization was really caused by solar radiation, as suggested by Heaviside.

## Consequences

The prediction and discovery of the ionosphere were important in stimulating new scientific understanding and advances in technology. In 1931, systematic experiments were begun to determine the variation of electron densities in the ionosphere, revealing an increase in ionization as the sun was rising and low ionization at night, except for sporadic increases possibly caused by meteoric activity. Noon ionization was found to increase as the sunspot maximum of 1937 was approached, suggesting a correlation between sunspots and increases in the ultraviolet radiation that ionizes the upper layers of the atmosphere. This made it possible to measure the ultraviolet radiation from the sun even though little of it reaches the ground. During the sunspot maximum of 1957-1958, the International Geophysical Year was established to study geophysical phenomena on a worldwide scale, including their relation to ionospheric variations. Thus, study of the ionosphere has contributed to developments in other sciences, such as astronomy, meteorology, and geophysics.

The early development of radar was closely associated with the studies of the ionosphere. The most powerful radar systems use the over-the-horizon technique of reflecting radar signals from the ionosphere to cover distances up to about 3,000 kilometers, about ten times farther than conventional radar. Over-the-horizon radar systems depend on computers to chart the constantly changing intensity and thickness of the ionosphere and to determine where conditions are best and which frequencies are needed for maximum performance. Thus, the ionosphere has become an indispensable tool for both communications and national security.

*Joseph L. Spradley*

# Zsigmondy Invents Ultramicroscope

> *Richard Zsigmondy developed a device called the "ultramicroscope," which allowed scientists to measure and identify individual particles in colloid solutions.*

**What:** Chemistry
**When:** 1902
**Where:** Göttingen, Germany
**Who:**
RICHARD ZSIGMONDY (1865-1929), an Austrian-born German organic chemist who won the 1925 Nobel Prize in Chemistry
H. F. W. SIEDENTOPF (1872-1940), a German physicist-optician
MAX VON SMOULUCHOWSKI (1879-1961), a German organic chemist

## Accidents of Alchemy

Richard Zsigmondy's invention of the ultramicroscope grew out of his interest in colloidal substances. Colloids consist of tiny particles of a substance that are dispersed throughout a solution of another material or substance (for example, salt in water). Zsigmondy first became interested in colloids while working as an assistant to the physicist Adolf Kundt at the University of Berlin in 1892. Although originally trained as an organic chemist, in which discipline he took his Ph.D. at the University of Munich in 1890, Zsigmondy became particularly interested in colloidal substances containing fine particles of gold that produce lustrous colors when painted on porcelain. For this reason, he abandoned organic chemistry and devoted his career to the study of colloids.

Zsigmondy began intensive research into his new field of interest in 1893, when he returned to Austria to accept a post as lecturer at a technical school at Graz. Zsigmondy became especially interested in gold-ruby glass, the accidental invention of the seventeenth century alchemist Johann Kunckle. Kunckle, while pursuing the alchemist's pipe dream of transmuting base sub-stances (such as lead) into gold, discovered instead a method of producing glass with a beautiful, deep red luster by suspending very fine particles of gold throughout the liquid glass before it was cooled. Zsigmondy also began studying a colloidal pigment called "purple of Cassius," the discovery of another seventeenth century alchemist, Andreas Cassius.

Zsigmondy soon discovered that purple of Cassius was a colloidal solution and not, as most chemists believed at the time, a chemical compound. This fact allowed him to develop techniques for glass and porcelain coloring with great commercial value, which led directly to his 1897 appointment to a research post with the Schott Glass Manufacturing Company in Jena, Germany. With the Schott Company, Zsigmondy concentrated on the commercial production of colored glass objects. His most notable achievement during this period was the invention of Jena milk glass, which is still prized by collectors throughout the world.

## Brilliant Proof

While studying colloids, Zsigmondy devised experiments that proved that purple of Cassius was colloidal. When he published the results of his research in professional journals, however, they were not widely accepted by the scientific community. Other scientists were not able to replicate Zsigmondy's experiments and consequently denounced them as flawed. The criticism of his work in technical literature stimulated Zsigmondy to make his greatest discovery, the ultramicroscope, which he developed to prove his theories regarding purple of Cassius.

The problem with proving the exact nature of purple of Cassius was that the scientific instruments available at the time were not sensitive enough for direct observation of the particles suspended in a colloidal substance. Using the facilities and assisted by the staff (especially H. F. W. Siedentopf, an expert in optical lens grinding) of the Zeiss Glass Manufacturing Company of Jena, Zsigmondy developed an ingenious device that permitted direct observation of individual colloidal particles.

This device, which its developers named the "ultramicroscope," made use of a principle that already existed. Sometimes called "dark-field illumination," this method consisted of shining a light (usually sunlight focused by mirrors) through the solution under the microscope at right angles to the observer, rather than shining the light directly from the observer into the solution. The resulting effect is similar to that obtained when a beam of sunlight is admitted to a closed room through a small window. If an observer stands back from and at right angles to such a beam, many dust particles suspended in the air will be observed that otherwise would not be visible.

Zsigmondy's device shines a very bright light through the substance or solution being studied. From the side, the microscope then focuses on the light shaft. This process enables the observer using the ultramicroscope to view colloidal particles that are ordinarily invisible even to the strongest conventional microscope. To a scientist viewing purple of Cassius, for example, colloidal gold particles as small as one ten-millionth of a millimeter in size become visible.

## Consequences

After Zsigmondy's invention of the ultramicroscope in 1902, the University of Göttingen appointed him professor of inorganic chemistry and director of its Institute for Inorganic Chemistry. Using the ultramicroscope, Zsigmondy and his associates quickly proved that purple of Cassius is indeed a colloidal substance.

That finding, however, was the least of the spectacular discoveries that resulted from Zsigmondy's invention. In the next decade, Zsigmondy and his associates found that color changes in colloidal gold solutions result from coagulation—that is, from changes in the size and number of gold particles in the solution caused by particles bonding together. Zsigmondy found that coagulation occurs when the negative electrical charge of the individual parti-

*Richard Zsigmondy.*

The Nobel Foundation

cles is removed by the addition of salts. Coagulation can be prevented or slowed by the addition of protective colloids.

These observations also made possible the determination of the speed at which coagulation takes place, as well as the number of particles in the colloidal substance being studied. With the assistance of the organic chemist Max von Smouluchowski, Zsigmondy worked out a complete mathematical formula of colloidal coagulation that is valid not only for gold colloidal solutions but also for all other colloids. Colloidal substances include blood and milk, which both coagulate, thus giving Zsigmondy's work relevance to the fields of medicine and agriculture. These observations and discoveries concerning colloids—in addition to the invention of the ultramicroscope—earned for Zsigmondy the 1925 Nobel Prize in Chemistry.

*Paul Madden*

# Sutton Predicts That Chromosomes Carry Hereditary Traits

> *Walter S. Sutton determined that inherited traits of organisms are physically located on chromosomes.*

**What:** Biology; Genetics
**When:** 1902
**Where:** New York, New York
**Who:**
WALTER S. SUTTON (1877-1916), an American geneticist and surgeon
THEODOR BOVERI (1862-1915), a German biologist
GREGOR JOHANN MENDEL (1822-1884), an Austrian monk and scientist
CARL ERICH CORRENS (1864-1933), a German geneticist
THOMAS HUNT MORGAN (1866-1945), an American geneticist and 1933 Nobel laureate in Physiology or Medicine

## Farming as a Scientific Endeavor

Beginning in 1856, an Austrian monk named Gregor Johann Mendel initiated a series of experiments with garden peas that would revolutionize twentieth century biology. Mendel cross-pollinated different lines of garden peas that had been bred for certain characteristics (purple or white flower color, wrinkled or smooth seeds, and the like). From his extensive experiments, he concluded that each garden pea plant carries two copies of each characteristic trait (flower color, seed texture, and so forth). These traits would later come to be known as "genes."

Diploid organisms have two copies of each gene; the copies may or may not be identical. Different forms of the same gene are called "alleles." For example, the gene for flower color may have two alleles, one conferring purple flower color and the other conferring white flower color. In most cases, one allele is dominant over other alleles for the same gene. For example, a plant having two purple flower-color alleles will have purple flowers, while a plant having two white flower-color alleles will have white flowers. A plant having one purple and one white flower-color allele, however, will have purple flowers; the dominant purple allele will mask the white allele.

Since most plants and animals reproduce sexually by the means of fusion of pollen (or sperm) with eggs, Mendel discovered the pattern of transmission of these genetic traits (that is, inheritance) from parents to offspring. While each individual has two copies of every gene, each individual can transmit only one copy of each gene (that is, only one of two alleles) to each of its offspring. If a parent has two different alleles for a given gene, only one of the two alleles can be passed on to each of its children. There is a fifty-fifty chance of either allele being transmitted for each of thousands of different genes conferring different characteristic traits. Before Mendel's findings were rediscovered in 1900, most investigators were concluding that the mechanism of heredity was a blending of characteristics, similar to the mixing of different colors of paint. According to Mendel's results, the mechanism was more similar to combining various colored balls.

Mendel summed up the random chance inheritance of different alleles of a gene in two principles. The first principle, allelic segregation, maintains that the different alleles for a given gene separate from each other during the formation of germ-line cells (that is, sperm and egg). The second principle, independent assortment, maintains that different alleles of different genes arrange themselves randomly during germ-cell production. When Mendel published his results in 1866, his work was scarcely noticed. Twenty years would pass before the importance of his research was understood.

**61**

## The Chromosomal Theory of Inheritance

From 1885 to 1893, the German biologist Theodor Boveri researched the chromosomes of the roundworm *Ascaris*. Chromosomes are molecules composed mostly of deoxyribonucleic acid (DNA) and protein. They are located within the nuclei of the cells of all living organisms. In the late 1890's, Walter S. Sutton studied chromosomes of the grasshopper *Brachyostola magna*. Sutton constructed detailed diagrams of *Brachyostola* chromosomes during mitosis (chromosome doubling and separating prior to cell division) and meiosis (chromosome dividing and splitting in sperm and egg production). Both Sutton and Boveri were independently attempting to understand chromosomal structure and function.

The breakthrough came in 1900, when the German biologist Carl Erich Correns rediscovered Mendel's work with garden peas. Correns boldly proposed that the chromosomes of living organisms carried the organisms' inherited traits. Unfortunately, he did not provide an exact mechanism to support his hypothesis. Nevertheless, Correns's hypothesis eventually caught the attention of both Sutton and Boveri. With Correns's hypothesis and their own research on chromosome behavior, Sutton and Boveri began to derive a mechanism for the chromosomal transmission of inherited traits. Together, the results of the two scientists culminated in the chromosomal theory of inheritance, one of the basic tenets of genetics and modern biology.

The chromosomal theory of inheritance makes four assertions. First, the fusion of sperm and egg is responsible for reproduction—the formation of a new individual. Second, each sperm or egg cell carries one-half of the genes for the new individual, or one copy of each chromosome. Third, chromosomes carry genetic information and are separated during meiosis. Finally, meiosis is the mechanism that best explains Mendel's principles of allelic segregation and independent assortment. Sutton reported his conclusions in a 1902 article in *The Biological Bulletin*.

## Consequences

The Sutton and Boveri chromosomal theory of inheritance represented a landmark in the history of biological thought. It reestablished Mendelism and provided a definite physical mechanism for inheritance. It demonstrated the molecular basis of life and thereby launched two successive waves of biological revolution: the pre-World War II genetic and biochemical revolution and the postwar molecular biology revolution, which continues today. It has also been very useful for the study of human genetic disorders.

Thomas Hunt Morgan and his associates generated hundreds of mutations in the fruit fly *Drosophila melanogaster* and mapped these mutations to specific chromosome locations, thereby verifying Sutton and Boveri's theory. Mutations were generated using either chemicals or radiation such as ultraviolet light and X rays. This work demonstrated that exposing living organisms to radiation and certain chemicals can cause chromosome and gene damage, often resulting in severe abnormalities, and sometimes death, in the exposed individuals and their descendants.

Chromosome studies also proved useful as a tool for understanding evolution. Evolution consists of mutational changes that occur in organisms over time, thereby giving rise to new types of organisms and new species that are better adapted to their environments. The chromosomal theory of inheritance helped to explain the processes by which evolution takes place.

*David Wason Hollar, Jr.*

# Pavlov Introduces Concept of Reinforcement in Learning

*Pavlov advanced the concept of reinforcement, which provided a physiochemical explanation for learning.*

**What:** Biology; Education
**When:** 1902-1903
**Where:** St. Petersburg, Russia
**Who:**
Ivan Petrovich Pavlov (1849-1936), a Russian physiologist

## From Eating to Learning

In April, 1903, Ivan Petrovich Pavlov delivered a surprising address to an International Medical Conference in Madrid. It was thought that the noted Russian physiologist would discuss his research on digestion; instead, he described new investigations into the links between mental and physical processes. In particular, he focused on the experience that begins, maintains, or eliminates a given kind of behavior. While he did not formally name that concept until the following year, he was referring to "reinforcement," the key to understanding the physical aspects of the learning process.

Scientists interested in the concept of learning generally fell into two groups. One group, variously described as "mentalists," "vitalists," or "subjectivists," held the opinion that thoughts and emotions were not subject to physical laws and that experimental attempts to penetrate the boundary between mind and body were useless. Their opponents, generally called "monists," insisted that explanations of human behavior were to be found in the laws of physics and chemistry. Before Pavlov's work on reinforcement, however, monists lacked a clearly explainable theory and enough experimental evidence to back it up. They needed a clear approach to the problem of learning combined with ways to measure the physical reactions of healthy subjects. Pavlov provided both.

The experimental technique Pavlov used to address learned behavior was an offshoot of his 1889 through 1897 investigations into the effect of the nervous system on digestion. In that project, which won for him the 1904 Nobel Prize in Physiology or Medicine, he diverted the duct of a dog's salivary gland (the parotid gland) to the outside of its muzzle so that the saliva could be collected from a funnel. This procedure became a major part of his research into reinforcement when he realized that the release of saliva was at least partly triggered by factors involving the brain. While doing research on the digestive system, he noticed an intriguing occurrence: When the presentation of food to a dog was regularly and closely preceded by an alerting signal such as approaching footsteps, the signal alone caused salivation. Pavlov directed preliminary investigations into that type of behavior (the conditional reflex) as early as 1897. Working in the well-equipped physiological laboratory of the St. Petersburg Institute of Experimental Medicine, he issued preliminary findings at Madrid in 1903 and presented formal results of his research in 1905.

## How Dogs Learn

Pavlov believed that all animals have an inborn neural capacity (the unconditional reflex) to react to events necessary for survival (the unconditional stimulus). For example, the sight of food normally produced a certain type and amount of salivation in a dog. As long as the dog was allowed to eat food presented to it, the unconditional reflex endured. Eating the food reinforced the unconditional reflex, because the act of eating maintained (excited) a neural association between the unconditional stimulus and the unconditional reflex. Yet, if a dog was repeatedly shown food it was not allowed to eat, salivation weakened and eventually disappeared. Pavlov interpreted this to mean that, on one hand, the association between the sight of food and salivation was actually a temporary neural connection that could be broken through lack of reinforcement. On the other hand, a broken connection could be reactivated by reestablishing the link between salivation and the sight of food. The link could also be redirected. If an unrelated stimulus (for example, the sound made by a bell) immediately preceded the delivery of food over a long enough period, it could provoke salivation. Moreover, a dog was able to discriminate between similar types of reinforced and unreinforced food. For example, it would stop salivating at the sight of white bread it was forbidden to

eat but continued salivating at the sight of brown bread that was regularly fed to it.

Thus, reinforcement is the notion that holds that animals can associate any two occurrences if the first one is promptly and reliably followed by the second. Reinforcement has a clearly adaptive function, because it allows an animal to change its behavior in reaction to changes in its environment. Learning takes place because of the forging and breaking of neural connections in reaction to changing circumstances.

**Consequences**

The importance of reinforcement was first recognized in Russia, where the monists and the subjectivists continued to debate the merits of Pavlov's new ideas. Yet although the state govern-

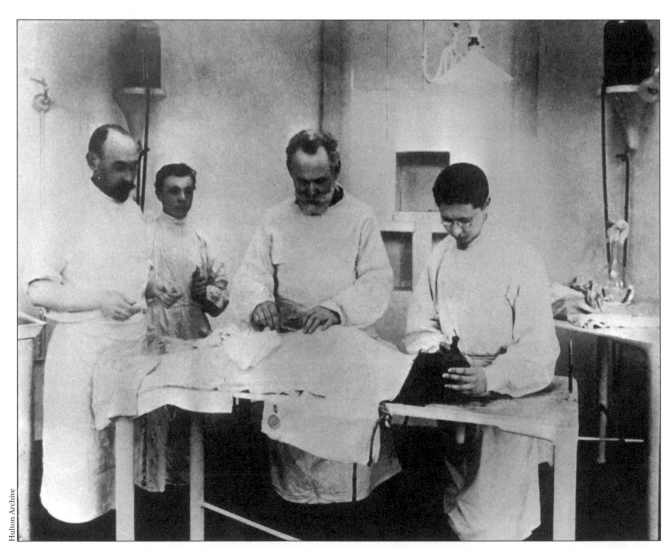

Hulton Archive

*Ivan Petrovich Pavlov (second from right) and his assistants demonstrate the theory of conditional reflex on a dog.*

ment (during the later years of the czar) backed subjectivism but opposed monism, private funding allowed Pavlov to expand his experiments.

The Revolutionary period disrupted Pavlov's work, but after 1921, the Soviet leadership vigorously supported his research. They did so for several reasons: Pavlov's emphasis on physical stimulation fit the materialistic orientation of the new Marxist regime, his doctrine of reinforcement strengthened the arguments of nurture over nature at a time when the government wanted to reeducate its citizenry, and his scientific triumphs made Soviet socialism look good to the rest of the world. With state sponsorship, Pavlov and his successors directed large research projects on complex aspects of association through reinforcement.

Largely because of the language barrier, non-Russians were unable to absorb Pavlov's concept of reinforcement until the 1920's. European reaction was unenthusiastic, and in the United States it was often confused with the idea of rewards and punishments. Generally, Americans and Western Europeans admire Pavlov's exacting investigations and associate his original concept with a specific kind of simple learning (classical, or passive, learning). Some aspects of Pavlovian reinforcement have been incorporated into theories of language formation and information processing. Nevertheless, it is only one of many influences affecting the course of Western theories of human behavior.

*Michael James Fontenot*

# Mathematics Responds to Crisis of Paradox

> *The logical paradox discovered by Bertrand Russell challenged the long-accepted belief in the consistency of mathematics.*

**What:** Mathematics
**When:** 1902
**Where:** Northern Europe
**Who:**
GOTTLOB FREGE (1848-1925), a German logician
BERTRAND RUSSELL (1872-1970), an English philosopher
DAVID HILBERT (1862-1943), a German mathematician
KURT GÖDEL (1906-1978), an Austrian mathematician
LUITZEN E. J. BROUWER (1881-1966), a Dutch mathematician

## Mathematics Looks Inward

The late nineteenth century and the early twentieth century were characterized by self-reflection within various intellectual domains. For example, Impressionism and later schools of art investigated the very methods of creating art, focusing on "art for art's sake." Sigmund Freud and others founded the field of psychology, which consists of the human psyche looking at itself. Literature, music, architecture, and science (which paid particular attention to the "scientific method") also turned inward.

Mathematics was no exception to this trend; the methods of mathematics were themselves being scrutinized. For example, Gottlob Frege was intensely investigating mathematical logic, the method of mathematical thinking.

Central to Frege's work was the mathematically pervasive concept of the "set"—a collection of objects, real or abstract, that could be defined by either listing its elements or providing a property that characterized only those elements. For example, one set might be defined as {2,4,6} or as those positive even numbers lower than 8. The property-based mode of definition must be used to define infinite sets (such as the set of integers), because infinite sets cannot be listed.

## Russell's Paradox Causes Crisis

Frege's work was well advanced when Bertrand Russell encountered a peculiar problem in his definition of sets by properties. Unable to see the solution, he wrote to Frege in 1902 to inquire about the problem. The older logician replied with one of the most gracious responses to bad news ever written, stating that he had never noticed the problem and that he could not see a solution for it. Thus, it was discovered that the foundation of Frege's life work was seriously flawed.

This problem, now called Russell's paradox, is deceptively simple to delineate. Because sets are well defined, they may be collected into other sets. For example, the set A may be defined as consisting of all sets with more than two elements. A would therefore contain the set of planets, the set of negative numbers, the set of polygons, and so forth. Since these are three sets collected by A, then A itself has more than two elements. Therefore, A is a member of itself. This fact may seem strange, but the defining property is absolutely unambiguous: "sets with more than two members." Thus, sets may be elements of themselves.

Russell then considered the set D, which consists of those sets that do not contain themselves. Then he asked, "Does D contain itself?" If it does, it is one of those sets that it must not contain. Therefore, D must not contain itself, but D is also one of those sets that it must contain. D contains itself only if it does not contain itself!

Paradoxes of ordinary language are well known. Two examples are the Cretan Epimenides' remark that "all Cretans are liars" (the liar paradox) and the sentence "This sentence is false." Both statements are true only if they are false. Russell's observation, however, represented the first time that the specter of paradox had arisen within mathematical thought. The seriousness of Russell's paradox stems from the assumption that mathematics embodied a higher truth and was therefore free from error, consistent, and unambiguous. Russell demonstrated that this assumption was false.

## Consequences

Many mathematicians, philosophers, and computer scientists regarded Russell's paradox as an assault on the very foundations of mathematics. If inconsistency could arise in an area as rigorous as set theory, how could consistency be guaranteed in more common areas of mathematics?

Russell and the philosopher Alfred North Whitehead set out to improve upon Frege's work (the Logicist school of thought). If mathematics could be derived from basic, self-evident axioms, no inconsistency would be possible. Russell and Whitehead's *Principia Mathematica* (1910) led to new uses of logic, but its means of avoiding paradox was too arbitrary for all mathematics, since it states that sets cannot contain both objects and other sets.

The Formalist school of David Hilbert, however, sought to establish the foundations of mathematics in the realm of symbol manipulation. The rules that governed such manipulation would be very simple and precise. Such a "proof-theoretic" or "metamathematical" analysis of proof was expected to confirm the consistency of mathematical systems. Much that was useful in mathematics and computer science came out of this work, but in 1931, the young Kurt Gödel astounded the world of mathematics by proving that the Formalist ideal was unreachable—that most mathematical systems could not be proved, by noncontroversial means, to be fully adequate and consistent.

The Intuitionist school of Luitzen E. J. Brouwer grew out of the work of Leopold Kronecker and therefore was not a response to Russell's paradox, but the Intuitionists believed that their insistence on meaning in mathematics would avoid paradox. Intuitionists limit mathematics to what actually can be "constructed" by the human mind. Therefore, infinite sets are ruled out, as is automatic acceptance of the "law of the excluded middle" (which states, basically, that any statement must be either true or false). In this school, the truth or falsity of any statement must be demonstrated. Both of these objections apply to Russell's paradox: D is an infinite set, and the assumption that "D contains D or it does not" is an application of the law of the excluded middle.

The Intuitionist/Constructivist school has not found wide acceptance, because it is viewed as too restrictive by many mathematicians. It is very important, however, in the field of computer science, in which most results must be demonstrated constructively.

It is possible that, before the discovery of Russell's paradox, an easily understood problem had never caused such a major crisis in a scientific field. Russell's paradox had this effect because it forced mathematicians and philosophers to re-examine traditional assumptions about mathematical truth.

*J. Paul Myers, Jr.*

# French Archaeologists Decipher Hammurabi's Code

> *The discovery of the code of laws attributed to the Babylonian king Hammurabi demonstrated that Judeo-Christian concepts of justice were much older than had been thought.*

**What:** Archaeology; Law
**When:** January, 1902
**Where:** Susa, Iran (formerly Persia)
**Who:**

M. J. DE MORGAN (1857-1924), a French archaeologist who supervised the excavations at Susa

VINCENT SCHEIL (1858-1940), a professor at the French School of Advanced Studies (École des Hautes Études)

GEORG FRIEDRICH GROTEFEND (1775-1853), a teacher in Göttingen, Germany

GEORGE SMITH (1840-1876), a self-trained assistant in the Egyptian-Assyrian section of the British Museum

## Plundered Laws

In January, 1902, French archaeologists, working under the supervision of M. J. de Morgan in the ruins of the ancient Elamite capital of Susa, unearthed several pieces of a large block of stone. This block, once reassembled, reached a height of nearly 4 meters and was covered on several sides by carved inscriptions. One side bore a representation of the Babylonian king Hammurabi, who reigned over the lands stretching from the Mesopotamian Tigris and Euphrates river valleys to the coast of the Mediterranean between approximately 1792 B.C.E. and 1750 B.C.E. King Hammurabi was depicted receiving laws from the seated Babylonian sun god Shamash. Other sides of the massive stone were inscribed, in the cuneiform script typical of ancient Mesopotamia, with the law code that would become known as the Code of Hammurabi.

The fact that the stone block was found in the ruins of Susa, and not in Babylon or a dependent city of the great Babylonian kings, can perhaps be explained by the fact that Babylon and Susa were frequently at war with each other. It is possible that Elamites carried the stone monument to their capital of Susa as a result of one or another of the military campaigns pitting Elamites against Babylonians sometime after the middle of the eighteenth century B.C.E.

On the obverse sides of Hammurabi's monuments, opposite the side bearing the inscription of the king and Babylon's sun god, were columns of writing that apparently represented a list of laws. One side contained sixteen columns of writing with a total of 1,114 lines. Five columns had been obliterated from the face of the stone monument, presumably by the Elamite "victors" over Babylon who intended to inscribe (but never did) a dedication to their own glory. An additional twenty-five hundred lines in twenty-eight columns were carved into the remaining sections of the surviving monument.

Even before the work of translation could be undertaken, it was apparent that an important section of the text (containing about seven hundred lines) was devoted to a description of Hammurabi himself, his titles, his qualities as a ruler, his dedication to the gods of Babylon, and—most important for archaeologists' records—the cities and districts under his rule. The text of Hammurabi's Code was most important, not only in terms of numbers of columns and lines of writing but also for its impact on archaeologists' knowledge of Babylonian civilization. The main concerns of the code were contracts of sale or business, farming and animal husbandry practices, and dowry and marriage contracts.

The French expedition to Susa discovered Hammurabi's Code in January of 1902, and a translation, by Father Vincent Scheil of the École des Hautes Études, was published in Paris in October of 1902. As a result, archaeologists' knowledge of ancient Babylonian and later Mesopotamian societies was, if not revolutionized, at least substantially broadened.

## The Importance of Textual Finds

Hieroglyphics from the Babylonian era were first "decodified" in 1802 by Georg Friedrich Grotefend, a schoolteacher from Göttingham, Germany. Grotefend worked without the advantages of word-for-word comparative language texts such as the ancient Rosetta Stone, which the French archaeologist Jacques-Joseph Champollion had used when he deciphered Egyptian hieroglyphics in the same generation. Until 1857, however, archaeologists tended to give little attention to textual finds such as those translated by Grotefend.

This changed when a debate was held in the Royal Asiatic Society in London over conflicting scholarly interpretations of cuneiform writings from the library of Tiglath-pileser III, king of Assyria. This rather esoteric event generated so much interest in the importance of textual details that portions of the Assyrian King Ashurbanipal's royal library at Nineveh were carefully reexamined. The reexamination was done by George Smith, an assistant in the Egyptian-Assyrian section of the British Museum, who in 1872 thus discovered the so-called Gilgamesh epic (2000 B.C.E.; *Gilgamesh Epic*, 1917) a mixture of Babylonian religious mythology and history. The most striking reference, inscribed long before comparable sections in the Old Testament, had to do with the "Great Deluge," or flood, and the role of Utnapishtim, whom Smith identified with the biblical Noah. This opened the possibility that some biblical stories had their roots much further back in history than had previously been thought.

## Consequences

Smith's findings helped to heighten the impact of the later discovery of the Code of Hammurabi. As a consequence of such discoveries, scholars came to believe that the biblical Mosaic code, as well as a number of other elements of legal practice associated with the Old Testament, might be traceable to ancient Mesopotamian history—and thus were much older than had previously been suspected.

*Byron D. Cannon*

# Bayliss and Starling Discover Hormones

*Sir William Maddock Bayliss and Ernest Henry Starling proved that chemical integration can occur without assistance from the nervous system.*

**What:** Biology
**When:** April-June, 1902
**Where:** Cambridge, England
**Who:**
SIR WILLIAM MADDOCK BAYLISS (1860-1924), an English physiologist
ERNEST HENRY STARLING (1866-1927), an English physiologist who coined the term "hormone"
SIR FREDERICK GRANT BANTING (1891-1941), a Canadian physiologist who was the cowinner of the 1923 Nobel Prize in Physiology and Medicine
ARNOLD BERTHOLD (1803-1861), a German physiologist
IVAN PETROVICH PAVLOV (1849-1936), a Russian physiologist

## Bypassing the Brain

In the human body, food is digested by being dissolved and broken down chemically into simple chemical compounds that can be easily absorbed and used for nourishment.

The process begins the moment food is chewed and swallowed and continues all the way down through the stomach and the small and large intestines. As the food reaches the stomach, the gastric glands in the stomach lining secrete gastric juices. These juices contain substances, such as enzymes and hydrochloric acid, that break down food and aid digestion. The food then enters the small intestine, triggering the action of other glands such as the pancreas. The pancreas also secretes digestive substances such as enzymes. It also produces insulin, the hormone that enables the body to store and use sugar.

At the beginning of the twentieth century, two English physiologists, Sir William Maddock Bayliss and Ernest Henry Starling, were interested in discovering what triggered the pancreas to release digestive juices as soon as food arrived in the small intestine. In 1902, they set up an experiment to find out whether a nerve signal from the intestine was ordering the pancreas to release these juices. The investigators took an animal and cut the nerve system controlling its small intestine. They then injected stimulating material such as food from the stomach into the intestine. To their astonishment, pancreatic juices poured promptly into the intestine. With all the nerves cut, some mysterious signal must have reached the pancreas and roused it to action. Bayliss and Starling discovered that the signal was chemical in nature, not nerve-related. Arrival of hydrochloric-acid-laden food from the stomach had caused the intestinal wall to secrete a substance called "secretin," which oozed into the bloodstream, eventually stimulating the pancreas.

## The Role of Hormones

Starling first used the word "hormone" (Greek *hormon*, exciting, setting in motion) in 1905 with reference to secretin. Today, physiologists know that hormones may inhibit as well as excite, and that they do not initiate metabolic transformation but merely alter the rate at which these changes occur.

Other things are known, as well. The hormone disappears rapidly from circulation owing to the destructive action of an enzyme called "secretinase." Small amounts of the hormone are excreted in the urine. Many materials other than hydrochloric acid stimulate the release of secretin: water, alcohol, fatty acids, partially hydrolyzed protein, and certain amino acids are all effective.

The discovery by Bayliss and Starling of how hormones trigger the operation of other bodily systems influenced the research of others. Ivan Petrovich Pavlov, a Russian scientist and great pioneer in the study of conditioned reflexes, repeated the work of Bayliss and Starling in 1910 and obtained similar results. Subsequently, S. Kopec demonstrated in 1917 that a hormone from the brain controlled pupation in certain invertebrates (insects), which illustrated for the first time that central nervous structures could perform hormonal roles.

In other areas, important medical research focused on the "islands of Langerhans." These are the small, scattered endocrine glands in the pancreas that produce insulin. They are named after Paul Langerhans, the German pathologist, who discovered them in 1869. If the islands fail to release enough insulin into the bloodstream during digestion, diabetes may result.

The bodies of people who have diabetes are unable to process properly the sugar in the food they eat. Since insulin regulates the body's ability to process sugar, diabetics must take in additional dosages of it. The condition can be fatal if not carefully controlled.

In 1920, Sir Frederick Grant Banting was intrigued by the possibility that the operation of the pancreas might somehow be related to the onset of diabetes. Previous medical evidence seemed to suggest that the islands of Langerhans were important in directly releasing into the bloodstream something that prevented the disease. He set out to study these islands in the hope of finding what the "something" was.

By 1921, a team led by Banting had succeeded in extracting a quantity of insulin from the embryos of animals. The insulin extract was next injected experimentally into dogs and then humans. It was found to be effective in relieving the symptoms of diabetes. For his discovery of insulin, Banting shared the 1923 Nobel Prize.

## Consequences

Other scientists built on the work of Bayliss and Starling. In one set of experiments, Arnold Berthold castrated six young cockerels, then returned a single testicle to the body cavity of each of the birds. Berthold observed that the host birds continued to exhibit the sexual behavior of normal young roosters. At autopsy, he found that the nerve supply of the grafted testes had not been reestablished. Hence, Berthold concluded that, since maintenance of sexual behavior and appearance could not have been accomplished

*Sir William Maddock Bayliss.*

by the nerves (which were severed), the results must have been caused by a contribution of the testes to the blood and then by the action of the added substance throughout the body.

In November, 1962, Donald G. Cooley, an American physiologist, published a manuscript entitled, "Hormones: Your Body's Chemical Rousers," in which he reviewed the experiments of Bayliss and Starling and presented an updated, salient summary concerning the mechanism of hormone action. The article appeared in the November, 1962, edition of *Today's Health.*

Strong evidence that a virus can cause juvenile-onset diabetes was reported in May, 1979, by scientists at the National Institute of Dental Research in Bethesda, Maryland. Ji-won Yoon, Marchall Austen, and Takashi Orodern isolated the virus, called "Cox-sackie B4," from the pancreas of a ten-year-old boy who had died of a sudden and severe case of diabetes. The researchers grew the virus in cultures and injected it into mice. Some strains of mice then developed diabetes. This evidence indicated that the Coxsackie virus can be a causal factor in some cases of diabetes, somehow interfering with the pancreas's ability to produce enough insulin.

*Nathaniel Boggs*

**71**

# Anthracite Coal Miners Strike

*Intervening to settle a coal miners' strike, President Theodore Roosevelt set a new precedent and increased the power of the United States presidency.*

**What:** Labor; National politics
**When:** May 12, 1902
**Where:** Pennsylvania and
   Washington, D.C.
**Who:**
THEODORE ROOSEVELT (1858-1919),
   president of the United States from
   1901 to 1909
JOHN MITCHELL (1870-1919), president of
   the United Mine Workers
GEORGE F. BAER (1842-1914), a mine
   operator and an owner of the Reading
   and Philadelphia Railroad
ELIHU ROOT (1845-1937), U.S. secretary
   of war under Roosevelt
E. E. CLARK (1856-1930), grand chief of
   the order of railway conductors
WILLIAM A. STONE (1846-1920), governor
   of Pennsylvania

## The Strike Is Launched

In 1902, anthracite coal miners were probably treated more unfairly than any other group of workers in the United States. They earned an average wage of $560 a year and had no guarantee of regular employment. Work in the mines was dangerous, and most of the miners lived in company-owned towns, so that their home life as well as their work was regulated by the mine owners.

United Mine Workers president John Mitchell served as a spokesman for these struggling workers. Threatening a strike in 1900, Mitchell had won a 10 percent increase in their wages. Though further changes were desperately needed, in 1902 negotiations between owners and the union failed. On May 12, 147,000 members of the United Mine Workers walked out of the anthracite mines of Pennsylvania. Their demands included recognition of their union, a workday that would not exceed nine hours, more accurate weighing of the coal, and a 20 percent increase in pay.

At that time, it was not known that another readily available form of coal, bituminous, could be substituted for anthracite. As the strike dragged on toward winter, residents of northern cities began to fear that they would have to endure the cold months with no fuel to warm their houses. In September, the price of anthracite coal, usually five dollars a ton, reached fourteen dollars. Poor people, who bought in smaller quantities, paid a penny a pound, which added up to twenty dollars a ton.

By October, schools began to close, and the small stores of coal that were left sold for thirty and thirty-five dollars a ton. In the West, mobs began to seize coal cars from passing trains. Leaders in business feared that the whole country would break out in riots. Mayors across the nation appealed to the president for help.

## The President Steps In

In response to their desperation, President Theodore Roosevelt arranged a conference between labor and management leaders at the White House on October 1, 1902. George F. Baer, the owner of the Reading and Philadelphia Railroad, represented the coal operators, while John Mitchell spoke for the striking miners. Many people across the country were already angry at Baer, for he had declared publicly that the rights of working men would be protected best "not by the labor agitators, but by the Christian men to whom God in his infinite wisdom has given the control of the property interests in this country." Baer was seen as proud and selfish, while Mitchell's manner was calm and polite.

The day-long conference between these two men failed, and it seemed that the strike could not be resolved peacefully. Mine owners claimed

**72**

that the workers wanted to return to the mines but feared violence from the union. Yet, when Pennsylvania governor William A. Stone called out the state militia to protect anyone who wished to work, most of the miners remained on strike. Roosevelt began making a plan for federal troops to occupy the mines, but at the same time Elihu Root, secretary of war and a friend of the business community, worked to arrange a less drastic settlement.

On October 12, Root met with J. P. Morgan, a New York millionaire whose railroads crisscrossed the coal fields. Together they hammered out a possible compromise. The next day, George Baer was summoned to talk with Morgan, and by October 15, the mine operators had ratified the agreement. The Root-Morgan proposal was to establish a five-man independent commission that would be given authority to resolve the dispute. Presented with the proposal, the United Mine Workers insisted that a union representative and a Roman Catholic priest be appointed to the five-man panel. The mine operators agreed to the priest but stubbornly opposed inclusion of a union member.

Roosevelt himself broke the deadlock by naming E. E. Clark, grand chief of the Order of Railway Conductors, as a sixth member of the commission. In order to satisfy the mine operators, Roosevelt publicly labeled Clark "an eminent sociologist" rather than "a labor leader." The commission proceeded to find a compromise solution to the strike. The miners' union was not recognized, but the miners won a nine-hour day and a 10 percent pay raise. The weight dispute was settled, and a board of conciliation was created to help resolve future difficulties.

Library of Congress

*In Hazelton, Pennsylvania, miners' wives search for coal on a dump during the strike.*

## Consequences

President Roosevelt later spoke of the coal-strike settlement as a turning point in his administration. He was right. The settlement increased his personal popularity and added to the power of the U.S. presidency. Americans began to expect their presidents to be negotiators, able to help resolve both standoffs between business and labor, and serious conflicts elsewhere in the world.

*Rex O. Mooney*
*Updated by Lewis L. Gould*

# Tsiolkovsky Proposes Using Liquid Oxygen for Rocket Fuel

> *Konstantin Tsiolkovsky used mathematics to show that liquid fuel could be used to launch rockets into space.*

**What:** Space and aviation
**When:** 1903
**Where:** Kaluga, Russia
**Who:**
KONSTANTIN TSIOLKOVSKY (1857-1935), a
   Russian scientist, one of the founders
   of modern astronautics

## Dedicated to Spaceflight

When Konstantin Tsiolkovsky was young, he was nicknamed "Bird" because of his lightheartedness and his flitting movements. When he was ten, however, scarlet fever made him deaf. From then on, he began to bury himself in books. He was one of the three most important people in the early history of rockets. The other two were Robert H. Goddard, a physicist from the United States, and Hermann Oberth, a scientist from Germany. Both Tsiolkovsky and Goddard suffered a severe childhood disease, both became teachers, and both were entirely dedicated to spaceflight.

Tsiolkovsky went to Moscow to study mathematics, physics, and later, astronomy. He believed even then that spaceflight was possible. Tsiolkovsky did not have enough money to do many experiments, so he had to prove most of his theories with mathematics. He kept journals of every step in his thinking. When he did perform experiments, the money came out of his food allowance. After a while, he got sick and had to be sent home. Before he left, however, he caught a vision of how spaceflight would be possible. He was sitting in a city park, watching some teenagers jumping off a hay wagon. As each one jumped off, the wagon would lurch forward slightly. At that moment, the English physicist and mathematician Sir Isaac Newton's law of action and reaction came alive for Tsiolkovsky, and he understood what would be needed to boost a rocket into space.

In 1885, Tsiolkovsky dedicated himself to aviation. For two years he spent all of his time working on his aerostat, a metal-skinned, piloted balloon. In 1887, at the Polytechnical Museum in Moscow, he presented his first public lecture, called "The Theory of the Aerostat." After his lectures and the two years of constant work, he came down with a serious illness and lost his voice for a year. Then his house burned, and he lost his library and models. The only work that was saved was the lecture he had given in Moscow.

In 1892, Tsiolkovsky moved to Kaluga, where he taught at two schools. For the first time since his illness, he returned to his scientific research. In 1894, he presented a design for a "birdlike flying machine" (a monoplane), something that would not be built for another twenty years. He also built the first Soviet wind tunnel. In 1898, he presented "Exploration of Space with Reactive Devices." This study was not published until 1903, partly because it was very technical. Not many people took notice of it, though Goddard's paper "A Method of Reaching Extreme Altitudes," published at about the same time, was received with great enthusiasm.

## Rocket Fuel

Between 1903 and 1933, Tsiolkovsky continued to devise different theories about spaceflight. One of the most important ones was that only reactive devices would work both within the atmosphere and in space. The aircraft of that time were pulled through the atmosphere solely by means of propellers. In the atmosphere, moreover, the wings of an aircraft are supported by the air itself. Airplanes would not work in space, Tsiolkovsky believed, since there is no air to hold them aloft or for the propellers to use. This had been a major problem for spaceflight until Tsiolkovsky watched the teenagers jumping off the wagon. In a reactive device, an explosion would push against the spacecraft, much as the legs of the teenagers had pushed against the wagon, making it lurch forward. Tsiolkovsky also showed that the black powder rockets that some people were using would not be enough to carry a rocket into space. He suggested that liquid fuel propellants be used instead.

Pointing out the problems with either hydrogen or oxygen, Tsiolkovsky suggested that hydrocarbons would be the best solution. Hydrogen evaporates quickly and therefore is hard to store.

Oxygen has a number of advantages, since it could be used as a coolant as well as to provide air for the pilots. Yet because the tanks holding hydrocarbons could be much lighter, Tsiolkovsky thought that hydrocarbons would be the best fuel, even though they do not explode with as much force and thus provide less thrust.

Tsiolkovsky described how the "explosion tube" of a rocket should look, as well as how it should be controlled. He also described the three rudders that should be included in the design. In short, Tsiolkovsky described, in 1903, the basic ideas that were used many years later for early rocket spaceflight. When one considers that 1903 was the year in which the Wright brothers flew the first successful airplane, Tsiolkovksy's achievement is truly amazing.

*Konstantin Tsiolkovsky stands before a model of one of the first rockets.*

## Consequences

In spite of the importance of Tsiolkovsky's theories, scientists did not pay much attention to him until 1924, when Oberth republished his work in both German and Russian. It was then that Tsiolkovsky was called "the father of space travel." Without his dedication and genius, the world might never have considered the possibility of space travel. Because his theories about rockets made such good sense, some of his more fantastic ideas have not been completely rejected. For example, his papers "Will the Earth Ever Be Able to Inform the Living Beings on Other Planets of the Existence on It of Intelligent Beings?" and "The Unknown Intelligent Forces" have led scientists to look for other living beings in the universe. At least two major projects have been set up to do that.

Tsiolkovsky also predicted the idea of a space station. He thought that it could serve as a sort of service station for rockets on their way to other planets. He included plans for how to build one, using rockets to send pieces into space, where they would be put together by astronauts. He suggested that plants would grow faster on a space station, since there would be no gravity and they would be closer to the sun. Tsiolkovsky had many other theories that are used in rocketry. His ideas of a multistage rocket and a gliding reentry were both used in the design of the U.S. space shuttles.

The memorial built in Tsiolkovsky's honor after his death quotes his belief that "mankind will not remain bound to the earth." Only three years before his death, Tsiolkovsky finally became recognized as a Soviet national hero. His determination that humans should explore space and his dedication to proving that space travel was possible helped lay the foundation of modern space technology.

*Ellen F. Mitchum*

# Benedictus Develops Laminated Glass

*Edouard Benedictus developed the idea and the practical means of making laminated glass, which consists of two sheets of glass with a thin layer of plastic sandwiched between them.*

**What:** Materials
**When:** 1903-1909
**Where:** Paris, France
**Who:**
EDOUARD BENEDICTUS (1879-1930), a
    French artist

## The Quest for Unbreakable Glass

People have been fascinated for centuries by the delicate transparency of glass and the glitter of crystals. They have also been frustrated by the brittleness and fragility of glass. When glass breaks, it forms sharp pieces that can cut people severely. During the 1800's and early 1900's, a number of people demonstrated ways to make "unbreakable" glass. In 1855 in England, the first "unbreakable" glass panes were made by embedding thin wires in the glass. The embedded wire grid held the glass together when it was struck or subjected to the intense heat of a fire. Wire glass is still used in windows that must be fire resistant. The concept of embedding the wire within a glass sheet so that the glass would not shatter was a predecessor of the concept of laminated glass.

A series of inventors in Europe and the United States worked on the idea of using a durable, transparent inner layer of plastic between two sheets of glass to prevent the glass from shattering when it was dropped or struck by an impact. In 1899, Charles E. Wade of Scranton, Pennsylvania, obtained a patent for a kind of glass that had a sheet or netting of mica fused within it to bind it. In 1902, Earnest E. G. Street of Paris, France, proposed coating glass battery jars with pyroxylin plastic (celluloid) so that they would hold together if they cracked. In Swindon, England, in 1905, John Crewe Wood applied for a patent for a material that would prevent automobile windshields from shattering and injuring people when they broke. He proposed cementing a sheet of material such as celluloid between two sheets of glass. When the window was broken, the inner material would hold the glass splinters together so that they would not cut anyone.

## Remembering a Fortuitous Fall

In his patent application, Edouard Benedictus described himself as an artist and painter. He was also a poet, musician, and philosopher who was descended from the philosopher Baruch Benedictus Spinoza; he seemed an unlikely contributor to the progress of glass manufacture. In 1903, Benedictus was cleaning his laboratory when he dropped a glass bottle that held a nitrocellulose solution. The solvents, which had evaporated during the years that the bottle had sat on a shelf, had left a strong celluloid coating on the glass. When Benedictus picked up the bottle, he was surprised to see that it had not shattered: It was starred, but all the glass fragments had been held together by the internal celluloid coating. He looked at the bottle closely, labeled it with the date (November, 1903) and the height from which it had fallen, and put it back on the shelf.

One day some years later (the date is uncertain), Benedictus became aware of vehicular collisions in which two young women received serious lacerations from broken glass. He wrote a poetic account of a daydream he had while he was thinking intently about the two women. He described a vision in which the faintly illuminated bottle that had fallen some years before but had not shattered appeared to float down to him from the shelf. He got up, went into his laboratory, and began to work on an idea that originated with his thoughts of the bottle that would not splinter.

Benedictus found the old bottle and devised a series of experiments that he carried out until the next evening. By the time he had finished, he had made the first sheet of Triplex glass, for which he applied for a patent in 1909. He also founded the Société du Verre Triplex (The Triplex Glass Society) in that year. In 1912, the Triplex Safety Glass Company was established in England. The company sold its products for military equipment in World War I, which began two years later.

Triplex glass was the predecessor of laminated glass. Laminated glass is composed of two or more sheets of glass with a thin layer of plastic (usually polyvinyl butyral, although Benedictus used pyroxylin) laminated between the glass sheets using pressure and heat. The plastic layer will yield rather than rupture when subjected to loads and stresses. This prevents the glass from shattering into sharp pieces. Because of this property, laminated glass is also known as "safety glass."

## Consequences

Even after the protective value of laminated glass was known, the product was not widely used for some years. There were a number of technical difficulties that had to be solved, such as the discoloring of the plastic layer when it was exposed to sunlight; the relatively high cost; and the cloudiness of the plastic layer, which ob-

scured vision—especially at night. Nevertheless, the expanding automobile industry and the corresponding increase in the number of accidents provided the impetus for improving the qualities and manufacturing processes of laminated glass. In the early part of the century, almost two-thirds of all injuries suffered in automobile accidents involved broken glass.

Laminated glass is used in many applications in which safety is important. It is typically used in all windows in cars, trucks, ships, and aircraft. Thick sheets of bullet-resistant laminated glass are used in banks, jewelry displays, and military installations. Thinner sheets of laminated glass are used as security glass in museums, libraries, and other areas where resistance to break-in attempts is needed. Many buildings have large ceiling skylights that are made of laminated glass; if the glass is damaged, it will not shatter, fall, and hurt people below. Laminated glass is used in airports, hotels, and apartments in noisy areas and in recording studios to reduce the amount of noise that is transmitted. It is also used in safety goggles and in viewing ports at industrial plants and test chambers. Edouard Benedictus's recollection of the bottle that fell but did not shatter has thus helped make many situations in which glass is safer for everyone.

*Gary W. Siebein*

# United States Acquires Panama Canal Zone

> *In order to build a canal that would connect the Atlantic and Pacific oceans, the United States supported Panama's secession from Colombia and then purchased rights to a ten-mile "canal zone."*

**What:** International relations; Economics
**When:** November 18, 1903
**Where:** Washington, D.C.; Bogotá, Colombia; and Panama
**Who:**
THEODORE ROOSEVELT (1858-1919), president of the United States from 1901 to 1909
JOHN MILTON HAY (1838-1905), U.S. secretary of state from 1898 to 1905
JOSÉ MANUEL MARROQUÍN (1827-1908), president of Colombia from 1900 to 1904
WILLIAM NELSON CROMWELL (1854-1948), a lawyer for the New Panama Canal Company
PHILIPPE JEAN BUNAU-VARILLA (1860-1940), a French engineer who became minister of Panama

## The Need for a Canal

Even in the time of Christopher Columbus and other early explorers of the New World, it was evident that a canal cutting through Central America would be very helpful to nations that engaged in world trade. For the United States, it became apparent during the Spanish-American War in 1898 that a canal would provide military advantages as well. To get from Puget Sound on the American West Coast to the Cuban war zone, the battleship *Oregon* had to journey all the way around Cape Horn, the southernmost tip of South America. The trip was so long that the *Oregon* almost missed the war.

The Clayton-Bulwer Treaty of 1850 between Great Britain and the United States, however, stipulated that neither nation could build a canal without the participation of the other. Yet Great Britain, eager to maintain its friendship with the United States, agreed to an all-American canal in 1900, in the first Hay-Paunceforte Treaty. The U.S. Senate refused to ratify this agreement, which specified that the canal would be neutral; the Senate insisted that the United States should have the right to defend the canal and keep a military presence there. An amendment to that effect was added, and Great Britain agreed reluctantly in the second Hay-Paunceforte Treaty of 1901.

## Negotiations and Revolution

Yet the battle for a canal was only beginning. The major debate was over the route. Most members of Congress—and, in fact, most Americans—preferred a route that would go through Panama. President William McKinley's Isthmian Canal Commission supported this view, as did Senator John T. Morgan of the Senate Canals Committee. On January 9, 1902, the House approved a Nicaraguan route, and the Senate seemed ready to agree.

At this point, however, William Nelson Cromwell, lawyer for the New Canal Company, appeared on the scene. The New Canal Company had bought the assets and rights of the de Lesseps Company, which had proposed a canal through Panama but had gone out of business without building it. Cromwell was supported by Philippe Jean Bunau-Varilla, an engineer who was committed to constructing a canal through Panama. Cromwell offered the United States the rights to the Panamanian route for $40 million.

Cromwell was a skilled lobbyist, and the Panamanian route actually proved to have certain engineering advantages that would make a canal easier to build. Moreover, the proposed route through Nicaragua lay rather close to an erupting volcano. In view of these factors, President Theodore Roosevelt, the Canal Commission, and Congress all changed their minds. The re-

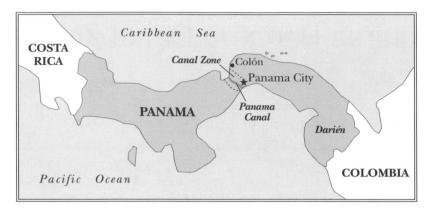

sult was the Spooner Act, which directed the president to build a canal through Panama if he could obtain the consent of Colombia, which owned Panama at that time. If Colombia did not agree, the Nicaraguan route was to be chosen.

The nation of Colombia was racked by severe internal conflict at the turn of the century. A coup had just deposed the president, and Vice President José Manuel Marroquín had been elevated to the president's office. In this tense situation the United States began to put great pressure on Colombia to sign a canal treaty. John Milton Hay, U.S. secretary of state, pushed through the Hay-Herrán Treaty, which was signed on January 22, 1903. Under this agreement, Colombia was to receive a onetime payment of $10 million and annual rents of $250,000, beginning in nine years, for rights to a canal zone six miles wide. The Colombian negotiators unwisely agreed not to ask for any of the $40 million being paid to the French canal company.

The treaty was quickly approved by the U.S. Senate on March 17, 1903, and the U.S. government settled back to wait for the expected approval by the Colombian legislature. President Roosevelt and members of the State Department failed, however, to understand the serious objections that the proud Colom-

bians had to the treaty. Colombia was being asked to give up some of its sovereignty at a time when the United States was expanding its power worldwide. The treaty did not offer enough compensation for a route that had many engineering advantages and a profitable railroad. Furthermore, it did not take account of the fact that the French company's rights would expire in 1904 in any case, so that the United States would be freed from its obligation to pay the $40 million. With only three dissenting votes, the Colombian senate rejected the treaty.

President Roosevelt reacted angrily to Colombia's decision and prepared to gain the canal zone by other means. According to State Department officials, an 1846 treaty between the United States and Colombia (then New Grenada) could be read to suggest that Americans had the right

*Construction of the canal, around 1909.*

to guarantee the neutrality of and free passage through the Panamanian isthmus. The State Department and the president began to argue that the Colombian vote against the canal treaty was a denial of free passage.

There were those in Panama who had long wanted to separate from Colombia and, in fact, had been prevented when the United States came to Colombia's aid. Bunau-Varilla now led the way in encouraging a Panamanian revolution and kept Roosevelt informed of its progress. Claiming to be guaranteeing free transit in the isthmus, Roosevelt sent the cruiser *Nashville* to the scene.

Panama City, on the Pacific side of the country, was taken by the revolutionaries on November 2, 1903. On the Atlantic side, the *Nashville* landed American forces at Colón to keep Colombian troops from passing across the isthmus and defeating the rebels. Bunau-Varilla bribed the Colombian general to leave Colón, and the revolution succeeded without lives being taken.

On November 18, 1903, Panama's new minister, Bunau-Varilla, joined Secretary Hay in signing the Hay-Bunau-Varilla Treaty. The United States agreed to pay $10 million cash and an annual rent of $250,000 in return for a canal zone ten miles wide that was to remain under permanent American control.

## Consequences

The Americans took possession of the canal site on May 4, 1904, and built the canal at a cost of more than $300 million. The Panama Canal was opened for shipping on August 15, 1914, two weeks after World War I began in Europe. Throughout the twentieth century the canal was extremely important for international shipping and as a U.S. military outpost.

The United States continued to operate the canal under the 1903 treaty provisions until the 1960's, when growing Panamanian nationalism led to pressure for the canal's return to Panamanian control. Throughout Latin America, the canal had become a symbol for North American "imperialism"—the apparent U.S. readiness to ignore the rights of other nations in order to enrich itself. On the last day of 1999, the United States finally turned over full ownership and control of the canal to Panama.

# Wright Brothers Fly First Successful Airplane

*The Wright brothers invented, built, and flew the world's first engine-powered airplane.*

**What:** Space and aviation
**When:** December 17, 1903
**Where:** Kitty Hawk, North Carolina
**Who:**
WILBUR WRIGHT (1867-1912), an
   American inventor
ORVILLE WRIGHT (1871-1948), an
   American inventor
OCTAVE CHANUTE (1832-1910), a French-
   born American civil engineer

## A Careful Search

Although people have dreamed about flying since the time of the ancient Greeks, it was not until the late eighteenth century that hot-air balloons and gliders made human flight possible. It was not until the late nineteenth century that enough experiments had been done with kites and gliders that people could begin to think seriously about powered, heavier-than-air flight. Two of these people were Wilbur and Orville Wright.

The Wright brothers were more than just tinkerers who accidentally found out how to build a flying machine. In 1899, Wilbur wrote the Smithsonian Institution for a list of books to help them learn about flying. They used the research of people such as George Cayley, Octave Chanute, Samuel Langley, and Otto Lilienthal to help them plan their own experiments with birds, kites, and gliders. They even built their own wind tunnel. They never fully trusted the results of other people's research, so they repeated the experiments of others and drew their own conclusions. They shared these results with Octave Chanute, who was able to offer them lots of good advice. They were continuing a tradition of excellence in engineering that began with careful research and avoided dangerous trial and error.

## Slow Success

Before the brothers had set their minds to flying, they had built and repaired bicycles. This was a great help to them when they put their research into practice and actually built an airplane. From building bicycles, they knew how to work with wood and metal to make a lightweight but sturdy machine. Just as important, from riding bicycles, they got ideas about how an airplane needed to work.

They could see that both bicycles and airplanes needed to be fast and light. They could also see that airplanes, like bicycles, needed to be kept under constant control to stay balanced, and that this control would probably take practice. This was a unique idea. Instead of building something solid that was controlled by levers and wheels like a car, the Wright brothers built a flexible airplane that was controlled partly by the movement of the pilot, like a bicycle.

The result was the 1903 *Wright Flyer.* The Flyer had two sets of wings, one above the other, which were about 12 meters from tip to tip. They made their own 12-horsepower engine, as well as the two propellers the engine spun. The craft had skids instead of wheels. On December 14, 1903, the Wright brothers took the *Wright Flyer* to the shores of Kitty Hawk, North Carolina, where Wilbur Wright made the first attempt to fly the airplane.

The first thing Wilbur found was that flying an airplane was not as easy as riding a bicycle. One wrong move sent him tumbling into the sand only moments after takeoff. Wilbur was not seriously hurt, but a few more days were needed to repair the *Wright Flyer.*

On December 17, 1903, at 10:35 A.M., after eight years of research and planning, Orville Wright took to the air for a historic twelve seconds. He covered 37 meters of ground and 152 meters of air space. Both brothers took two flights that morning. On the fourth flight, Wil-

*First Orville Wright, then his brother Wilbur, fly their engine-powered plane at Kitty Hawk, North Carolina.*

bur flew for fifty-nine seconds over 260 meters of ground and through more than 800 meters of air space. After he had landed, a sudden gust of wind struck the plane, damaging it beyond repair. Yet no one was able to beat their record for three years.

### Consequences

Those first flights in 1903 got little publicity. Only a few people, such as Octave Chanute, understood the significance of the Wright brothers' achievement. For the next two years, they continued to work on their design, and by 1905 they had built the *Wright Flyer III*. Although Chanute tried to get them to enter flying contests, the brothers decided to be cautious and try to get their machine patented first, so that no one would be able to steal their ideas.

News of their success spread slowly through the United States and Europe, giving hope to others who were working on airplanes of their own. When the Wright brothers finally went public with the *Wright Flyer III*, they inspired many new advances. By 1910, when the brothers started flying in air shows and contests, their feats were matched by another American, Glen Hammond Curtiss. The age of the airplane had arrived.

Later in the decade, the Wright brothers began to think of military uses for their airplanes. They signed a contract with the U.S. Army Signal Corps and agreed to train military pilots.

Aside from these achievements, the brothers from Dayton, Ohio, set the standard for careful research and practical experimentation. They taught the world not only how to fly but also how to design airplanes. Indeed, their methods of purposeful, meaningful, and highly organized research had an impact not only on airplane design but also on the field of aviation science in general.

*Harry J. Eisenman*

**83**

# Edison Develops First Alkaline Storage Battery

> *Thomas Edison developed the nickel-iron alkaline battery as a lightweight, inexpensive portable power source for vehicles with electric motors.*

**What:** Energy
**When:** 1904
**Where:** West Orange, New Jersey
**Who:**
THOMAS ALVA EDISON (1847-1931), American chemist, inventor, and industrialist
HENRY FORD (1863-1947), American inventor and industrialist
CHARLES F. KETTERING (1876-1958), American engineer and inventor

## A Three-Way Race

The earliest automobiles were little more than pairs of bicycles harnessed together within a rigid frame, and there was little agreement at first regarding the best power source for such contraptions. The steam engine, which was well established for railroad and ship transportation, required an external combustion area and boiler. Internal combustion engines required hand cranking, which could cause injury if the motor backfired. Electric motors were attractive because they did not require the burning of fuel, but they required batteries that could store a considerable amount of energy and could be repeatedly recharged. Ninety percent of the motorcabs in use in New York City in 1899 were electrically powered.

The first practical storage battery, which was invented by the French physicist Gaston Planté in 1859, employed electrodes (conductors that bring electricity into and out of a conducting medium) of lead and lead oxide and a sulfuric acid electrolyte (a solution that conducts electricity). In somewhat improved form, this remained the only practical rechargeable battery at the beginning of the twentieth century. Edison considered the lead acid cell (battery) unsuitable as a power source for electric vehicles because using lead, one of the densest metals known, resulted in a heavy battery that added substantially to the power requirements of a motorcar. In addition, the use of an acid electrolyte required that the battery container be either nonmetallic or coated with a nonmetal and thus less dependable than a steel container.

### The Edison Battery

In 1900, Edison began experiments aimed at developing a rechargeable battery with inexpensive and lightweight metal electrodes and an alkaline electrolyte so that a metal container could be used. He had already been involved in manufacturing the nonrechargeable battery known as the Lalande cell, which had zinc and copper oxide electrodes and a highly alkaline sodium hydroxide electrolyte. Zinc electrodes could not be used in a rechargeable cell because the zinc would dissolve in the electrolyte. The copper electrode also turned out to be unsatisfactory. After much further experimentation, Edison settled on the nickel-iron system for his new storage battery. In this system, the power-producing reaction involved the conversion of nickel oxide to nickel hydroxide together with the oxidation of iron metal to iron oxide, with both materials in contact with a potassium hydroxide solution. When the battery was recharged, the nickel hydroxide was converted into oxide and the iron oxide was converted back to the pure metal.

Although the basic ingredients of the Edison cell were inexpensive, they could not readily be obtained in adequate purity for battery use. Edison set up a new chemical works to prepare the needed materials. He purchased impure nickel

**84**

alloy, which was then dissolved in acid, purified, and converted to the hydroxide. He prepared pure iron powder by using a multiple-step process. For use in the battery, the reactant powders had to be packed in pockets made of nickel-plated steel that had been perforated to allow the iron and nickel powders to come into contact with the electrolyte. Because the nickel compounds were poor electrical conductors, a flaky type of graphite was mixed with the nickel hydroxide at this stage.

Sales of the new Edison storage battery began in 1904, but within six months it became apparent that the battery was subject to losses in power and a variety of other defects. Edison took the battery off the market in 1905 and offered full-price refunds for the defective batteries. Not a man to abandon an invention, however, he spent the next five years examining the failed batteries and refining his design. He discovered that the repeated charging and discharging of the battery caused a shift in the distribution of the graphite in the nickel hydroxide electrode. By using a different type of graphite, he was able to eliminate this problem and produce a very dependable power source.

The Ford Motor Company, founded by Henry Ford, a former Edison employee, began the large-scale production of gasoline-powered automobiles in 1903 and introduced the inexpensive, easy-to-drive Model T in 1908. The introduction of the improved Edison battery in 1910 gave a boost to electric car manufacturers, but their new position in the market would be short-lived. In 1911, Charles Kettering invented an electric starter for gasoline-powered vehicles that eliminated the need for troublesome and risky hand cranking. By 1915, this device was available on all gasoline-powered automobiles, and public interest in electrically powered cars rapidly diminished. Although the Kettering starter required a battery, it required much less capacity than an electric motor would have and was almost ideally suited to the six-volt lead-acid battery.

## Consequences

Edison lost the race to produce an electrical power source that would meet the needs of automotive transportation. Instead, the internal combustion engine developed by Henry Ford

*Thomas Alva Edison.*

became the standard. Interest in electrically powered transportation diminished as immense reserves of crude oil, from which gasoline could be obtained, were discovered first in the southwestern United States and then on the Arabian peninsula. Nevertheless, the Edison cell found a variety of uses and has been manufactured continuously throughout most of the twentieth century much as Edison designed it.

Electrically powered trucks proved to be well suited for local deliveries, and some department stores maintained fleets of such trucks into the mid-1920's. Electrical power is still preferable to internal combustion for indoor use, where exhaust fumes are a significant problem, so forklifts in factories and passenger transport vehicles at airports still make use of the Edison-type power source. The Edison battery also continues to be used in mines, in railway signals, in some communications equipment, and as a highly reliable source of standby emergency power.

*Donald R. Franceschetti*

**85**

# Elster and Geitel Devise First Practical Photoelectric Cell

> *Julius Elster and Hans Friedrich Geitel's pioneering work on the photoelectric effect and photoelectric cells was of decisive importance in the electron theory of metals.*

**What:** Energy
**When:** 1904
**Where:** Wolfenbüttel, Germany
**Who:**
JULIUS ELSTER (1854-1920), a German experimental physicist
HANS FRIEDRICH GEITEL (1855-1923), a German physicist
WILHELM HALLWACHS (1859-1922), a German physicist

## Early Photoelectric Cells

The photoelectric effect was known to science in the early nineteenth century when the French physicist Alexandre-Edmond Becquerel wrote of it in connection with his work on glass-enclosed primary batteries. He discovered that the voltage of his batteries increased with intensified illumination and that green light produced the highest voltage. Since Becquerel researched batteries exclusively, however, the liquid-type photocell was not discovered until 1929, when the Wein and Arcturus cells were introduced commercially. These cells were miniature voltaic cells arranged so that light falling on one side of the front plate generated a considerable amount of electrical energy. The cells had short lives, unfortunately; when subjected to cold, the electrolyte froze, and when subjected to heat, the gas generated would expand and explode the cells.

What came to be known as the "photoelectric cell," a device connecting light and electricity, had its beginnings in the 1880's. At that time, scientists noticed that a negatively charged metal plate lost its charge much more quickly in the light (especially ultraviolet light) than in the dark. Several years later, researchers demonstrated that this phenomenon was not an "ionization" effect because of the air's increased conductivity, since the phenomenon took place in a vacuum but did not take place if the plate were positively charged. Instead, the phenomenon had to be attributed to the light that excited the electrons of the metal and caused them to fly off: A neutral plate even acquired a slight positive charge under the influence of strong light. Study of this effect not only contributed evidence to an electronic theory of matter—and, as a result of some brilliant mathematical work by the physicist Albert Einstein, later increased knowledge of the nature of radiant energy—but also further linked the studies of light and electricity. It even explained certain chemical phenomena, such as the process of photography. It is important to note that all the experimental work on photoelectricity accomplished prior to the work of Julius Elster and Hans Friedrich Geitel was carried out before the existence of the electron was known.

## Explaining Photoelectric Emission

After the English physicist Sir Joseph John Thomson's discovery of the electron in 1897, investigators soon realized that the photoelectric effect was caused by the emission of electrons under the influence of radiation. The fundamental theory of photoelectric emission was put forward by Einstein in 1905 on the basis of the German physicist Max Planck's quantum theory (1900). Thus, it was not surprising that light was found to have an electronic effect. Since it was known that the longer radio waves could shake electrons into resonant oscillations and the shorter X rays could detach electrons from the atoms of gases, the intermediate waves of visual light would have been expected to have some effect upon electrons—such as detaching them from metal plates and therefore setting up a difference of potential. The photoelectric cell, developed by Elster and Geitel in 1904, was a practical device that made use of this effect.

In 1888, Wilhelm Hallwachs observed that an electrically charged zinc electrode loses its charge when exposed to ultraviolet radiation if the charge is negative, but is able to retain a positive charge under the same conditions. The following year, Elster and Geitel discovered a photoelectric effect caused by visible light; however, they used the alkali metals potassium and sodium for their experiments instead of zinc.

The Elster-Geitel photocell (a vacuum emission cell, as opposed to a gas-filled cell) consisted of an evacuated glass bulb containing two electrodes. The cathode consisted of a thin film of a rare, chemically active metal (such as potassium) that lost its electrons fairly readily; the anode was simply a wire sealed in to complete the circuit. This anode was maintained at a positive potential in order to collect the negative charges released by light from the cathode. The Elster-Geitel photocell resembled two other types of vacuum tubes in existence at the time: the cathode-ray tube, in which the cathode emitted electrons under the influence of a high potential, and the thermionic valve (a valve that permits the passage of current in one direction only), in which it emitted electrons under the influence of heat. Like both of these vacuum tubes, the photoelectric cell could be classified as an "electronic" device.

The new cell, then, emitted electrons when stimulated by light, and at a rate proportional to the intensity of the light. Hence, a current could be obtained from the cell. Yet Elster and Geitel found that their photoelectric currents fell off gradually; they therefore spoke of "fatigue" (instability). It was discovered later that most of this change was not a direct effect of a photoelectric current's passage; it was not even an indirect effect but was caused by oxidation of the cathode by the air. Since all modern cathodes are enclosed in sealed vessels, that source of change has been completely abolished. Nevertheless, the changes that persist in modern cathodes often are indirect effects of light that can be produced independently of any photoelectric current.

*Julius Elster (left) and Hans Friedrich Geitel.*

## Consequences

The Elster-Geitel photocell was, for some twenty years, used in all emission cells adapted for the visible spectrum, and throughout the twentieth century, the photoelectric cell has had a wide variety of applications in numerous fields. For example, if products leaving a factory on a conveyor belt were passed between a light and a cell, they could be counted as they interrupted the beam. Persons entering a building could be counted also, and if invisible ultraviolet

**87**

rays were used, those persons could be detected without their knowledge. Simple relay circuits could be arranged that would automatically switch on street lamps when it grew dark. The sensitivity of the cell with an amplifying circuit enabled it to "see" objects too faint for the human eye, such as minor stars or certain lines in the spectra of elements excited by a flame or discharge. The fact that the current depended on the intensity of the light made it possible to construct photoelectric meters that could judge the strength of illumination without risking human error—for example, to determine the right exposure for a photograph.

A further use for the cell was to make talking films possible. The early "talkies" had depended on gramophone records, but it was very difficult to keep the records in time with the film. Now, the waves of speech and music could be recorded in a "sound track" by turning the sound first into current through a microphone and then into light with a neon tube or magnetic shutter; next, the variations in the intensity of this light on the side of the film were photographed. By reversing the process and running the film between a light and a photoelectric cell, the visual signals could be converted back to sound.

*Genevieve Slomski*

# Hartmann Discovers Interstellar Matter

*Johannes Franz Hartmann discovered the first indications that matter exists in the vast space between the stars that had long been thought to be an empty vacuum.*

**What:** Astronomy
**When:** 1904
**Where:** Potsdam, Germany
**Who:**
JOHANNES FRANZ HARTMANN (1865-1936), a German astronomer
HENRI-ALEXANDRE DESLANDRES (1853-1948), a French astronomer
VESTO MELVIN SLIPHER (1875-1969), an American astronomer
EDWIN BRANT FROST (1866-1935), an American astronomer
EDWARD EMERSON BARNARD (1857-1923), an American astronomer
SIR ARTHUR STANLEY EDDINGTON (1882-1944), an English astronomer

## The Study of Starlight

The general approach to astronomy changed in the late nineteenth and early twentieth centuries. Rather than studying only the motions and positions of stars, planets, and so forth, astronomers began using new tools that allowed them to learn about the physical makeup of these objects. The English mathematician and physicist Sir Isaac Newton was the first to realize that white light (direct sunlight) is made up of light of many colors, or wavelengths, that blend together and appear white.

By the end of the nineteenth century, spectroscopes had been added to telescopes in order to put this knowledge to use in astronomy. The spectroscope is an instrument that breaks starlight down into its spectrum (band of colors) by passing it through a prism or reflecting it from a finely ruled grating. Astronomers can use the colors and dark bands of an object's spectrum to determine its chemical and other properties. In particular, the spectrum of a star reveals its velocity along a line of sight to the observer (the rate at which the distance between the star and the observer is changing) and the velocity with which it turns on its axis. The dark lines appearing in a star's spectrum reveal the presence of particular elements, each of which absorbs certain wavelengths of light. Thus, it is possible to identify the elements in a star's atmosphere, or in any other cloud of gas in between the observer and the star, by measuring the wavelength at which a band appears.

In 1900, Henri-Alexandre Deslandres discovered that lines in the spectrum of the star Theta Orionis were moving rapidly compared to the rest of the spectrum. Johannes Franz Hartmann followed up this observation at the Potsdam Astrophysical Observatory, using state-of-the-art spectrographs of the star Delta Orionis. Hartmann confirmed the movement, but found that the shifts were slower than Deslandres had described. He also discovered that some of the lines shifted differently in relation to one another.

## The Doppler Effect

The reason for all the motion turned out to be what is called the Doppler effect. As an object that is emitting waves (light waves, for example, or sound waves) approaches an observer, the wavelengths appear shorter than they really are. Light looks bluer and sounds seem higher-pitched. As the object moves away, however, the wavelengths appear longer—light looks redder and sounds seem lower-pitched. The amount of change or shift has to do with the velocity of the moving object. This effect can appear to shift the lines in the star's spectrum either redward or blueward, and it is the tool used in measuring the velocity of an individual star moving directly toward or away from the observer.

The Doppler effect can be observed in double-star systems such as Delta Orionis. (A double-star

**89**

is described as two stars so close together that they look like one star to the naked eye.) The stars in the system, as they orbit each other, are moving alternately toward or away from the earth, and thus their wavelengths appear to be either shorter or longer.

The motion of the double stars explained the type of shift (redward and then blueward) observed for most of the spectral lines in Delta Orionis. It did not, however, explain the lines that had a different Doppler motion, one that did not move the lines from redward to blueward and back but moved the lines in the same direction and by the same amount in a constant manner. Hartmann called these lines "stationary." He at first thought that the unchanging Doppler shift represented the motion of the entire system with respect to Earth. Hartmann later carefully ruled out this explanation, however, as well as the possibility that Earth's atmosphere was responsible for producing the lines. The remaining explanation was that there was a fixed cloud of matter that was producing the lines. Because the lines were of the right wavelength for the element calcium, Hartmann concluded that a cloud of calcium gas was causing the lines. Yet Hartmann was not sure where the calcium was located. He believed it could be found somewhere between Earth and the double-star system and had nothing to do with the double stars. Or it could be part of the double-star system.

## Consequences

The discovery of the stationary lines was only one step in a long story. Hartmann's work was confirmed by Vesto Melvin Slipher at the Naval Observatory in Flagstaff, Arizona, and by Edwin Brant Frost at Yerkes Observatory in Wisconsin. Slipher made more observations of Delta Ori-

onis and was the first to suggest that it was truly interstellar matter that was responsible for the lines. Frost, in 1909, observed other stars whose spectra also displayed stationary lines.

Some astronomers disagreed with Hartmann, Slipher, and Frost about the nature of the lines and used several observations to argue that the lines were, in fact, connected with the stars in question. Reynold Kenneth Young noted in 1920 that stationary lines appeared only in the spectra of relatively young stars and argued that there must be some connection between the star and the calcium gas causing the lines.

In the late 1920's, astronomers Sir Arthur Stanley Eddington and Otto Struve finally settled the question of interstellar matter. Struve wrote several papers on the calcium lines in the spectra of stars that led scientists to believe that interstellar matter was responsible for the stationary lines. Eddington explained why the lines appeared only in relatively young stars by showing that older stars have spectra that would not reveal the presence of the stationary lines easily, even when they were present.

The presence of interstellar matter affected various aspects of astronomy. For example, when interstellar matter appears in the form of dust, it absorbs some of the light from stars and makes stars appear dimmer (and more distant) than they really are. This realization forced astronomers to rewrite their star maps to include changes in distance that arose during further study. Also, interstellar matter plays an important role in theories of how stars form because it provides the material for star formation. Because these clouds move about galactic centers, their motion, when measured, can give indications of the speed and direction of galactic rotation.

*Mary Hrovat*

# Kapteyn Determines Rotation of Galaxy

*Jacobus Cornelis Kapteyn discovered that the proper motions of the stars tended in two opposite directions, implying the rotation of the galaxy.*

**What:** Astronomy
**When:** 1904
**Where:** Groningen, The Netherlands
**Who:**
JACOBUS CORNELIS KAPTEYN (1851-1922), a Dutch astronomer who devoted much of his life to the detailed study of the Milky Way Galaxy
HARLOW SHAPLEY (1885-1972), an American astronomer who disputed the Kapteyn view of the universe
HEBER DOUST CURTIS (1872-1942), an American astronomer who supported the Kapteyn view

## Measuring the Galaxy

Following several years of study of the structure of the Milky Way, Jacobus Cornelis Kapteyn concluded, on the basis of exhaustive star counts in sampled portions of the sky, that the "proper" motions (motions across the sky, as opposed to motions directly toward, or directly away from, the earth) of the stars were not random. Instead, the stars move in opposite directions in different parts of the sky. Kapteyn's discovery had major cosmological implications, including the implication that the Milky Way was a rotating galaxy.

Kapteyn studied at the University of Utrecht and was awarded a doctorate of physics in 1875. Following completion of his studies, he pursued a position at the Leiden Observatory, where he became a successful astronomer. By 1878, he had won a professorship in astronomy at the University of Groningen. Upon arrival at Groningen, however, he was unable to raise funds to equip an observatory. Undaunted, he became a leader in arranging collaboration with other astronomers, notably with Sir David Gill at the Cape of Good Hope in South Africa. From 1885 to 1899, Kapteyn analyzed photographic plates taken by Gill and assisted in producing a star catalog with nearly a half million entries. He later continued studies as a visiting astronomer at the Mount Wilson Observatory in Pasadena, California.

If the galaxy were essentially a static collection of stars, then, theoretically, their movements across the sky (proper motions) ought to be random. Only those stars relatively close to Earth have measurable proper motions, but there were enough of these that Kapteyn could analyze their behavior. He found consistent movement in two opposite directions for neighboring stars. The motions, one toward the constellation Orion and one toward the constellation Sagittarius, implied what Kapteyn called the "streaming of stars," or a pattern of movement that indicated that stars were moving in opposite directions.

In 1904, Kapteyn made the announcement of his discovery at the International Congress of Science at the World's Fair in St. Louis. Desiring to follow up these early discoveries, Kapteyn devoted much of his time after 1904 to organizing the worldwide astronomical community in cooperative efforts to photograph and analyze 206 portions of the sky. This collaborative effort was intended to study such features of stars as their magnitude (brightness), color, proper motion, radial velocity (velocity directly toward or away from the observer), and spectral type. The outbreak of World War I in 1914, however, largely ended such international cooperation. Kapteyn did, however, generate enough information to convince him that the galaxy was ellipsoidal (lens-shaped), its major axis measuring approximately 52,000 light-years. His main concern about the accuracy of his model had to do with whether the sun was near the center of the Milky Way, a point that also caused other astronomers to question the model.

Kapteyn was also deeply concerned about how the absorption of light by gas and dust might have affected his model. He realized that appreciable absorption in the direction of the galactic center would have caused his central position for the sun to be wrong and, more important, would have caused him to underestimate the size of the galaxy. As a result, he continued to look for some means of measuring the amount of absorption. Having failed to measure any sizable amounts by 1918, he convinced himself that it was negligible and that his model was accurate.

## Debate

Kapteyn's inability to measure the effects of the absorption of light did, however, cause him to underestimate the size of the galaxy. Harlow Shapley, who was studying many of the same problems at Mount Wilson in California, proposed a much larger universe. By 1914, Shapley, using superior equipment, was confident in his distance measurements, and he argued that the

Yerkes Observatory, University of Chicago

*Jacobus Cornelis Kapteyn.*

streaming of the stars, which was firmly established, would have to be reinterpreted.

Shapley debated his views, including his much larger estimate of the scale of the universe, with Heber Doust Curtis at the annual meeting of the National Academy of Sciences on April 26, 1920. Shapley presented his position, while Curtis defended the perspectives of Kapteyn's followers. Both sides, on the basis of later discoveries, were partially right and partially wrong. Kapteyn's view of "nebulas" as island universes was eventually established, and Shapley's estimation of the dimensions of the Milky Way was found to be more accurate. While Kapteyn's interpretation of the cause of the streaming effect turned out to be only partially correct, his recognition of the phenomenon remained extremely important to cosmology.

## Consequences

The efforts of Kapteyn in the early twentieth century demonstrate clearly that careful observation and experimentation can have exceptional value, even if the subsequent interpretation of such efforts is erroneous. Kapteyn's view of the universe dominated the first two decades of the twentieth century. His work significantly affected the cosmology of the day, especially his view of nebulas as "island universes" beyond the bounds of the Milky Way. The Belgian astrophysicist Georges Lemaître's cosmology of the mid-1920's followed this view, which, in turn, led to the currently accepted big bang theory.

By the late 1920's, as the result of work by astronomers Bertil Lindblad, Jan Hendrik Oort, and Edwin Powell Hubble, a more accurate picture emerged of Earth's galaxy as a spiral rotating in the same fashion as the other exterior galaxies. In 1927, Oort, a former student of Kapteyn, demonstrated (once the sun's position near the edge of the galaxy was established) that the apparent motion of stars in one direction was the result of the sun lagging behind stars nearer the center. The apparent motion in the opposite direction was the result of slower outer stars falling behind the sun. Kapteyn's view of the universe was thus superseded by the research that it had stimulated.

*Ivan L. Zabilka*

# Gorgas Controls Yellow Fever in Panama Canal Zone

> *Recognizing that mosquitoes spread malaria and yellow fever, William Crawford Gorgas applied strict sanitary controls within the Panama region, enabling construction of the Panama Canal.*

**What:** Medicine
**When:** 1904-1905
**Where:** Panama Canal Zone
**Who:**
WILLIAM CRAWFORD GORGAS (1854-1920), an American army surgeon and sanitarian
SIR RONALD ROSS (1857-1932), an English physician who was awarded the 1902 Nobel Prize in Physiology or Medicine
WALTER REED (1851-1902), a bacteriologist
JOHN FRANK STEVENS (1853-1943), the chief engineer of the Isthmanian Canal Commission

## Canal Efforts Blocked by Insects

With the discovery of gold in California in 1848, considerable interest quickly developed in the United States concerning the construction of a transoceanic canal through Central America that would shorten the time and distance necessary to travel between the East and West coasts. Following his success in building the Suez Canal during the 1860's, the French diplomat Ferdinand de Lesseps initiated efforts in the 1870's to construct such a canal.

The challenges facing the highly skilled French engineers arriving in Panama at the start of the project, in 1881, were formidable. They included the taming of the unpredictable Chagres River and excavation difficulties at Culebra, where geological formations resulted in a continual problem with slides. Yet the greatest obstacle was the high incidence of various diseases, especially malaria and yellow fever. After eight years in Panama, the French had lost an estimated two thousand workers to yellow fever and more than fifty-five hundred to other illnesses. A second brief attempt at continuing excavations ensued during the 1890's; in the end, however, de Lesseps and the French gave up. Despite these failures, American interest in the canal intensified after the Spanish-American War. Beginning in 1904, the United States embarked on a canal construction program that succeeded where the French had not. Much of the American success can be credited to advances in tropical medicine related to the eradication of malaria and yellow fever.

During the late 1890's, the English physician Sir Ronald Ross, working in India, showed that malaria is spread by the *Anopheles* mosquito. It became obvious to Ross that in order to stamp out malaria, one had to prevent mosquitoes from infecting humans. His ideas were systematically outlined in his *Mosquito Brigades and How to Organise Them*, published in 1901.

## Stopping Mosquitoes Where They Live

Concurrent with the publication of *Mosquito Brigades*, a team of U.S. Army doctors, including Walter Reed and William Crawford Gorgas, were eliminating yellow fever in Havana, Cuba, using similar techniques. Reed, borrowing from the work of Cuban physician Carlos Finley, had concluded that only one type of mosquito, *Stegomyia fasciata*, was responsible for transmission of the dreaded yellow fever. To eliminate yellow fever, the doctors realized, they had to prevent the spread of the insect by keeping the female *Stegomyia fasciata* from laying its eggs.

Gorgas would employ this theoretical understanding of yellow fever in his practical public-health efforts in Havana. He learned that fresh

**93**

Library of Congress

*William Crawford Gorgas.*

Leland O. Howard, water left standing was covered with a thin film of kerosene or oil wherever this method was feasible. In addition, adult mosquitoes were killed by the fumigation of every house in Havana where a case of yellow fever had appeared. Since the insect had only a ten-day life cycle, the *Stegomyia* population diminished rapidly.

With this significant accomplishment, the stage was set for Gorgas's work in Panama. A deeply religious man, Gorgas had long thought of his life experiences as ordered by God with the purpose of readying him for this task. In 1904, he was sent to Panama as the Isthmanian Canal Commission's chief sanitary officer, and he was charged with the elimination of malaria and yellow fever there. Despite his success in Cuba, however, he met with resistance from the political leadership in Washington and from key members of the Isthmanian Canal Commission who did not believe the mosquito theory. Gorgas's arguments fell on deaf ears until a deadly epidemic hit the Canal Zone in the spring and early summer of 1905 and the subsequent appointment of John Frank Stevens as chief engineer.

Stevens understood that before construction could begin effectively, diseases such as yellow fever and malaria had to be eliminated, and he gave Gorgas unequivocal support. In 1905, therefore, Stevens's engineering department gave Gorgas first priority in terms of men and materials. By the fall of 1905, Gorgas had more than four thousand men engaged in sanitation work, and his budget was increased dramatically. Cases of yellow fever in the Canal Zone fell from sixty-two in June, 1905, to twenty-seven in August and one in December, with no further outbreaks in 1906. Similar techniques were used to combat malaria; that disease was reduced but not eradicated, since the *Anopheles* that carried it had a much broader range of flight and bred in a more widespread area than the *Stegomyia* did.

water had often been left standing inside the sick rooms of French engineers and workers at the Panama Canal Company hospital at Ancón during the 1880's; tragically, the water had served as a mosquito breeding ground. Under Gorgas's direction, such water was immediately disposed off or sealed off with wooden lids or screens. Further, following a suggestion by entomologist

## Consequences

Gorgas's measures to control the spread of mosquito-transmitted diseases in Panama had both short-term and long-term significance. Although malaria proved to be much more difficult to contain than yellow fever, construction on the Panama Canal progressed steadily to its completion and the opening of the waterway to commercial traffic on January 1, 1915. In 1907, Gorgas was appointed by President Theodore Roosevelt as a member of the Isthmanian Canal Commission; the following year, he was elected president of the American Medical Association.

Promoted to the rank of brigadier general and named surgeon general of the U.S. Army in 1914, Gorgas played an influential role in sanitation work during World War I, retiring in 1918; he subsequently served as director of the yellow fever program for the International Health Board of the Rockefeller Foundation. His pioneering work in the Canal Zone is regarded both as an important contribution to a historic engineering achievement and as a model of practical public-health work.

*John A. Heitmann*

# Russia and Japan Battle over Territory

> *The conflicting imperial ambitions of Russia and Japan led to a war over Manchuria and Korea.*

**What:** War
**When:** February 9, 1904-September 5, 1905
**Where:** Manchuria, Korea, and the Yellow Sea
**Who:**
NICHOLAS II (1868-1918), czar of Russia from 1894 to 1917
SERGEI YULIEVICH WITTE (1849-1915), Russian minister of finance from 1892 to 1903
ALEKSEI NIKOLAEVICH KUROPATKIN (1848-1921), Russian minister of war and supreme commander of Russian forces in the Far East from 1904 to 1905
ZINOVI PETROVICH ROZHDESTVENSKI (1848-1909), commander of the Russian Baltic fleet
WILLIAM II (1859-1941), emperor of Germany from 1888 to 1918
THEODORE ROOSEVELT (1858-1919), president of the United States from 1901 to 1909

## War Begins

Influenced by the Western nations' drive to expand their control over other territories, both Russia and Japan had, by the end of the nineteenth century, become imperialist powers themselves in the Far East. When both of them attempted to expand their holdings in Manchuria and Korea, they came into conflict.

Between 1901 and 1904, Japanese and Russian officials negotiated to try to agree on the exact limits of their countries' control in these regions. Among the Russians, however, there were two distinct groups that made negotiations between the two countries difficult. Count Sergei Yulievich Witte, minister of finance, and Count Vladimir Nikolaevich Lamsdorff, minister of foreign

affairs, joined General Aleksei Nikolaevich Kuropatkin, minister of war, in favoring accommodation with Japan.

The other group, led by Aleksandr Bezobrazov, persuaded Czar Nicholas II that the Russians needed to respond to Japan with a policy of military strength.

In 1903, there was an increase in revolutionary activity in Russia. In times of war, feelings of patriotism tend to increase. Russian government officials thought that if Russia were threatened by a foreign foe, the Russian people would rally around the government, and revolution would become less attractive.

Meanwhile, the Japanese shored up their position by making an alliance with Great Britain, a traditional foe of Russia, in 1902. With increased confidence, the Japanese representatives became just as stubborn as the Russians during the 1903 negotiations. The talks between Russia and Japan reached a deadlock in early 1904.

On February 9, 1904, the Japanese mounted a surprise attack on the Russian fleet at Port Arthur, in the Yellow Sea, and decimated it. Russian national pride was dealt a major blow; after all, Russia was a huge country, while Japan was tiny. As the Russians scrambled to respond to the attack, Japanese troops were able to disembark on the mainland without opposition.

Because the Trans-Siberian Railway had not been completed, Russian reinforcements were slow to arrive, and the Japanese forces defeated Russian armies in Manchuria. In a series of major battles along the Yalu River on the border between Korea and Manchuria, and at Liaoyang and Mukden in Manchuria itself, the Japanese defeated Russia again and again. A standoff was finally reached near Mukden in October, 1904.

Meanwhile, the Russian government tried desperately to balance Japan's naval forces by sending its Baltic Fleet, commanded by Admiral Zinovi Petrovich Rozhdestvenski, to the Far East.

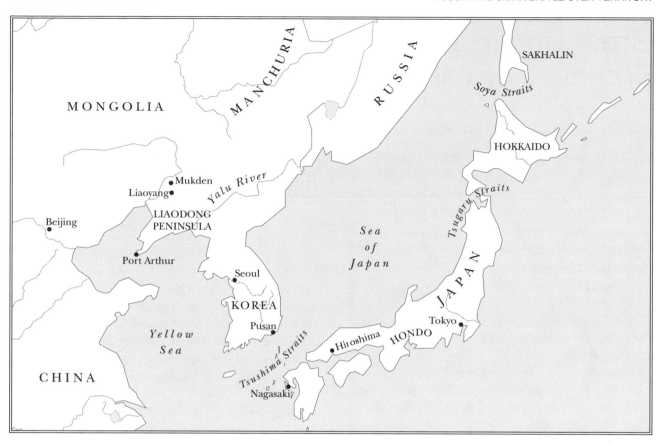

This fleet finally arrived in the China Sea after narrowly avoiding a conflict with Great Britain, but by then Port Arthur had surrendered. Rozhdestvenski set sail for Vladivostok. On May 24, 1905, the Baltic Fleet was waylaid by a Japanese flotilla in the Tsushima Straits, between Korea and Japan. By the end of the battle, the Russian fleet had been practically annihilated.

## Peace Negotiations

Such major defeats on land and sea, made worse by the revolutionary activity within Russia, made Nicholas II grateful to accept the help of American president Theodore Roosevelt and German emperor William II. They arranged for peace negotiations to be held in Portsmouth, New Hampshire.

By the time of the negotiations, though Japan had been winning the war, its resources were seriously depleted. Witte, who represented Russia at the peace conference, was able to take advantage of this situation, so that Russia did not suffer heavy losses under the terms of the agreement. By a treaty signed on September 5, 1905, Russia

surrendered to Japan the areas already lost during the fighting. Japan was to keep control of the Liaodong Peninsula, the naval base of Port Arthur, and Dalny, a commercial port. Russia also surrendered control over the South Manchurian Railway, and the southern portion of the island of Sakhalin was transferred to Japan.

## Consequences

The Russo-Japanese War had repercussions around the world. Russia had lost some of its influence in the southern European Balkan states, and Germany would move in to fill the gap. France began to lose confidence in Russia, while Great Britain and Russia began to let go of their old rivalry in Asia.

Within Russia, the war and defeat added fuel to the revolutionary movement, which would eventually culminate in the 1917-1918 Bolshevik Revolution. In Japan, the terms of peace, which were seen as too soft on Russia, brought about a mood of frustration among the people and the leaders, and a spirit of nationalism began to arise.

*Hans Heilbronner*

**97**

# Great Britain and France Form Entente Cordiale

> *Realizing that both countries had much to gain by leaving behind their old rivalry, leaders of France and Great Britain entered into a new relationship of cooperation.*

**What:** International relations
**When:** April 8, 1904
**Where:** London and Paris
**Who:**
Sir Thomas Barclay (1853-1941), a British official active in international relations
Henry Charles Keith Petty-Fitzmaurice Marquis of Lansdowne (1845-1927), foreign secretary of Great Britain from 1900 to 1905
Edward VII (1841-1910), king of Great Britain from 1901 to 1910
Théophile Delcassé (1852-1923), French minister of foreign affairs from 1898 to 1906
Émile Loubet (1838-1929), president of France from 1899 to 1906
Pierre-Paul Cambon (1848-1924), French ambassador at London

## Need for Good Relations

As colonial powers, Great Britain and France had been rivals through the eighteenth and nineteenth centuries. As the twentieth century began, however, leaders of both countries began to realize that much could be gained through a new friendship between them.

There were several reasons that better relations were needed. In the Far East, tension was growing between Japan (an ally of Great Britain) and Russia (an ally of France). In 1904, the Russo-Japanese War broke out, and if Great Britain and France were to avoid being drawn into war against each other, they needed to mend their hostile relations.

The 1898 Fashoda crisis, a struggle for control of Sudan, had already brought France and Great Britain into conflict in Africa. More conflict over African colonies seemed possible. For more than a quarter century, France had claimed certain political and financial rights in Egypt, a region Great Britain was attempting to control. At the same time, France wanted control of Morocco. Groups of Moroccans often carried out raids against French settlements in neighboring Algeria; if France could dominate Morocco, then, it would be in a much better position to protect its other North African holdings. Great Britain's large share of Morocco's trade, however, was an obstacle to France's aims. Great Britain and France also continued to dispute rights to fisheries in Newfoundland.

Within Europe, Great Britain made numerous attempts between 1898 and 1901 to enter into an alliance with Germany. Yet the Boer War of 1899-1902 increased anti-British feeling in Germany. With Germany refusing to come into alliance, and with the buildup of the German navy, which began to threaten British domination of the seas, British leaders looked increasingly to France. At the same time, France had been disappointed by its alliance with Russia, which had been financially costly and had not brought much reward to the French.

## Steps Toward Friendship

Both the British and the French governments appointed new officials to improve relations between their countries. The new minister of foreign affairs in France, Théophile Delcassé, tried to persuade the British to agree to France's domination over Morocco. In return,

Hulton Archive

*This cartoon depicts the formation of the Entente Cordiale between France and Great Britain.*

he was willing to avoid a confrontation with British forces at Fashoda in 1898. On March 21, 1899, he reached an agreement with Great Britain about how the two colonial powers would divide their influence in the Upper Nile (Egypt) and the Congo.

A year later, Sir Thomas Barclay and Delcassé cooperated to bring British chambers of commerce to visit the great Paris Exposition. Barclay also arranged a return visit of French chambers of commerce to England.

On October 14, 1903, both states signed the Anglo-French Treaty of Arbitration, promising to submit most of their disputes to the Permanent Court of Arbitration at The Hague, in the Netherlands. Barclay was quite involved in negotiating this treaty, and he also helped bring about an exchange of visits between King Edward VII and President Émile Loubet in 1903.

When Loubet went to London, Delcassé accompanied him and took the opportunity to begin serious discussions with Henry Charles Keith Petty-Fitzmaurice—Marquis of Lansdowne and British foreign secretary—to settle the conflicts that remained between Great Britain and France. The result was that on April 8, 1904, Lansdowne and French ambassador Pierre-Paul Cambon signed the Entente Cordiale.

The Entente Cordiale was not a military pact, but it resolved a number of disputes having to do with the Newfoundland fisheries, West African boundaries, Siam, Madagascar, and the New Hebrides Islands. Most important, France agreed to allow Great Britain a free hand in Egypt in exchange for being allowed a free hand in Morocco. A secret provision of the Entente Cordiale was that Morocco would eventually be divided between Spain and France: Spain would control the coastal area, opposite Gibraltar, while the French occupied the inland area.

**99**

## Consequences

Delcassé and other French leaders were pleased to have gained greater control over Morocco. Germany, however, also had interests in Morocco, and German leaders, especially those in the Foreign Office, were insulted that Delcassé apparently thought he could disregard them.

Since France was already allied with Russia, the balance of power in Europe was beginning to make the Germans uneasy. Germany attempted to defend its Moroccan interests in 1905 and 1911, and two serious crises resulted. The effect was to strengthen bonds between Great Britain and France even further, and to push Germany further away. The overall effect was to divide Europe into two armed camps, which would move against each other in 1914 and set off World War I.

*Edward P. Keleher*

# Fleming Patents First Vacuum Tube

*Sir John Ambrose Fleming used the Edison effect to detect the presence of radio waves, starting the electronics industry.*

**What:** Energy
**When:** November 16, 1904
**Where:** London, England
**Who:**

SIR JOHN AMBROSE FLEMING (1849-1945), an English physicist and professor of electrical engineering

THOMAS ALVA EDISON (1847-1931), an American inventor

LEE DE FOREST (1873-1961), an American scientist and inventor

ARTHUR WEHNELT (1871-1944), a German inventor

## A Solution in Search of a Problem

The vacuum tube is a sealed tube or container from which almost all the air has been pumped out, thus creating a near vacuum. When the tube is in operation, currents of electricity are made to travel through it. The most widely used vacuum tubes are cathode-ray tubes (television picture tubes).

The most important discovery leading to the invention of the vacuum tube was the Edison effect by Thomas Alva Edison in 1884. While studying why the inner glass surface of lightbulbs blackened, Edison inserted a metal plate near the filament of one of his lightbulbs. He discovered that electricity would flow from the positive side of the filament to the plate, but not from the negative side to the plate. Edison offered no explanation for the effect.

Edison had, in fact, invented the first vacuum tube, which was later termed the "diode"; at that time there was no use for this device. Therefore, the discovery was not recognized for its true significance. A diode converts electricity that alternates in direction (alternating current) to electricity that flows in the same direction (direct current). Since Edison was more concerned with producing direct current in generators, and not household electric lamps, he essentially ignored this aspect of his discovery. Like many other inventions or discoveries that were ahead of their time—such as the laser—for a number of years, the Edison effect was "a solution in search of a problem."

The explanation for why this phenomenon occurred would not come until after the discovery of the electron in 1897 by Sir Joseph John Thomson, an English physicist. In retrospect, the Edison effect can be identified as one of the first observations of "thermionic" emission, the freeing up of electrons by the application of heat. Electrons were attracted to the positive charges and would collect on the positively charged plate, thus providing current; but they were repelled from the plate when it was made negative, meaning that no current was produced. Since the diode permitted the electrical current to flow in only one direction, it was compared to a valve that allowed a liquid to flow in only one direction. This analogy is popular since the behavior of water has often been used as an analogy for electricity, and this is the reason that the term "valves" became popular for vacuum tubes.

## Same Device, Different Application

Sir John Ambrose Fleming, acting as adviser to the Edison Electric Light Company, had studied the lightbulb and the Edison effect starting in the early 1880's, before the days of radio. Many years later, he came up with an application for the Edison effect as a radio detector when he was a consultant for the Marconi Wireless Telegraph Company. Detectors (devices that conduct electricity in one direction only, just as the diode does, but at higher frequencies) were required to make the high-frequency radio waves audible by converting them from alternating current to direct current. Fleming was able to detect radio waves quite effectively by using the Edison effect. Fleming used essentially the same device that Edison had created, but for a different purpose. Fleming applied for a patent on his detector on November 16, 1904.

In 1906, Lee de Forest refined Fleming's invention by adding a zigzag piece of wire between the metal plate and the filament of the vacuum tube. The zigzag piece of wire was later replaced by a screen called a "grid." The grid allowed a small voltage to control a larger voltage between the filament and plate. It was the first complete vacuum tube and the first device ever constructed capable of amplifying a signal—that is, taking a small-voltage signal and making it much larger. He named it the "audion" and was granted a U.S. patent in 1907.

In 1907-1908, the American navy carried radios equipped with de Forest's audion in its goodwill tour around the world. While useful as an amplifier of the weak radio signals, it was not useful at this point for the more powerful signals of the telephone. Other developments were made quickly as the importance of the emerging fields of radio and telephony were realized.

## Consequences

With many industrial laboratories working on vacuum tubes, improvements came quickly. For example, tantalum and tungsten filaments quickly replaced the early carbon filaments. In 1904, Arthur Wehnelt, a German inventor, discovered that if metals were coated with certain materials such as metal oxides, they emitted far more electrons at a given temperature. These materials enabled electrons to escape the surface of the metal oxides more easily. Thermionic emission and, therefore, tube efficiencies were greatly improved by this method.

Another important improvement in the vacuum tube came with the work of the American chemist Irving Langmuir of the General Electric Research Laboratory starting in 1909, and Harold D. Arnold of Bell Telephone Laboratories. They used new techniques such as the mercury diffusion pump to achieve higher vacuums. Working independently, Langmuir and Arnold discovered that very high vacuum used with higher voltages increased the power these tubes could handle from small fractions of a watt to hundreds of watts. The de Forest tube was now useful for the higher-power audio signals of the telephone. This resulted in the introduction of the first speech transmission across the United States in 1914, followed by the first transatlantic communication in 1915.

The invention of the transistor in 1948 by the American physicists William Shockley, Walter H. Brattain, and John Bardeen ultimately led to the downfall of the tube. With the exception of the cathode-ray tube, transistors could accomplish the jobs of nearly all vacuum tubes much more efficiently. Also, the development of the integrated circuit allowed the creation of small, efficient, highly complex devices that would be impossible with radio tubes. By 1977, the major producers of the vacuum tube had stopped making it.

*Christopher J. Biermann*

# Einstein Describes Photoelectric Effect

*Albert Einstein explained how a metal surface releases electrons after exposure to light.*

**What:** Physics
**When:** 1905
**Where:** Bern, Switzerland
**Who:**

ALBERT EINSTEIN (1879-1955), a German-born American physicist who was awarded the 1921 Nobel Prize in Physics for his description of the photoelectric effect

MAX PLANCK (1858-1947), a German physicist who was awarded the 1918 Nobel Prize in Physics

HEINRICH HERTZ (1857-1894), a German physicist who discovered the photoelectric effect

SIR JOSEPH JOHN THOMSON (1856-1940), an English physicist who discovered the electron, an achievement for which he won the 1906 Nobel Prize in Physics

## Puzzling Effects

The photoelectric effect is the process by which electrons are ejected from a metal surface when light is shined on that surface. Since it requires energy to remove an electron from a metal, it is clear that this energy comes from the incident light (that is, the light that falls upon, or strikes, the surface). In 1887, Heinrich Hertz discovered that light striking a metal surface can produce visible sparks if that surface is in the presence of an electric field.

In 1888, the German physicist Wilhelm Hallwachs showed that shining light on a surface can cause an uncharged body to become positively charged (that is, to lose electrons). In 1899, Sir Joseph John Thomson, who had discovered the electron two years earlier, stated that the photoelectric effect involved the emission of electrons from the surface of the metal. This explained Hertz's observation (the sparks were created by electrons that were accelerated by the electric field) and also explained Hallwachs's results (the emitted electrons were carrying negative charge away from the metal body, thus leaving it with a net positive charge).

In 1902, the German physicist Philipp P. Lenard showed that the energy of the ejected electrons—or, equivalently, their speed—did not depend on the intensity, or brightness, of the incident light. It was shown in 1904 that the energy of the ejected electrons depended on the frequency, or color, of the light: The higher the frequency of the incident light, the greater the speed of the escaping electrons.

These two discoveries contradicted the classical theories of physics, which held that light was an electromagnetic wave that carried energy based on its intensity. When this energy struck a certain surface, the electrons on the surface would gain energy gradually, or "heat up," until eventually they became energetic enough to escape from the surface. If the light was very bright, or intense, the electrons should escape with a large supply of energy—in contradiction to the findings of Lenard. Scientists began to look for a solution to the problem.

## A Revolutionary Paper

In 1905, Albert Einstein published three revolutionary papers. The most famous was on relativity, one was on Brownian motion as evidence for the existence of atoms, and the third was on the photoelectric effect. Einstein suggested that the photoelectric effect could be understood by discarding certain key concepts from classical physics and replacing them with radical new ideas—ideas that would form the basis of modern physics. One of these radical ideas was the concept of light "quanta" (parcels), which had been proposed by Max Planck in 1900.

**103**

As an aid to understanding, Einstein suggested that the incident light of the photoelectric effect should not be viewed as a classical wave but rather as a collection of particles—light quanta, later to be renamed "photons." The amount of energy that these photons carry depends on their frequency, not their intensity.

By viewing the incident light as a collection of photons, Einstein was able to explain the photoelectric effect as follows: When a photon strikes a metal surface, there is a strong chance that it will encounter "free electrons," which are electrons that are detached from atoms and can move from atom to atom, conducting electricity or heat. When a photon encounters an electron, it will usually transfer all of its energy to the electron. In general, an electron can absorb only one photon, but it will always absorb this photon in its entirety. After the absorption, the electron, which had very little energy before it absorbed the photon, has the added energy of the photon. If this energy is high enough, the electron will be able to escape from the surface of the metal.

Einstein was able to make several predictions: that the energy of a photoejected electron can never exceed the energy of the photon, and that if the photon's energy is less than the energy needed for the electron to escape, no electrons will be ejected no matter how bright the incident light.

Einstein's explanation for the photoelectric effect came at a time when classical ideas were still strong and the notion of light quanta seemed radical and mysterious. Even Planck and Einstein had reservations about the concepts they had put forward, but they believed that the concepts were helpful in describing what they observed and useful in predicting the outcome of future observations. It would be nearly two decades before these important ideas were universally accepted.

## Consequences

The light-quanta hypothesis became an important part of several larger theories. In 1911, the Danish physicist Niels Bohr began to use the idea of light quanta to account for the emission spectra of atoms. It was known that atoms, when excited, gave off light with certain characteristic frequencies that differed from one atom to the

*Albert Einstein.*

next. The famous "Bohr model" of the atom stated that these frequencies could be understood as the frequency of the light quantum, or photon, emitted by an atom when an electron jumped from a large orbit to a smaller one. Since electrons generally are limited to specific orbits within an atom, the frequency of the emitted photon would depend on the difference in energy levels between the two orbits.

Later, the French physicist Louis de Broglie recognized that light, which had been demonstrated to behave like a wave, also behaved like a particle at times. If light indeed had a "dual character," should not electrons, which had always been understood as behaving like particles, also behave like waves? De Broglie then proposed his famous theory of wave-particle duality, which stated that light and matter had both wave and particle characteristics. These radical notions would have been unthinkable without the concept of photons.

*Karl W. Giberson*

# Hertzsprung Notes Relationship Between Stellar Color and Luminosity

> *Ejnar Hertzsprung discovered that the color of a star is related to its luminosity, or brightness, which led to the presentation of the first Hertzsprung-Russell diagram.*

**What:** Astronomy; Photography
**When:** 1905
**Where:** Denmark
**Who:**
EJNAR HERTZSPRUNG (1873-1967), a
    Danish astronomer and photographer
HENRY NORRIS RUSSELL (1877-1957), an
    American astronomer

## Giants and Dwarfs

With the advent of spectroscopy and photography in the late nineteenth century, astronomers began for the first time to study the nature of stars and other bodies rather than merely their motions. Spectroscopy is the study of the spectra of stars. A star's "spectrum" is the rainbow band of colored light and dark lines that result when the star's light is passed through a prism or grating. The patterns of lines can be used to classify the stars and provide clues to the star's properties, such as what it is made of, how hot it is, whether it is rotating and at what speed, and more. Photography was used to record these spectra so that they could be studied and compared with the spectra of other stars. As data were gathered for large numbers of stars, scientists could begin to search for orders and patterns in the data. Ejnar Hertzsprung's work was a part of this quest to understand the nature of stars as revealed by their light.

Hertzsprung began the study of astronomy in 1902, after several years of working as a chemist. He studied stellar spectra photographed in Denmark, examined star classification work done at Harvard College Observatory, and wrote two papers, in 1905 and 1907, both entitled "The Radiation of Stars." He had discovered that there are two differ-

ent types of stars: giants and dwarfs. Hertzsprung had studied the colors, brightnesses, motions, and distances of stars to arrive at this discovery, which led to his development of a diagram plotting the actual, or absolute, brightness against the temperature for a group of stars. This type of plot, developed independently in the United States by Henry Norris Russell and presented in 1913, is known as the Hertzsprung-Russell, or H-R, diagram and is a basic tool of astrophysics today.

## Star Plots

Hertzsprung studied the colors of stars in several clusters whose spectra he had photographed at the Urania Observatory in Copenhagen. He was able to measure the wavelength from individual stars and then use this wavelength as an index to the stars' colors. The "wavelength" is the distance from crest to crest of each electromagnetic wave making up the light. These wavelengths are very small. Red light has a longer wavelength than green light, which, in turn, has a longer wavelength than blue light; the wavelength is related to the color.

Once Hertzsprung had indexed the stars' colors, he constructed a diagram in which he plotted the color versus the brightness for the stars in two star clusters, the Pleiades and the Hyades.

In 1911, Hertzsprung published his diagrams, the first of their kind to be published, when he was senior staff astronomer at the Astrophysical Observatory at Potsdam, East Germany. Hertzsprung made the important observation that stars do not appear in all possible combinations of color and brightness; there is a relationship between how bright the star is and what color it is. The brightnesses and colors of stars in the two clusters form a narrow diagonal band across the plot. The band stretches from bright blue stars in the upper left

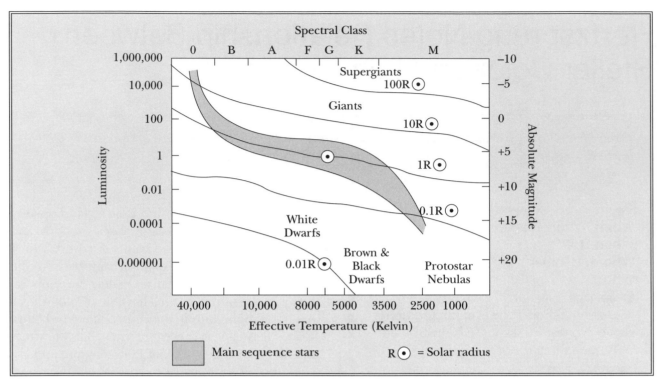

*A typical Hertzsprung-Russell diagram.*

corner to dim red stars in the lower right corner. This band was called the "main" sequence, or sometimes the "dwarf" sequence, to differentiate it from the "giant" sequence of large stars, which was revealed on other H-R diagrams. The discovery of this relationship provided astronomers with valuable information about celestial bodies. Astronomers would be busy for years studying the relationship between color and brightness.

## Consequences

Hertzsprung continued his work on the color-luminosity relationship by comparing the H-R diagrams for the Pleiades, Hyades, and Praesepe star clusters, and he noticed differences in the types of stars in the three clusters. Astronomers today interpret these differences as indications that the Pleiades are younger than the other two clusters; the H-R diagrams are a useful way of measuring the ages of clusters. It is now known that the brighter and bluer a star is, the shorter its lifetime is; therefore, a cluster with bright blue stars still left in it must be younger than the maximum age such stars would reach.

In addition, a method was later developed in which the brightness-color diagram for a cluster was compared with that of another cluster at a known distance; this comparison can yield an estimate of the distance of the first cluster. In working on the Pleiades cluster, Hertzsprung did related work on their spectra that led to estimates of the total mass of the cluster. H-R diagrams thus provide a rich source of information about star clusters.

Hertzsprung's plot of stars on the main sequence, together with his demonstration that a giant sequence exists, provided support and information to Russell, who based his theory of stellar evolution on the H-R diagram. Russell's evolutionary picture turned out to be incorrect, but it was an important first step in the use of H-R diagrams.

Hertzsprung's work with stars' spectra and their colors yielded results that were vital for the further studies conducted by astrophysicists in the twentieth century. The recognition of a color-luminosity relationship was the first step toward understanding the reasons behind the relationship, reasons that have had profound implications for ideas on stellar formation and evolution.

*Mary Hrovat*

# Punnett Argues for Acceptance of Mendelism

> *Reginald Crundall Punnett published* Mendelism, *explaining how traits are inherited.*

**What:** Biology; Genetics
**When:** 1905
**Where:** Cambridge, England
**Who:**
REGINALD CRUNDALL PUNNETT (1875-1967), an English geneticist who crusaded for the acceptance of Mendel's theories
GREGOR JOHANN MENDEL (1822-1884), an Austrian monk and botanist who performed pioneering research into heredity

## The Heredity of Peas

In 1905, Reginald Crundall Punnett's landmark book *Mendelism* was published. This small book stated clearly the principles of heredity espoused by the Austrian botanist Gregor Johann Mendel in his classic paper of 1865. Mendel's original paper, long ignored because it had been published in an obscure journal, was independently rediscovered in 1900 by three botanists: Hugo de Vries, Carl Erich Correns, and Erich Tschermak von Seysenegg. Yet although Mendel's ideas were intriguing, early twentieth century biologists did not agree on the universality of his proposed laws of heredity, and they searched for an understanding of the significance of the Mendelian patterns. Punnett's work played a major role in earning acceptance of Mendel's theories, and his book became an excellent learning tool for the rediscovered Mendelism.

In 1856, Mendel began his experiments with garden peas in an effort to study the inheritance of individual characteristics. It was known at the time that the first generation reproduced from hybrids, or crossbred plants, tended to be uniform in appearance, but that the second generation reverted to the characteristics of the two original plants that had been crossbred. Such facts were observable, but the explanations remained unsatisfactory. Before Mendel, the concept of "blending" inheritance predominated; that is, it was assumed that offspring were typically similar to their parents because the essences of the parents were blended at conception. Mendel's work with pea plants suggested another theory, that of "particulate" inheritance; according to this theory, a gene passes from one generation to the next as a unit, without any blending.

Mendel laid the groundwork for his experiments by testing thirty-four varieties of peas to find the most suitable varieties for research. From these, he chose twenty-two to examine for two different traits, color and texture. He was then able to trace the appearance of green and yellow seeds, as well as round and wrinkled ones, in several generations of offspring. By counting the results of his hybridization, he found that the ratio of "dominant" genes to "recessive" ones was three to one. In effect, he demonstrated that there were rules governing the process of inheritance.

## Punnett Squares

Punnett's work served to explain Mendel's theories of sex determination, sex linkage, complementary factors and factor interaction, autosomal linkage, and mimicry. He also had a number of other notable scientific and practical achievements; in World War I, for example, when food was in short supply in Great Britain, he devised an ingenious scheme to distinguish the sex of very young chickens. He realized that by noting sex-linked color factors that would appear in the plumage only of male chicks, the unwanted males could be distinguished and de-

**107**

stroyed so that food supplies would not be wasted on them. His work with *The Journal of Genetics,* which he founded jointly with the British geneticist William Bateson in 1911, also contributed significantly to the advancement of genetics. Yet it was Punnett's invention and use in his book of a simple diagrammatic scheme that helped to assure the acceptance of Mendelism and most directly influenced the understanding of genetics.

In his book, Punnett included simple charts to illustrate the inherited characteristics that would appear in offspring when particular characteristics are crossbred. The diagrams are explicit, clear, and convincing. These "Punnett squares," as they came to be known, soon became a valuable tool for teaching genetics students the rudiments of the subject.

## Consequences

During the late nineteenth century, cytologists, or scientists who study cells, began to observe the behavior of chromosomes in cell division. Although some of these scientists suspected that chromosomes might play a part in heredity, they had no real understanding of how traits were distributed among offspring. Although Mendel had already determined some of these

relationships, his ideas remained generally unavailable to cytologists. Before the rediscovery of Mendel's work, therefore, many theories of heredity were competing for acceptance by the scientific community; no single theory could be agreed upon as adequately explaining all the observed facts. Even after Mendelism was rediscovered, it was by no means universally accepted; critics argued that rules that seemed to explain the inheritance of color traits in pea plants did not seem to apply to the inheritance of other traits in other species.

With the aid of his convincing diagrams, Punnett engaged in a crusade for Mendelism, showing how apparent exceptions to the rules proposed by Mendel could be explained. Mendel's theories thus won gradual acceptance and were combined with other ideas to explain the inheritance process more fully still. For example, it soon became understood that chromosomes were the physical basis for the ratios of inheritance that Mendel had observed. Punnett's work played a key role in making possible this fusion between Mendelian genetics and cytology and helped to establish the basis for modern genetic theory.

*Marilyn Bailey Ogilvie*

# Baekeland Produces First Artificial Plastic

> *Leo Hendrik Baekeland developed the first totally synthetic thermosetting plastic, which paved the way for modern materials science.*

**What:** Materials
**When:** 1905-1907
**Where:** Yonkers, New York
**Who:**
LEO HENDRIK BAEKELAND (1863-1944), a Belgian-born chemist, consultant, and inventor
CHRISTIAN FRIEDRICH SCHÖNBEIN (1799-1868), a German chemist who produced guncotton, the first artificial polymer
ADOLF VON BAEYER (1835-1917), a German chemist

### Exploding Billiard Balls

During the 1860's, the firm of Phelan and Collender offered a prize of ten thousand dollars to anyone producing a substance that could serve as an inexpensive substitute for ivory, which was somewhat difficult to obtain in large quantities at reasonable prices. Earlier, Christian Friedrich Schönbein had laid the groundwork for a breakthrough in the quest for a new material in 1846 by the serendipitous discovery of nitrocellulose, more commonly known as "guncotton," which was produced by the reaction of nitric acid with cotton.

An American inventor, John Wesley Hyatt, while looking for a substitute for ivory as a material for making billiard balls, discovered that the addition of camphor to nitrocellulose under certain conditions led to the formation of a white material that could be molded and machined. He dubbed this substance "celluloid," and this product is now acknowledged as the first synthetic plastic. Celluloid won the prize for Hyatt, and he promptly set out to exploit his product. Celluloid was used to make baby rattles, collars, dentures, and other manufactured goods.

As a billiard ball substitute, however, it was not really adequate, for various reasons. First, it is thermoplastic—in other words, a material that softens when heated and can then be easily deformed or molded. It was thus too soft for billiard ball use. Second, it was highly flammable, hardly a desirable characteristic. A widely circulated, perhaps apocryphal, story claimed that celluloid billiard balls detonated when they collided.

### Truly Artificial

Since celluloid can be viewed as a derivative of a natural product, it is not a completely synthetic substance. Leo Hendrik Baekeland has the distinction of being the first to produce a com-

*Leo Hendrik Baekeland.*

pletely artificial plastic. Born in Ghent, Belgium, Baekeland emigrated to the United States in 1889 to pursue applied research, a pursuit not encouraged in Europe at the time. One area in which Baekeland hoped to make an inroad was in the development of an artificial shellac. Shellac at the time was a natural and therefore expensive product, and there would be a wide market for any reasonably priced substitute. Baekeland's research scheme, begun in 1905, focused on finding a solvent that could dissolve the resinous products from a certain class of organic chemical reaction.

The particular resins he used had been reported in the mid-1800's by the German chemist Adolf von Baeyer. These resins were produced by the condensation reaction of formaldehyde with a class of chemicals called "phenols." Baeyer found that frequently the major product of such a reaction was a gummy residue that was virtually impossible to remove from glassware. Baekeland focused on finding a material that could dissolve these resinous products. Such a substance would prove to be the shellac substitute he sought.

These efforts proved frustrating, as an adequate solvent for these resins could not be found. After repeated attempts to dissolve these residues, Baekeland shifted the orientation of his work. Abandoning the quest to dissolve the resin, he set about trying to develop a resin that would be impervious to any solvent, reasoning that such a material would have useful applications.

Baekeland's experiments involved the manipulation of phenol-formaldehyde reactions through precise control of the temperature and pressure at which the reactions were performed. Many of these experiments were performed in a 1.5-meter-tall reactor vessel, which he called a "Bakelizer." In 1907, these meticulous experiments paid off when Baekeland opened the reactor to reveal a clear solid that was heat resistant, nonconducting, and machinable. Experimentation proved that the material could be dyed practically any color in the manufacturing process, with no effect on the physical properties of the solid.

Baekeland filed a patent for this new material in 1907. (This patent was filed one day before that filed by James Swinburne, a British electrical engineer who had developed a similar material in his quest to produce an insulating material.) Baekeland dubbed his new creation "Bakelite" and announced its existence to the scientific community on February 15, 1909, at the annual meeting of the American Chemical Society. Among its first uses was in the manufacture of ignition parts for the rapidly growing automobile industry.

## Consequences

Bakelite proved to be the first of a class of compounds called "synthetic polymers." Polymers are long chains of molecules chemically linked together. There are many natural polymers, such as cotton. The discovery of synthetic polymers led to vigorous research into the field and attempts to produce other useful artificial materials. These efforts met with a fair amount of success; by 1940, a multitude of new products unlike anything found in nature had been discovered. These included such items as polystyrene and low-density polyethylene. In addition, artificial substitutes for natural polymers, such as rubber, were a goal of polymer chemists. One of the results of this research was the development of neoprene.

Industries also were interested in developing synthetic polymers to produce materials that could be used in place of natural fibers such as cotton. The most dramatic success in this area was achieved by Du Pont chemist Wallace Carothers, who had also developed neoprene. Carothers focused his energies on forming a synthetic fiber similar to silk, resulting in the synthesis of nylon.

Synthetic polymers constitute one branch of a broad area known as "materials science." Novel, useful materials produced synthetically from a variety of natural materials have allowed for tremendous progress in many areas. Examples of these new materials include high-temperature superconductors, composites, ceramics, and plastics. These materials are used to make the structural components of aircraft, artificial limbs and implants, tennis rackets, garbage bags, and many other common objects.

*Craig B. Lagrone*

# Boltwood Uses Radioactivity to Date Rocks

*Bertram Borden Boltwood pioneered the radiometric dating of rocks, leading to the use of nuclear methods in geology and establishing a new chronology of Earth.*

**What:** Earth science
**When:** 1905-1907
**Where:** New Haven, Connecticut
**Who:**
BERTRAM BORDEN BOLTWOOD (1870-1927), the first American scientist to study radioactive transformations
ERNEST RUTHERFORD (1871-1937), an English physicist who won the 1908 Nobel Prize in Chemistry
SIR WILLIAM THOMSON, LORD KELVIN (1824-1907), an English physicist

## Not Enough Time

Radioactivity is the property exhibited by certain chemical elements that, during spontaneous nuclear decay, emit radiation in the form of alpha particles, beta particles, or gamma rays. Related to this property is the process of nuclear disintegration. In this process, an atomic nucleus, through its emission of particles or rays, undergoes a change in structure. To take an example, the presence of helium in rocks is the result of radioactive elements in the rocks emitting alpha particles during disintegration.

Bertram Borden Boltwood was fascinated by the theory of radioactive disintegrations proposed in 1903 by McGill University scientists Ernest Rutherford and Frederick Soddy. According to that theory, radioactivity is always accompanied by the production of new chemical elements on an atom-by-atom basis. In 1904, Boltwood impressed Rutherford by demonstrating that all uranium minerals contain the same number of radium atoms per gram of uranium. This confirmation of the theory of radioactivity marked the beginning of a close collaboration between the two scientists.

The significance of Boltwood's work can be seen in the light of a chronological controversy raging at that time between geologists and physicists. It had been accepted generally that the earth, at some time in its history, was a liquid ball and that its solid crust was formed when the temperature was reduced by cooling. The geological age of Earth was, thus, defined as the period of time necessary to cool it down from the melting point to its present temperature. Using these guidelines, several estimates of that time were made by the famous physicist Sir William Thomson, Lord Kelvin, who in 1877 claimed that the age of the earth was probably close to 20 million years and certainly not as large as 40 million. His calculations were mathematically correct but did not take into account radioactivity, which had been discovered the year before.

Geologists believed that 40 million years simply was not enough time to create continents, to erode mountains, or to supply oceans with minerals and salts. Studies of sequences of layers (stratigraphy) and of fossils (paleontology) led them to believe that the earth was older than 100 million years, but they were not able to prove it.

## Earth Clocks

Boltwood and Rutherford proved that geologists were right. This came as a by-product of their research on the nature of radioactivity. Rutherford knew that helium was always present in natural deposits of uranium, and this led him to believe that in radioactive minerals, alpha particles somehow were turned into ordinary atoms of helium. Accordingly, each rock of a radioactive mineral is a generator of helium. The accumulation of the gas proceeds more or less uniformly so that the age of the rock can be determined from the amount of trapped helium. In that sense, radioactive rocks are natural clocks.

Knowing how much helium is produced from each gram of uranium per billion years, Rutherford and his collaborators were able to see that naturally radioactive rocks are often older than 100 million years. The ages of some samples exceeded 500 million years. Moreover, Rutherford was aware that in assigning ages he would have to account for helium that was escaping from the rocks.

Impressed by these results, and trying to eliminate the uncertainties associated with the leakage of helium, Boltwood decided to work on another method of dating. This decision stemmed from his earlier attempts to demonstrate that, as in the case of radium, all uranium minerals contain the same number of atoms of lead per gram of uranium. Chemical data, however, did not confirm this expectation—the measured lead-to-uranium ratios were found to be different in minerals from different locations.

According to Rutherford and Soddy's theory, a spontaneous transformation of uranium into a final product proceeds through a set of steps, in which alpha particles and electrons are emitted, one after another. Boltwood realized that lead

must be the final product and that its accumulation could be used to date minerals. By focusing on lead rather than helium, he hoped to reduce the uncertainties associated with the leakage. Lead, he argued, is less likely to escape from rocks than helium because, once trapped, lead becomes part of a solid structure.

Motivated by these ideas, Boltwood proceeded with the development of the uranium-lead method of dating. To accomplish this, he had to determine the rate at which lead is produced from uranium. Having achieved this, Boltwood started his investigations in 1905, and before the end of the year, he had analyzed twenty-six samples. One of them was identified as 570 million years old. In a formal publication, which appeared in 1907, he described forty-six minerals collected in different locations; their reported ages were between 410 and 2,200 million years old.

Similar results had been reported earlier by Rutherford from his laboratory in Montreal and by the English physicist Robert John Strutt, Lord

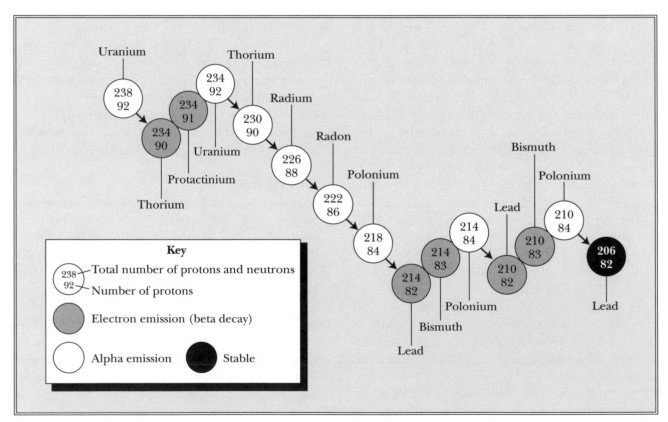

*Uranium decays naturally, over a predictable period of time, to form lead. Boltwood realized that the accumulation of lead in rocks could be used to determine their age.*

Rayleigh, from the Imperial College in London, both of whom had used helium methods. Although there was a wide variation in dates, it became clear that many rocks were at least ten times older than what had been calculated by Lord Kelvin.

## Consequences

The main results of the pioneering work of Boltwood and his successors was the realization that geological times must be expressed in hundreds and thousands of millions of years, rather than in tens of millions of years, as advocated by Lord Kelvin. This was particularly significant for the acceptance of the theory of evolution by biologists and, in general, for a better understanding of many long-term processes on Earth.

Geochronology, for example, has been used in investigations of reversals of the terrestrial magnetic field. Such reversals occurred many times during the geological history of Earth.

They were discovered and studied by dating pieces of lava, naturally magnetized during solidification. The most recent reversal took place approximately 700,000 years ago.

It is clear, in retrospect, that the discovery of radioactivity affected geochronology in two ways: by providing tools for radiometric dating and by invalidating the thermodynamic calculations of Lord Kelvin. These calculations were based on the assumption that the geothermal energy lost by Earth is not replenished. The existence of radioactive heating, discovered in 1903 in France, contradicted that assumption and prepared scientists for the acceptance of Boltwood's findings. Lord Kelvin died in the same year in which these findings were published, but he knew about Rutherford's findings as early as spring, 1904. He was very interested in radioactive heating but never came forth with a public retraction of his earlier pronouncements.

*Ludwik Kowalski*

# Russian Workers and Czarist Troops Clash on Bloody Sunday

> *A peaceful demonstration by workers in Saint Petersburg ended when czarist troops killed about one hundred protesters.*

**What:** Civil strife
**When:** January 22, 1905
**Where:** Saint Petersburg, Russia
**Who:**
NICHOLAS II (1868-1918), czar of Russia from 1894 to 1917
FATHER GEORGI APPOLLONOVICH GAPON (1870-1906), a Russian Orthodox priest

## Rising Discontent

A severe economic depression from 1900 to 1902 made life for Russian peasants and city workers even more difficult than it had been. Between 1900 and 1905, unemployment increased in large centers of industry such as Moscow and Saint Petersburg. Workers went on strike over low wages and long hours.

Many Russians were angry about their government's seeming indifference to their poverty. Czar Nicholas II and a number of his closest advisers, including Konstantin Petrovich Pobedonostsev, Procurator of the Holy Synod, were seen as stubbornly conservative and resistant to changes that might benefit the common people of Russia. Faced with the need for social and political reform, and frustrated at the government's inaction, various revolutionary reformers were driven to desperate acts.

Unrest in the cities spread to the countryside, where peasants began to revolt with the encouragement of the Social Revolutionary Party. In the cities, the workers were supported by the Social Democrats and an organization—known as the Society for the Mutual Help of the Workers in the Engineering Industry—founded in Moscow in 1902 by Sergei Zubatov, head of the city's security police. This movement actually received some support from the government, for it was designed to make sure that the workers would not adopt a program of political reform. This movement of "police socialism" was mostly staffed by police officials. It soon collapsed in Moscow but reappeared in Saint Petersburg under the leadership of a Russian Orthodox priest, Father Georgi Appollonovich Gapon.

Father Gapon's organization also received government recognition as being nonpolitical. Actually, it was nonpolitical at first, but by 1904, when its membership had grown to eight thousand, many of its members were attracted to the more revolutionary socialist approach of the Social Democrats and Social Revolutionaries. Laborers were becoming more and more disgusted with their working conditions. Russia's war losses to Japan, such as the fall of Port Arthur in December, 1904, were another reason for discontent among the Russian people.

## Petition and Protest

In January, 1905, a large factory in the capital dismissed many workers. Fellow workers began a protest strike, which within a few days had spread across the city. Gapon's organization met on January 20, 1905, and its members angrily decided that he should lead a march to the czar's Winter Palace in Saint Petersburg and present Nicholas II with a petition.

The workers and the radical intellectuals who helped to draft the petition reaffirmed their loyalty to the czar. Yet the petition called on him to lift from their shoulders the yoke of oppression that his corrupt officials had placed on them. The petitioners demanded better treatment from officials and factory owners, and an end to the war with Japan. They also asked for major civil and political reforms, including freedom of speech, freedom of the press, and freedom of association; broader voting rights for local elections; equality under the law; and, above all, the creation of a representative assembly to make laws for the country.

For Russia at that time in its history, these demands were truly revolutionary. Father Gapon believed that some of the petition's demands went too far, but the bitterly cold Sunday morning of January 22, 1905, found him leading his followers to the Czar's Winter Palace.

Wearing his priestly robes, Father Gapon led long columns of workers from the outskirts of Saint Petersburg toward the center of the city and the Winter Palace. The march was intended as a peaceful demonstration, and some marchers were accompanied by their families. Others carried religious images and portraits of the czar. Nevertheless, if the czar rejected the petition, the marchers were prepared to use force to achieve their demands. Gapon had assured them that part of the army would side with them and provide them with arms if fighting was necessary.

Police spies within Gapon's organization, however, had alerted the government to have military forces prepared for action. When the workers approached the Winter Palace and refused to break up their demonstration, the army troops, who remained loyal to the czar, fired on them. About a hundred workers were killed, and Gapon was unable to deliver the petition. The workers had not realized that Nicholas had left the city a few days earlier.

## Consequences

The march and massacre of January 22, 1905, came to be known as "Bloody Sunday." This tragic episode triggered the first Russian revolution in the twentieth century. Father Gapon managed to escape from the capital and fled into exile. From his place of exile he issued an open letter to the czar, denouncing him for having refused to accept the petition. "Let all blood which

*This drawing depicts the clash between protesters and czarist troops on Bloody Sunday.*

has to be shed fall upon thee, hangman, and thy kindred." In this statement, Gapon expressed the real historical consequence of Bloody Sunday: The Russian people had now completely lost faith in the czar. The massacre made it clear to all that the Russian autocratic government was bent on protecting itself, not the Russian people.

Gapon's appeal for revolution was soon answered by strikes of workers throughout Russia. Although the Revolution of 1905 lasted less than a year and did not succeed in its aims, Nicholas II never felt secure afterward. His failure to respond to the needs of workers had led to the revolution, and the disasters Russia suffered in the war with Japan had added fuel to the flames. Yet an even greater disaster, World War I, was soon to come; it provided the occasion in 1917 for a successful revolution that finally toppled czarist autocracy in Russia.

*Edward P. Keleher*

# Lowell Predicts Existence of Planet Pluto

*The astronomer Percival Lowell set off a search for a trans-Neptunian planet.*

**What:** Astronomy
**When:** August, 1905
**Where:** Flagstaff, Arizona
**Who:**
PERCIVAL LOWELL (1855-1916), a
mathematician and astronomer and
the founder of the Lowell Observatory
VESTO MELVIN SLIPHER (1875-1969), the
director of the Lowell Observatory
during the final search for Pluto
CARL O. LAMPLAND (1873-1951), an
astronomer who assisted with Lowell's
search for Pluto
CLYDE W. TOMBAUGH (1906-1997), the
staff assistant at Lowell Observatory
who made the actual discovery of
Pluto

## The Search for Martians

In 1877, the Italian astronomer Giovanni Virginio Schiaparelli began an observational study of Mars that led to his announcement of complex and detailed surface patterns on the planet's surface. He called these features *canali*, meaning "channels" (the Italian term, however, was widely translated into English as "canals," a source of much subsequent confusion).

Percival Lowell followed the discoveries of Schiaparelli keenly and imagined many great Martian civilizations. By the early 1890's, Schiaparelli's eyesight had deteriorated, and he could no longer make telescopic observations. Determined to continue Schiaparelli's research, Percival founded the Lowell Observatory (opened on June 1, 1894) in Flagstaff, Arizona, and devoted it to a study of Mars and the other planets. Lowell made personal observations of the so-called Martian canals and imagined these canals as irrigation channels built by a desperate civilization striving to retain its water supply from the polar ice caps. His theories on Martian life were well received by science-fiction writers and the general public, but they were ridiculed by professional astronomers; Lowell, frustrated, sought ways to improve his credibility with his colleagues.

The mathematical prediction of the planet Neptune and its discovery in 1846 gave rise to the possibility of additional remote and undiscovered planets in the solar system. Lowell realized that if he could correctly predict the location of a ninth planet, beyond Neptune, the feat would improve his reputation as an astronomer. In 1905, Lowell inspired his colleagues at the Lowell Observatory to begin the first systematic photographic and visual search for "Planet X." He expected this distant planet would be similar to Neptune in density, size, and magnitude (brightness) and therefore easy to spot.

## The Search for Planet X

Lowell began his search in 1905. Each sky area was photographed, and Lowell personally examined the photographic plates with a handheld magnifying glass. His technique used the basic assumption that a comparison of two photographs of the same star fields would show movement of nonstellar objects such as comets, asteroids, and planets. Later, Lowell used a Zeiss Blink Comparator (a device that superimposes two photographic plates of the same region for direct comparison), which made comparison easier and faster.

By 1916, after more than a thousand photographs had been taken, Lowell realized that discovering the ninth planet by observation would be difficult. He then attempted to use statistical and mathematical methods to discover its orbit using information gathered from Uranus and Neptune. On January 13, 1915, Lowell presented his theory on Planet X to the American Academy of Arts and Sciences. His book, *Memoir on a Trans-Neptunian Planet*, was published in the spring of

**117**

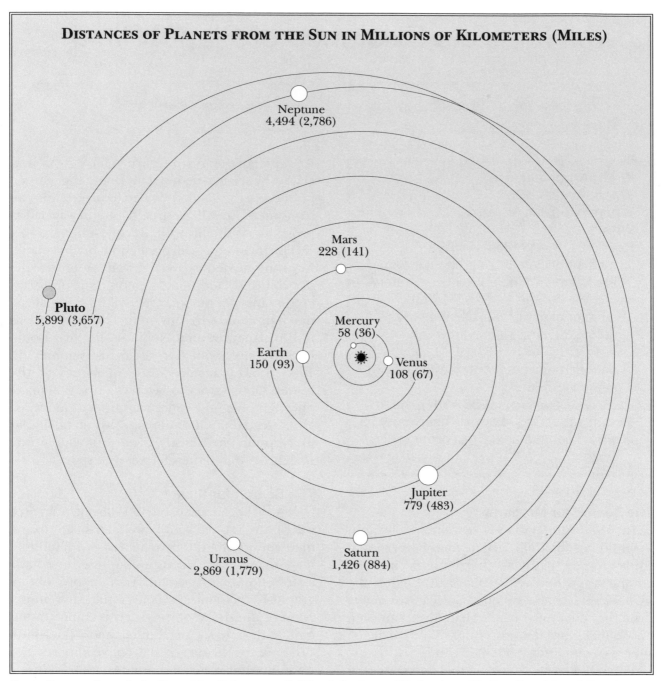

**DISTANCES OF PLANETS FROM THE SUN IN MILLIONS OF KILOMETERS (MILES)**

Neptune
4,494 (2,786)

Mars
228 (141)

Pluto
5,899 (3,657)

Mercury
58 (36)

Earth
150 (93)

Venus
108 (67)

Jupiter
779 (483)

Uranus
2,869 (1,779)

Saturn
1,426 (884)

*Source:* Data are from the Jet Propulsion Laboratory, California Institute of Technology. *The Deep Space Network.* Pasadena, Calif.: JPL, 1988, p. 17.

that year. Lowell had studied tiny disturbances in the orbits of Uranus and Neptune, and he had concluded that the disturbances were caused by a planet whose orbit intercepted that of Neptune. Neptune's orbit was not well known at the time (it had been observed for only fifty years), so its motion could not be accurately calculated. Also, any deviations in the motion of Uranus

were expected to be further minimized because of its great distance from Planet X; these errors were introduced into his calculations. Lowell also had studied comets, and he suggested that a distant planet could affect a cometary orbit.

Lowell's mathematics led him to assume that Planet X would have a mass seven times that of Earth, an inclination (with respect to the ecliptic, or plane of the solar system) of 10 degrees, and an average distance of forty-three times the distance between Earth and the sun. He also believed that the undiscovered planet would have a low density and a high albedo (ratio of sunlight received versus sunlight reflected), as is the case with the four largest planets. In addition, Lowell expected the planet to appear faint because of its great distance from Earth, and he could only generally predict its location to be in or near the constellation of Gemini. Originally, he had predicted the location to be within the constellation of Libra.

Lowell overworked himself in his attempt to locate the ninth planet, and he died of a stroke on November 12, 1916, at the age of sixty-one. This caused an interruption of thirteen years in the search for Planet X. In 1925, however, A. Lawrence Lowell, Percival's brother, provided funds for the purchase of a new telescope that began operation in 1929, at which time an assistant, Clyde W. Tombaugh, was employed. On February 18, 1930, Tombaugh, using the Zeiss Blink Comparator, discovered Planet X on photographic plates taken in January, 1930.

## Consequences

News of the discovery of a ninth planet was made at Lowell Observatory on the seventy-fifth anniversary of Percival Lowell's birth. The press and general public were intrigued by the search for and discovery of the planet; hundreds of telegrams and congratulatory letters were received by the observatory staff.

The planet's location and basic characteristics were so similar to those predicted by Lowell that its discovery could not have been by chance. In fact, Vesto Slipher, the director of the Lowell Observatory, stated that the discovery was simply the conclusion of a search begun in 1905 by Lowell and his theoretical work.

The discovery of a ninth planet prompted astronomers at other institutions to begin searches for still more distant planets. Lowell's early ideas of intelligent life on Mars and his search for a trans-Neptunian planet excited the general population. The discovery of Pluto (the name was suggested by an eleven-year-old English girl) produced a sense of mystery and fascination that would continue to prompt the public to learn more about astronomy and keep looking up toward the stars.

During the late 1990's, after astronomers had learned much more about the size and composition of Pluto, the idea arose that Pluto might be made up primarily of water ice and not be a true "planet" at all. Because of Pluto's apparent similarity to the ice moons of the Jovian planets and because of its highly eccentric orbit, some astronomers theorized that it might be nothing more than an "escaped" planetary moon, or perhaps simply a large piece of icy debris remaining from the solar system's formation. For these reasons, the astronomers proposed to stop including Pluto among the solar system's list of planets. However, this opinion was far from unanimous in the scientific community.

*Noreen A. Grice*

# Einstein Articulates Special Theory of Relativity

*Albert Einstein's special theory of relativity challenged the ideas of the day regarding space and time.*

**What:** Physics
**When:** Fall, 1905
**Where:** Bern, Switzerland
**Who:**
ALBERT EINSTEIN (1879-1955), a German American physicist who was awarded the 1921 Nobel Prize in Physics
SIR ISAAC NEWTON (1642-1727), an English mathematician
ALBERT A. MICHELSON (1852-1931), an American physicist who was awarded the 1907 Nobel Prize in Physics

## Waiting for the Light

Albert Einstein's special theory of relativity was first presented in an article in 1905. In it, he points out that time cannot be viewed as being "absolute." He says, rather, that time is "relative." He also questions the idea of "simultaneity"— that is, of being able to view an event, such as a man throwing a baseball, as it is occurring. Because it takes a certain (infinitesimal) amount of time for the light that illuminates an event to travel to an observer, the event can be seen only after it happens rather than as it happens.

Light always travels at the same speed, according to Einstein, and nothing can move faster than the speed of light. In 1882, Albert A. Michelson's experiments led him to measure the speed of light more accurately than anyone had been able to do previously (299,853 kilometers per second, according to his findings). The speed of light has now been measured as 299,792 kilometers per second. It was Michelson's measurement, however, that proved indispensable to Einstein as he moved toward devising his special theory of relativity.

Einstein further discovered that as a moving object gets closer to the speed of light, its mass begins to increase dramatically. That object's length also appears to change, depending on the rate at which it is moving. He also found that a moving clock appeared to run more slowly than an identical clock at rest. This observation has been proved to be true in various ways, notably by physicists who carried finely tuned atomic clocks with them on around-the-world commercial flights and, upon their return, checked them against comparable clocks in their laboratories, only to find that the clocks they had carried with them had, indeed, lost time as had been predicted by Einstein's theory. Other laboratory experiments with high-speed electrons (one of the three basic units that make up an atom) show that they often achieve ten thousand times their normal mass as their speed increases.

It was not until 1905 that Einstein published his famous formula, $E = mc^2$, which states that mass and energy are equivalent: The energy, $E$, of a particle of matter equals its mass, $m$, multiplied by the speed of light, $c$, squared. Matter itself is a form of concentrated energy. This equation would later suggest the possibility of creating very powerful explosions.

## Approaching the Speed of Light

If it were possible for an object to travel faster than the speed of light, that object would predictably move backward in time. Because objects do not move that fast, however, the physical changes that rapidly moving objects experience (in mass, length, and time) are very hard to spot. An object moving at a speed one-seventh that of the speed of light—42,827 kilometers, or slightly more than one trip around the world along the equator per second—would change

**120**

in mass, length, and time measurement by only 1 percent.

Drastic alterations would take place, however, if the object were to begin moving at nearly the speed of light. When the object's velocity reached six-sevenths of the speed of light, the mass would be twice what it was at rest, while the length and time measurements would be cut in half. Were the velocity of the object to equal the speed of light, its mass would become infinite. According to Einstein's theory of relativity, if a person were in a vehicle capable of coming within 99.995 percent of the speed of light, the mass of the person and the vehicle would be increased one thousand times. In addition, time would pass for that person at one one-thousandth of the normal rate; in other words, each year of such travel would be the same as one thousand years of time as it is normally measured.

The only major physical phenomenon that Einstein could not explain using his special theory of relativity was gravity. This problem led him, in 1915, to formulate his general theory of relativity. Gravity needed to be explained separately because space and time are curved. Therefore, when scientists are trying to draw up mathematical calculations having to do with space and time, their findings are affected by this curvature.

## Consequences

Einstein's special and general theories of relativity corrected some of Sir Isaac Newton's scientific views, which had, until that time, been highly respected by many physicists. Central to Einstein's special theory, in particular, is his shattering of the Newtonian principle that space exists in three dimensions and that time exists in only one. Rather than thinking of space and time separately, Einstein saw "space-time" as a single four-dimensional system in which neither space nor time can exist without the other. These new ideas, once they became accepted, led other scientists to pursue studies that resulted in the splitting of the atom, the development of nuclear energy, space exploration, the development of a theory of superconductivity, and countless other accomplishments that have helped humankind explore the universe and understand it in much greater detail.

Einstein's later research led him to conclude that such moving particles as protons and electrons (like the electron, the proton is a basic building block of the atom) travel in waves and are closely connected with photons (the basic particles of electromagnetic energy). This conclusion led to the development of the field of wave mechanics in physics and was a fundamental element in Einstein's work with the photoelectric effect, for which he was awarded the Nobel Prize. Although the prize did not come in recognition of Einstein's work in relativity (his best-known achievement), his special theory established him as the most compelling and influential physicist of his day. It led to a total rethinking of the entire field of physics.

*R. Baird Shuman*

# Norway Becomes Independent

*Norway dissolved its union with Sweden, gaining complete independence for the first time since 1380.*

**What:** Political independence
**When:** October 26, 1905
**Where:** Norway and Sweden
**Who:**
OSCAR II (1829-1907), king of Sweden from 1872 to 1907, and king of Norway from 1872 to 1905
JOHAN SVERDRUP (1816-1892), Norwegian Liberal leader, president from the Storting from 1871 to 1881
ERIK GUSTAF BOSTRÖM (1842-1907), prime minister of Sweden in 1904
PETER CHRISTIAN MICHELSEN (1857-1925), Norwegian radical leader, premier of Norway from 1905 to 1907
HAAKON VII (1872-1957), king of Norway from 1905 to 1957 (formerly Prince Carl of Denmark)

### Desire for Independence

In 1380, Norway had entered into a union with Denmark that brought the two countries under one ruler; this union lasted until 1814. After the Vienna Settlement ending the Napoleonic Wars, Finland was taken from Sweden and given to Russia. In return, Russia's czar, Alexander I, promised to support Sweden's annexation of Norway. By Sweden and Denmark's Treaty of Kiel in 1814, Norway was annexed to Sweden.

Yet the Norwegian people remained stubbornly independent. On May 17, 1814, a Norwegian assembly adopted a liberal constitution that provided for a parliament, the Storting, and severely restricted the rights of the executive branch of government, represented by a monarch. The constituent assembly then proceeded to elect a Danish prince as king.

In response, Sweden invaded Norway, which then had to submit to Swedish rule. Yet on No-vember 4, 1814, the Storting declared Norway to be "a free, independent, and indivisible kingdom, united with Sweden under one king." In 1815 Sweden accepted this declaration and set up a special agreement, the Riksakt, to define the relationship between the two countries. The Riksakt was not a very clear or consistent agreement; for example, it gave the Swedish government authority to conduct foreign affairs for both countries, but the consular service was a shared responsibility.

Norway and Sweden were very different in the makeup of their populations and in their traditions of government. Norway was made up mostly of free peasants and people who made their living from the sea, while in Sweden, there was an aristocracy that owned most of the land and kept the peasants dependent. By the mid-1800's, Sweden had developed considerable industry, but Norway remained primarily agricultural and commercial. Norway's liberal structure of government was quite different from Sweden's conservative system, which was run by the aristocracy.

A rising nationalism in Norway was fostered by a great interest in Norwegian language and folklore. Late in the nineteenth century, playwright Henrik Ibsen and writer Björnstjerne Björnson led a Norwegian literary revival. Young Norwegians became passionately committed to breaking off their country's union with Sweden. The Liberal Party emerged after 1870 as the champion of Norwegian democracy and greater freedom for Norway within the union. By 1890, however, the Radical Party had attracted many young nationalists with its call for a complete breaking of the union with Sweden.

### Steps Toward Separation

The Norwegian Liberal leader Johan Sverdrup issued a call for changing the terms of union to permit members of the Norwegian ministry, who were appointed by the Crown, to take part in the deliberations of the Storting. The Storting passed resolutions to this effect three times, but each time King Oscar II exercised his veto. In June, 1880, the Storting passed still another resolution making Sverdrup's proposal a part of the constitution. Because the Liberals had just won a major victory at the polls, the king finally had to accept the amendment. This marked the full establishment of parliamentary government in Norway.

Under the Military Service Act, supported by Radicals and passed by the Storting in 1885, Norway withdrew the largest and best part of its military forces from the common army of the union. This caused a further weakening of the monarchy and of the union.

In 1891, the Radicals won a majority of votes and took control of the ministry and the Storting. They then raised the issue that led to the end of the union between Sweden and Denmark. Contrary to the provisions of the Riksakt, they demanded a separate Norwegian consular service. On this matter the Radicals came up against the opposition of both the Swedish government and the Norwegian Conservative Party. These two groups agreed to try to resolve not only this question but also all other matters that had been sources of tension between the two nations. So it was that Sweden began what proved to be a long series of negotiations leading nowhere.

At first, the Swedes insisted on keeping a joint consular service but proposed giving Norway an equal share in the conduct of foreign policy. This proposal failed. In 1902, Sweden reopened negotiations and agreed to the Radical demand for separate consular services. The Radical government was defeated in 1903, and it seemed that the negotiations had a chance of success. Yet in 1904, the very stubborn, uncompromising Erik Gustaf Boström became prime minister of Sweden and began to insist that Norway's consular service could not be allowed real independence from Sweden's Foreign Ministry. On February 7,

*Scandinavia.*

1905, the two governments announced that negotiations were formally over.

Ironically, Boström had managed to unite the people of Norway as never before against Sweden. In March, 1905, the Radicals returned to power in Norway under the leadership of Peter Christian Michelsen. Along with the Radical-dominated Storting, Michelsen moved to destroy the union. In May, 1905, the Storting passed a bill calling for the establishment of a separate Norwegian consular service. When the king vetoed the bill, the entire Norwegian ministry resigned. Oscar refused to accept the resignations, and on June 7, the Storting declared that the royal power had ceased to function.

**Consequences**

In a vote on August 13, 1905, the Norwegian people approved ending the union with Sweden by the overwhelming majority of 368,208 to 184. The Swedish parliament reluctantly accepted the separation on September 24. A month later, on October 26, the two states signed a formal treaty dissolving the union. Norway then bestowed the crown on Prince Carl of Denmark (grandson of King Christian IX of Denmark), who ruled until 1957 as Haakon VII.)

*Edward P. Keleher*
*Updated by John Quinn Imholte*

**123**

# Russia's October Manifesto Promises Constitutional Government

> *Though Nicholas II's October Manifesto promised important freedoms under the guarantee of a new constitution, his later actions fell far short of his promises.*

**What:** National politics
**When:** October 30, 1905
**Where:** Saint Petersburg, Russia
**Who:**
NICHOLAS II (1868-1918), czar of Russia from 1894 to 1917
COUNT SERGEI YULIEVICH WITTE (1849-1915), first constitutional premier of Russia, from 1905 to 1906
ALEXANDER DUBROVIN (1855-1918), president of the Union of the Russian People, a reactionary group
LEON TROTSKY (1879-1940), leader of the leftist Soviet of Workers' Deputies

## Constitutional Manifesto

Growing discontent among Russian workers and peasants had led to a public call for government reform in January, 1905, in Saint Petersburg. Czarist troops had opened fire on the demonstrators, initiating a massacre that had come to be known as Bloody Sunday. This episode combined with Russia's losses in the Russo-Japanese War of 1904-1905 to touch off the Revolution of 1905.

Between January and October, 1905, there were strikes in industrial cities and peasant revolts in the countryside, along with mutinies in the army and navy. Seeking to calm the people, Czar Nicholas II proclaimed in August the establishment of the Duma, or Russian parliament. He did not propose giving the Duma much power: It would be elected only indirectly, and the vote would be given only to a few. Furthermore, it would not have power to make laws, only to advise the czar.

His proclamation did not have the effect he had hoped. In October, a general strike brought almost all activity in the country to a standstill for ten days. The pressure on Nicholas had increased. Count Sergei Yulievich Witte, his former finance minister and now premier, drafted a constitution, which Nicholas reluctantly agreed to publish as an imperial manifesto.

If fully carried out, the October Manifesto would have given Russia a constitutional monarchy. It had several important provisions to liberalize the government. The manifesto guaranteed civil liberties such as freedom of speech, assembly, and conscience; freedom of the press; freedom from arbitrary arrest; and the right to form trade unions. Citizens who had been excluded from voting under the August decree would receive the right to vote. The manifesto promised that no law would be enacted without the consent of the Duma, and it gave the Duma the right to decide whether actions of the czar's officials were legal.

## The People Respond

Government officials had hoped that publication of the October Manifesto would quiet the country. Instead, the Russian people responded with more disorder. Those who supported the manifesto demonstrated against those who opposed the new freedoms. The revolutionary movement split between rejecters and supporters of the new constitution. In the end, this worked in the government's favor.

On the extreme right, conservatives who were loyal to the government and the Russian Orthodox Church organized the Union of the Russian People under the presidency of Alexander Dubrovin. Czar Nicholas himself belonged to this group. Some of these conservatives led the Black Hundreds—gangs of tough young men—in demonstrations supporting the czar. During the week

**124**

*A Moscow crowd celebrates the freedoms promised by Czar Nicholas in the October Manifesto.*

after the manifesto's publication, the Black Hundreds launched many attacks on Jewish people, many of whom lost property or were killed.

The Octobrists were moderate rightists who were very pleased with the October Manifesto and called it the climax of a successful revolution. The Constitutional Democrats, also known as Cadets, also favored the manifesto but urged the government to move quickly on land reform and other issues.

Leftist parties rejected the manifesto completely, insisting that it did not do enough for the Russian people. They attempted to continue the revolution. The Soviet (Council) of Workers' Deputies in Saint Petersburg, made up of members of the Social Revolutionary, Bolshevik, and Menshevik parties, had been set up before the manifesto was published. Now its leaders, including Leon Trotsky, laid plans for new strikes, which they hoped would lead to an armed uprising.

Though peasant revolts and mutinies in the armed forces continued, the Soviet of Workers' Deputies began to lose the support of urban workers. The program of strikes did not succeed, and by mid-December the government had arrested the leaders. In Moscow the Soviet of Workers' Deputies took up arms against the czar, but by the end of the month the government was able to defeat them.

## Consequences

The Revolution of 1905 did not succeed; the czarist monarchy had survived. Nicholas II had lost some of his confidence, but he now began to take back his traditional autocratic way of ruling the nation. The October Manifesto had served its purpose, gaining some support for Nicholas among the people and dividing the revolutionary movement. Its provisions were not put into practice.

Though historians speak of the Revolution of 1905 in Russia, the uprisings of that year were actually only a revolt that prepared the way for the Revolution of 1917. At the end of 1905, the czar remained on the throne, most of the army remained loyal, and the promises of the October Manifesto had proved hollow. Yet the people's discontent had not been squelched, and it would rise again to end czarist rule for good in 1917.

*Edward P. Keleher*

**125**

# Crile Performs First Human Blood Transfusion

> *George Washington Crile performed the first direct human blood transfusion and developed a device that made the technique practical during surgery.*

**What:** Medicine
**When:** December, 1905
**Where:** Cleveland, Ohio
**Who:**
GEORGE WASHINGTON CRILE (1864-1943), an American surgeon, author, and brigadier general in the U.S. Army Medical Officers' Reserve Corps
ALEXIS CARREL (1873-1944), a French surgeon
SAMUEL JASON MIXTER (1855-1923), an American surgeon

## Nourishing Blood Transfusions

It is impossible to say when and where the idea of blood transfusion first originated, although descriptions of this procedure are found in ancient Egyptian and Greek writings. The earliest documented case of a blood transfusion is that of Pope Innocent VII. In April, 1492, the pope, who was gravely ill, was transfused with the blood of three young boys. As a result, all three boys died without bringing any relief to the pope.

In the centuries that followed, there were occasional descriptions of blood transfusions, but it was not until the middle of the seventeenth century that the technique gained popularity following the English physician and anatomist William Harvey's discovery of the circulation of the blood in 1628. In the medical thought of those times, blood transfusion was considered to have a nourishing effect on the recipient. In many of those experiments, the human recipient received animal blood, usually from a lamb or a calf. Blood transfusion was tried as a cure for many different diseases, mainly those that caused hemorrhages, as well as for

other medical problems and even for marital problems.

Blood transfusions were a dangerous procedure, causing many deaths of both donor and recipient as a result of excessive blood loss, infection, passage of blood clots into the circulatory systems of the recipients, passage of air into the blood vessels (air embolism), and transfusion reaction as a result of incompatible blood types. In the mid-nineteenth century, blood transfusions from animals to humans stopped after it was discovered that the serum of one species agglutinates and dissolves the blood cells of other species. A sharp drop in the use of blood transfusion came with the introduction of physiologic salt solution in 1875. Infusion of salt solution was simple and was safer than blood.

## Direct-Connection Blood Transfusions

In 1898, when George Washington Crile began his work on blood transfusions, the major obstacle he faced was solving the problem of blood clotting during transfusions. He realized that salt solutions were not helpful in severe cases of blood loss, when there is a need to restore the patient to consciousness, steady the heart action, and raise the blood pressure. At that time, he was experimenting with indirect blood transfusions by drawing the blood of the donor into a vessel, then transferring it into the recipient's vein by tube, funnel, and cannula, the same technique used in the infusion of saline solution.

The solution to the problem of blood clotting came in 1902 when Alexis Carrel developed the technique of surgically joining blood vessels without exposing the blood to air or germs, either of which can lead to clotting. Crile learned this technique from Carrel and used it to join the peripheral artery in the donor to a peripheral vein of the recipient. Since the transfused blood remained sealed in the inner lining of the vessels, blood clotting did not occur.

The first human blood transfusion of this type was performed by Crile in December, 1905. The patient, a thirty-five-year-old woman, was transfused by her husband but died a few hours after the procedure.

The second, but first successful, transfusion was performed on August 8, 1906. The patient, a twenty-three-year-old male, suffered from severe hemorrhaging following surgery to remove kidney stones. After all attempts to stop the bleeding were exhausted with no results, and the patient was dangerously weak, transfusion was considered as a last resort. One of the patient's brothers was the donor. Following the transfusion, the patient showed remarkable recovery and was strong enough to withstand surgery to remove the kidney and stop the bleeding. When his condition deteriorated a few days later, another transfusion was done. This time, too, he showed remarkable improvement, which continued until his complete recovery.

For his first transfusions, Crile used the Carrel suture method, which required using very fine needles and thread. It was a very delicate and time-consuming procedure. At the suggestion of Samuel Jason Mixter, Crile developed a new method using a short tubal device with an attached handle to connect the blood vessels. By this method, 3 or 4 centimeters of the vessels to be connected were surgically exposed, clamped, and cut, just as under the previous method. Yet, instead of suturing of the blood vessels, the recipient's vein was passed through the tube and then cuffed back over the tube and tied to it. Then the donor's artery was slipped over the cuff. The clamps were opened, and blood was allowed to flow from the donor to the recipient. In order to accommodate different-sized blood vessels, tubes of four different sizes were made, ranging in diameter from 1.5 to 3 millimeters.

## Consequences

Crile's method was the preferred method of blood transfusion for a number of years. Following the publication of his book on transfusion, a number of modifications to the original method were published in medical journals. In 1913, Edward Lindeman developed a method of transfusing blood simply by inserting a needle through the patient's skin and into a surface vein, making it for the first time a nonsurgical method. This method allowed one to measure the exact quantity of blood transfused. It also allowed the donor to serve in multiple transfusions. This development opened the field of transfusions to all physicians. Lindeman's needle and syringe method also eliminated another major drawback of direct blood transfusion: the need to have both donor and recipient right next to each other.

*Gershon B. Grunfeld*

# Anschütz-Kaempfe Perfects Gyrocompass

> *Hermann Anschütz-Kaempfe designed and manufactured the first practical gyroscopic compass, a navigational device that enables ships and submarines to stay on course.*

**What:** Transportation
**When:** 1906
**Where:** Kiel, Germany
**Who:**
HERMANN ANSCHÜTZ-KAEMPFE (1872-1931), a German inventor and manufacturer
JEAN-BERNARD-LÉON FOUCAULT (1819-1868), a French experimental physicist and inventor
ELMER AMBROSE SPERRY (1860-1930), an American engineer and inventor

## From Toys to Tools

A gyroscope consists of a rapidly spinning wheel mounted in a frame that enables the wheel to tilt freely in any direction. The amount of momentum allows the wheel to maintain its "attitude" even when the whole device is turned or rotated.

These devices have been used to solve problems arising in such areas as sailing and navigation. For example, a gyroscope aboard a ship maintains its orientation even while the ship is rolling. Among other things, this allows the extent of the roll to be measured accurately. Moreover, the spin axis of a free gyroscope can be adjusted to point toward true north. It will (with some exceptions) stay that way despite changes in the direction of the vehicle in which it is mounted. Gyroscopic effects were employed in the design of various objects long before the theory behind them was formally known. A classic example is a child's top, which balances, seemingly in defiance of gravity, as long as it continues to spin. Boomerangs and flying disks derive stability and accuracy from the spin imparted by the thrower. Likewise, the accuracy of rifles improved when barrels were manufactured with internal spiral grooves that caused the emerging bullet to spin.

In 1852, the French inventor Jean-Bernard-Léon Foucault built the first gyroscope, a measuring device consisting of a rapidly spinning wheel mounted within concentric rings that allowed the wheel to move freely about two axes. This device, like the Foucault pendulum, was used to demonstrate the rotation of the earth around its axis, since the spinning wheel, which is not fixed, retains its orientation in space while the earth turns under it. The gyroscope had a related interesting property: As it continued to spin, the force of the earth's rotation caused its axis to rotate gradually until it was oriented parallel to the earth's axis, that is, in a north-south direction. It is this property that enables the gyroscope to be used as a compass.

## When Magnets Fail

In 1904, Hermann Anschütz-Kaempfe, a German manufacturer working in the Kiel shipyards, became interested in the navigation problems of submarines used in exploration under the polar ice cap. By 1905, efficient working submarines were a reality, and it was evident to all major naval powers that submarines would play an increasingly important role in naval strategy.

Submarine navigation posed problems, however, that could not be solved by instruments designed for surface vessels. A submarine needs to orient itself under water in three dimensions; it has no automatic horizon with respect to which it can level itself. Navigation by means of stars or landmarks is impossible when the submarine is submerged. Furthermore, in an enclosed metal hull containing machinery run by electricity, a magnetic compass is worthless. To a lesser extent, increasing use of metal, massive moving parts, and electrical equipment had also rendered the magnetic compass unreliable in conventional surface battleships.

It made sense for Anschütz-Kaempfe to use the gyroscopic effect to design an instrument that would enable a ship to maintain its course while under water. Yet producing such a device would not be easy. First, it needed to be suspended in such a way that it was free to turn in any direction with as little mechanical resistance as possible. At the same time, it had to be able to resist the inevitable pitching and rolling of a vessel at sea. Finally, a continuous power supply was required to keep the gyroscopic wheels spinning at high speed.

The original Anschütz-Kaempfe gyrocompass consisted of a pair of spinning wheels driven by an electric motor. The device was connected to a compass card visible to the ship's navigator. Motor, gyroscope, and suspension system were mounted in a framc that allowed the apparatus to remain stable despite the pitch and roll of the ship.

In 1906, the German navy installed a prototype of the Anschütz-Kaempfe gyrocompass on the battleship *Undine* and subjected it to exhaustive tests under simulated battle conditions, sailing the ship under forced draft and suddenly reversing the engines, changing the position of heavy turrets and other mechanisms, and firing heavy guns. In conditions under which a magnetic compass would have been worthless, the gyrocompass proved a satisfactory navigational tool, and the results were impressive enough to convince the German navy to undertake installation of gyrocompasses in submarines and heavy battleships, including the battleship *Deutschland*.

Elmer Ambrose Sperry, a New York inventor intimately associated with pioneer electrical development, was independently working on a design for a gyroscopic compass at about the same time. In 1907, he patented a gyrocompass consisting of a single rotor mounted within two concentric shells, suspended by fine piano wire from a frame mounted on gimbals. The rotor of the Sperry compass operated in a vacuum, which enabled it to rotate more rapidly. The Sperry gyrocompass was in use on larger American battleships and submarines on the eve of World War I (1914-1918).

## Consequences

The ability to navigate submerged submarines was of critical strategic importance in World War I. Initially, the German navy had an advantage both in the number of submarines at its disposal and in their design and maneuverability. The German U-boat fleet declared all-out war on Allied shipping, and, although their efforts to blockade England and France were ultimately unsuccessful, the tremendous toll they inflicted helped maintain the German position and prolong the war. To a submarine fleet operating throughout the Atlantic and in the Caribbean, as well as in near-shore European waters, effective long-distance navigation was critical.

Gyrocompasses were standard equipment on submarines and battleships and, increasingly, on larger commercial vessels during World War I, World War II (1939-1945), and the period between the wars. The devices also found their way into aircraft, rockets, and guided missiles. Although the compasses were made more accurate and easier to use, the fundamental design differed little from that invented by Anschütz-Kaempfe.

*Martha Sherwood-Pike*

# Nernst Develops Third Law of Thermodynamics

*Walther Nernst showed that it is impossible to reach a temperature of absolute zero and that, close to that temperature, matter exhibits no disorder. Together, these two statements are known as the third law of thermodynamics.*

**What:** Physics
**When:** 1906
**Where:** Berlin, Germany
**Who:**
WALTHER NERNST (1864-1941), a German physicist
SIR JAMES DEWAR (1842-1923), an English physicist
HEIKE KAMERLINGH ONNES (1853-1926), a Dutch physicist

## Reaching Very Low Temperatures

Beginning in the 1870's, scientists were able to achieve temperatures lower than −100 degrees Celsius by allowing gases to expand rapidly. At such low temperatures, the gases themselves often became liquid and could be used to cool other materials to similar temperatures. Work in this area led to the idea of "absolute zero," which was the lowest temperature possible. That temperature was −273 degrees Celsius (0 degrees Celsius is the temperature of an ice-water mixture). During the next thirty years, physicists such as James Dewar tried to see how closely they could approach absolute zero. In trying to reach absolute zero, Dewar invented new equipment for cooling gas and for storing it after it had been liquefied.

Scientists also studied the properties of matter at such very low temperatures, and in doing so they came to two conclusions. First, the closer one approached absolute zero, the more difficult it became to go any lower. This was the result of a combination of practical problems. For example, the colder a sample of matter became, the more rapidly heat leaked into it and warmed it up. Also, at very low temperatures, the liquid gases that were used in the cooling process actually froze into solids, thereby becoming useless in any further cooling. The second conclusion was quite unexpected: At these extremely low temperatures, many properties of matter and energy changed in ways that seemed to contradict existing ideas.

## Approaching Absolute Zero

Physicists began asking what led to such dramatic changes, and they also wondered whether it was possible to achieve absolute zero. More than any other physicist of his generation, Walther Nernst was able to provide important answers to these questions. Together, some of these answers became known as the third law of thermodynamics.

To understand the third law of thermodynamics, it is necessary to consider the first two laws. Thermodynamics is the study of a number of forms of energy, including heat energy, mechanical energy, and the energy associated with the orderliness of things. This last kind of energy is called "entropy." More exactly, entropy reflects the degree of disorder that exists; a decrease in orderliness produces an increase in entropy. An example of order is the way in which molecules are arranged in a crystal. An example of disorder is the chaotic way that molecules are scattered in a gas. The first law of thermodynamics states that different forms of energy have to add up: If one kind of energy increases, another kind must decrease. The second law, which is concerned with entropy, states that, in any process, some energy is lost through an increase in disorder. The third law also says something about entropy: As temperature approaches absolute zero, entropy also approaches zero. Thus, at absolute zero, all disorder vanishes.

Walther Nernst was a great experimentalist and an even greater theoretical physicist, and his development of the third law enabled him to combine the two talents. He made ingenious heat measurements at different temperatures close to absolute zero. He also thought deeply about his results and reached a conclusion of great importance. He realized that, at lower and lower temperatures, the total energy of a system becomes more and more nearly equal to its heat energy. In such a case, the total energy is the sum of heat energy and entropic energy. Therefore, if the total energy and heat energy become equal, entropy cannot exist.

This conclusion was astonishing to most of Nernst's contemporaries, who believed that at absolute zero all energy, and not just entropy, became zero. In fact, a certain amount of energy remained at absolute zero. Also, if entropy was zero, then the system was completely ordered.

This makes sense, since at zero, even gases would exhibit a solid, crystalized state.

There was also, however, an additional consequence of zero entropy. Nernst realized that, in the process of achieving low temperatures by means of gas expansion, the drop in temperature resulted from converting heat energy into entropy. Therefore, the closer one approached zero (where entropy became zero), the more difficult it became to perform this conversion, and no number of steps could ever reach absolute zero. Thus, the third law makes two statements that seem to be different but are, in fact, closely related: First, as absolute zero is approached, entropy also approaches zero; second, it is impossible to reach absolute zero.

## Consequences

Nernst's work indicated that quite unexpected things happened at temperatures approaching absolute zero. During the next few years after Nernst's discoveries, additional surprises were in store, many of which were discovered in the laboratory of Kamerlingh Onnes in Holland. For one thing, it was discovered that the electrical resistance of certain metals falls suddenly to zero at a specific temperature, a result that had not been predicted by existing theories. Such "superconductivity" has led to a new understanding of the way in which matter is constructed.

Superconductivity has become important in such practical applications as the efficient transmission of electrical power and the construction of powerful magnets that do not dissipate their power in useless heat. Also, when helium gas is liquefied (at approximately 2 degrees Celsius), it proves to flow much more easily than other "normal" liquids do; this phenomenon is called "superfluidity."

The list of unexpected phenomena that manifest at very low temperatures is a long one, and it will probably grow longer. Some of the most significant advances in physics have come from studies of extremely low temperatures, and Walther Nernst's insights continue to provide important guidance to low-temperature physicists.

*John L. Howland*

Francis Simon/AIP Emilio Segrè Visual Archives

*Walther Nernst.*

# Oldham and Mohorovičić Determine Inner Structure of Earth

*Richard Dixon Oldham, Andrija Mohorovičić and other seismologists, using data from earthquakes, revealed the layered internal structure of the earth.*

**What:** Earth science
**When:** 1906 and 1910
**Where:** England and Croatia
**Who:**
RICHARD DIXON OLDHAM (1858-1936), an Irish seismologist
BENO GUTENBERG (1889-1960), a German American seismologist
ANDRIJA MOHOROVIČIĆ (1857-1936), a Croatian meteorologist and seismologist
INGE LEHMANN (1888-1993), a Danish seismologist

## Earth Tides

The scientific picture of the deep interior of the earth must rely upon indirect evidence. This picture has thus changed over time as new evidence becomes available. Only in the twentieth century, with the development of new measurement techniques, did a clear picture begin to emerge.

The preeminent geological text of the later half of the seventeenth century, German Jesuit and scholar Athanasius Kircher's *Mundus Subterraneus* (1664), described the earth with a fiery central core from which emerged a web of channels that carried molten material into fire chambers or "glory holes" throughout the "bowels" of the earth. The "contraction" theory, popular through the first half of the nineteenth century, held that the earth had formed from material from the sun and had since been slowly cooling, creating a solid crust. As the fluid cooled, it would contract, and the crust would collapse around the now smaller core, causing earthquakes and producing wrinkles that formed the surface features of mountains and valleys.

Physicists in the 1860's, however, pointed out several physical consequences of the contraction model that appeared to be violated. For example, the gravitational force of the moon, which creates tides in the ocean, would produce tides also in a fluid interior. The resulting effect should either cause the postulated thin crust to crack and produce earthquakes whenever the moon was overhead or, if the crust were sufficiently elastic, cause tides in the crust as well. Instruments were developed to detect crustal tides, but no such tides were discovered.

By the beginning of the twentieth century, geologists were forced to abandon the contraction theory and conclude that the earth probably was completely solid and "as rigid as steel." Thus, when the German geophysicist and meteorologist Alfred Lothar Wegener suggested the idea of continental drift in 1912, it was dismissed on this basis as being physically impossible. Ironically, the very measuring instruments that overturned the simple liquid interior contraction theory would, in turn, disprove the solid earth view that had challenged and replaced it. Seismographs—such as the inverted pendulum seismograph invented in 1900 by German seismologist Emil Wiechert— would provide a completely new source of information about the internal structure of the earth.

## An X Ray of the Earth

In 1900, Richard Dixon Oldham published a paper that established that earthquakes give rise to three separate forms of wave motion that travel through the earth at different rates and along different paths. When an earthquake occurs, it causes waves throughout the earth that fan out from the earthquake's point of origin or "focus." Primary (P) waves emanate like sound waves from the focus, successively compressing and expanding the surrounding material. While P waves can travel through gases, liquids, and solids, secondary (S) waves can travel only through solids. S waves travel at about two-thirds the speed of P waves. Surface waves, the third type of seismic wave, travel only near the earth's surface.

Seismologists now speak of seismic data as being able to provide an X-ray picture of the earth. In his groundbreaking 1906 article, "The Earth's Interior as Revealed by Earthquakes," Oldham analyzed worldwide data on fourteen earthquakes. He also observed that within certain interior earth zones, P waves behaved differently from what he expected. In 1912, Beno Gutenberg was able to establish that at a certain depth the velocity at which the seismic waves traveled changed sharply. He estimated this depth to be about 2,900 kilometers (almost half of the earth's radius).

Oldham had also recognized in his own data the suggestion of a thin outer crust, but his data were insufficient to determine its depth. This estimation was to be made by Andrija Mohorovičić, a professor at the University of Zagreb, from an analysis (published in 1910) of an earthquake that had hit Croatia's Kulpa Valley in late 1909. Mohorovičić had noticed that at any one seismic station, both P and S waves from the earthquake appeared in two sets, but the time between the sets varied according to how far away the station was from the earthquake's focus.

Based upon these different arrival times, Mohorovičić reasoned that waves from a single shock were taking two paths, one of which traveled for a time through a "faster" type of rock. He calculated the depth of this change in material. His estimate fell within the now-accepted figure of 20 to 70 kilometers (the crustal thickness varies under the continents and shrinks to between 6 and 8 kilometers under the ocean). This boundary between crust and mantle is now called the "Mohorovičić Discontinuity," or the Moho.

## Consequences

The use of seismic data to plumb the earth's interior produced an outpouring of research, both theoretical and experimental. By the mid-1920's, seismologists had learned to make subtler interpretations of their data and realized that no S waves had been observed to penetrate through the core. Since S waves cannot pass through liquid, there was reason to think that the core was liquid.

Independent support for this hypothesis came from rigidity studies. Seismic waves could be used to determine the rigidity of the mantle; it was revealed to be much greater than the average rigidity of the earth as a whole (the existence of a low-density fluid region would account for this discrepancy). In 1936, however, Inge Lehmann was able to show that P waves passing close to the earth's center changed velocity slightly, and she correctly inferred the existence of an inner core to account for this phenomenon. The current picture of the earth now includes a liquid outer core surrounding a solid inner core about 1,200 kilometers in radius.

*Robert T. Pennock*

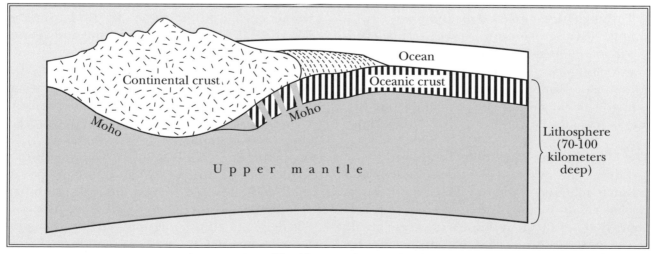

*A cross-section of Earth's crust, showing the Moho layer.*

# Coolidge Patents Tungsten Filament

*William David Coolidge discovered how to create tungsten wire for use in incandescent lightbulbs.*

**What:** Materials
**When:** 1906-1911
**Where:** Schenectady, New York
**Who:**
WILLIAM DAVID COOLIDGE (1873-1975), an
    American electrical engineer
THOMAS ALVA EDISON (1847-1931), an
    American inventor

## The Incandescent Lightbulb

The electric lamp developed along with an understanding of electricity in the latter half of the nineteenth century. In 1841, the first patent for an incandescent lamp was granted in Great Britain. A patent is a legal claim that protects the patent holder for a period of time from others who might try to copy the invention and make a profit from it. Although others tried to improve upon the incandescent lamp, it was not until 1877, when Thomas Alva Edison, the famous inventor, became interested in developing a successful electric lamp, that real progress was made. The Edison Electric Light Company was founded in 1878, and in 1892, it merged with other companies to form the General Electric Company.

Early electric lamps used platinum wire as a filament. Because platinum is expensive, alternative filament materials were sought. After testing many substances, Edison finally decided to use carbon as a filament material. Although carbon is fragile, making it difficult to manufacture filaments, it was the best choice available at the time.

## The Manufacture of Ductile Tungsten

Edison and others had tested tungsten as a possible material for lamp filaments but discarded it as unsuitable. Tungsten is a hard, brittle metal that is difficult to shape and easy to break, but it possesses properties that are needed for lamp filaments. It has the highest melting point (3,410 degrees Celsius) of any known metal; therefore, it can be heated to a very high temperature, giving off a relatively large amount of radiation without melting (as platinum does) or decomposing (as carbon does). The radiation it emits when heated is primarily visible light. Its resistance to the passage of electricity is relatively high, so it requires little electric current to reach its operating voltage. It also has a high boiling point (about 5,900 degrees Celsius) and therefore does not tend to boil away, or vaporize, when heated. In addition, it is mechanically strong, resisting breaking caused by mechanical shock.

William David Coolidge, an electrical engineer with the General Electric Company, was assigned in 1906 the task of transforming tungsten from its natural state into a form suitable for lamp filaments. The accepted procedure for producing fine metal wires was (and still is) to force a wire rod through successively smaller holes in a hard metal block until a wire of the proper diameter is achieved. The property that allows a metal to be drawn into a fine wire by means of this procedure is called "ductility." Tungsten is not naturally ductile, and it was Coolidge's assignment to make it into a ductile form. Over a period of five years, and after many failures, Coolidge and his workers achieved their goal. By 1911, General Electric was selling lamps that contained tungsten filaments.

Originally, Coolidge attempted to mix powdered tungsten with a suitable substance, form a paste, and squirt that paste through a die to form the wire. The paste-wire was then sintered (heated at a temperature slightly below its melting point) in an effort to fuse the powder into a solid mass. Because of its higher boiling point, the tungsten would remain after all the other components in the paste boiled away. At about 300 degrees Celsius, tungsten softens sufficiently to be hammered into an elongated form. Upon cooling, however, tungsten again becomes brittle, which prevents it from being shaped further into filaments. It was suggested that impurities in the tungsten caused the brittleness, but specially purified tungsten worked no better than the unpurified form.

Many metals can be reduced from rods to wires if the rods are passed through a series of rollers that are successively closer together. Some success was achieved with this method when the rollers were heated along with the metal, but it was still not possible to produce sufficiently fine wire. Next, Coolidge tried a procedure called "swaging," in which a thick wire is repeatedly and rapidly struck by a series of rotating hammers as the wire is drawn past them. After numerous failures, a fine wire was successfully produced using this procedure. It was still too thick for lamp filaments, but it was ductile at room temperature.

Microscopic examination of the wire revealed a change in the crystalline structure of tungsten as a result of the various treatments. The individual crystals had elongated, taking on a fiberlike appearance. Now the wire could be drawn through a die to achieve the appropriate thickness. Again, the wire had to be heated, and if the temperature was too high, the tungsten reverted to a brittle state. The dies themselves were heated, and the reduction progressed in stages, each of which reduced the wire's diameter by one thousandth of an inch.

Finally, Coolidge had been successful. Pressed tungsten bars measuring $\frac{1}{4} \times \frac{3}{8} \times 6$ inches were hammered and rolled into rods $\frac{1}{8}$ inch, or $\frac{125}{1000}$ inch, in diameter. The unit $\frac{1}{1000}$ inch is often called a "mil." These rods were then swaged to approximately 30 mil and then passed through dies to achieve the filament size of 25 mil or smaller, depending on the power output of the lamp in which the filament was to be used. Tungsten wires of 1 mil or smaller are now readily available.

## Consequences

Ductile tungsten wire filaments are superior in several respects to platinum, carbon, or sintered tungsten filaments. Ductile filament lamps can withstand more mechanical shock without breaking. This means that they can be used in, for example, automobile headlights, in which jarring frequently occurs. Ductile wire can also be coiled into compact cylinders within the lamp bulb, which makes for a more concentrated source of light and easier focusing. Ductile tungsten filament lamps require less electricity than do carbon filament lamps, and they also last longer. Because the size of the filament wire can be carefully controlled, the light output from lamps of the same power rating is more reproducible. One 60-watt bulb is therefore exactly like another in terms of light production.

Improved production techniques have greatly reduced the cost of manufacturing ductile tungsten filaments and of lightbulb manufacturing in general. The modern world is heavily dependent upon this reliable, inexpensive light source, which turns darkness into daylight.

*Gordon A. Parker*

# Willstätter Discovers Composition of Chlorophyll

> *Richard Willstätter unified theories concerning the material responsible for photosynthesis in plants and traced this activity to two basic chlorophyll molecules.*

**What:** Biology
**When:** 1906-1913
**Where:** Zurich, Switzerland
**Who:**
RICHARD WILLSTÄTTER (1872-1942), a German organic chemist and winner of the 1915 Nobel Prize in Chemistry
ARTHUR STOLL (1887-1971), a Swiss chemist

## A Laborious Process

When Richard Willstätter fled from the Gestapo in Nazi Germany to Switzerland in 1939, one of the few items that he took with him was his 1915 Nobel Prize certificate honoring him for chemical research. His work had established the composition and partial structure of chlorophyll. The certificate was decorated with a border of green leaves, blue cornflowers, scarlet geranium blossoms, and red berries. It was the presence of these and other colors in the plant kingdom that had led to the initial observations resulting in his productive research.

Willstätter knew and practiced the techniques then available for analyzing the molecular structure of plants. The general method at the time involved a series of chemical reactions in which compounds of known composition were reacted with the original material (or a material derived from the original), with the goal of breaking down the original into smaller fragments. These reactions were followed by tedious purification of these fragments until a simple molecule of known structure resulted. Then the series of reactions was traced backward to determine the molecular structure of the starting material.

For the most part, the reconstructed trail was neither direct nor clear, and many possibilities had to be ruled out by performing further reactions before a definitive answer could be found. The answers were also very open to challenge, and unless the utmost care were exercised at each step of the way, they could be easily refuted. The heart of Willstätter's research for the remainder of his active career was modifying, adapting, and refining these basic tools in order to establish a clearer trail leading to the structure of molecules.

*Richard Willstätter.*

## Diversity from Simplicity

At the time Willstätter began studying photosynthesis, very little was known about the source of color in plants; however, the generally accepted view was that each variation of green observed was the expression of a unique chemical molecule. Therefore, it was thought that there must be a vast multitude of molecules responsible. These were, as a group, referred to as "chlorophylls" and were said to be where photosynthesis occurred.

Willstätter worked in an area that was just beginning to emerge. Building on the meager knowledge available, he created techniques to extract plant materials and to recover, unchanged, the chemical substances that formed these materials. The raw material for developing these processes was the dried leaves of plants. With the raw material in hand, he was able to separate and purify the extracted substances. Some of these methods involved the use of enzymes, a then-poorly-understood class of biological catalysts; others used the technique of adsorption chromatography, which was just developing.

Of particular interest is his finding that there are but two chlorophylls, a blue-green, or alpha, form, and a yellow-green, or beta, form, rather than the myriad forms previously projected. All the hues of green observed resulted from only these two molecules. Willstätter demonstrated this simplicity in the presence of diversity by extracting and analyzing the green parts of more than two hundred species of plants. Only the two types of chlorophyll were found.

The culmination of Willstätter's work was the establishment of the chemical composition of these two forms of chlorophyll. Magnesium, previously believed by others to be an impurity, was shown to be an integral part of the molecule.

Phosphorus, thought by others to be a part of the molecule, was shown to be an impurity. Moreover, each chlorophyll form is a very large molecule.

Willstätter also established the similarity of these molecules to hemoglobin by showing how the magnesium of chlorophyll and the iron of hemoglobin play similar structural roles. This similarity had been conceived as early as 1851, but Willstätter's work was the first to establish the idea as fact. In collaboration with Arthur Stoll, he published a book explaining his work on chlorophyll in 1913.

## Consequences

Willstätter's work opened the door to understanding photosynthesis. His research became the basis for a knowledge of the structures of large, biologically important molecules.

Recognizing the uniqueness of the chlorophyll molecule, its composition, and the role that it plays in photosynthesis has significantly improved human life. Willstätter's work, along with its extension by others, has guided agriculturists in their search for greater plant productivity. Of most direct impact was the understanding of the important role of magnesium as a plant nutrient. This essential element, along with others, must be supplied by fertilization in soils that lack magnesium. As a result, land previously thought to be useless has produced crops, increasing the food supply.

Willstätter was a member of a small group of chemists who pioneered the field of organic chemistry by breaking new ground and by devising new methods of experimentation. For this effort, Willstätter is often referred to as one of the fathers of organic chemistry.

*Kenneth H. Brown*

# First Russian Duma Convenes

*Established by Czar Nicholas II in an attempt to quiet those who criticized his domineering way of governing the country, the new Russian parliament, or Duma, was not given a chance to exercise any real authority.*

**What:** National politics
**When:** May 10, 1906
**Where:** Saint Petersburg, Russia
**Who:**
NICHOLAS II (1868-1918), czar of Russia
   from 1894 to 1917
COUNT SERGEI YULIEVICH WITTE (1849-
   1915), first constitutional premier of
   Russia, from November, 1905, to May,
   1906
PAUL MILIUKOV (1859-1943), leader of the
   Constitutional Democrats, or Cadets
ALEKSANDR IVANOVICH GUCHKOV (1862-
   1936), leader of the Octobrists
IVAN LOGINOVICH GOREMYKIN (1839-1917),
   premier of Russia from May to July,
   1906
PËTR ARKADEVICH STOLYPIN (1862-1911),
   premier of Russia in July, 1906

## Limits Set for the Duma

In August, 1905, as workers' and peasants' revolts were breaking out across Russia, Czar Nicholas II issued a proclamation that a Duma, or parliament, would be established for the country. In October, he followed that announcement with a draft constitution called the October Manifesto. Before the end of the year, there were further uprisings in Saint Petersburg, Moscow, and elsewhere, but the czar and his army were able to suppress them.

As 1906 began, there was continued debate in Russia over whether a constitutional monarchy should be established or whether the absolute monarchy of the czar should be continued and strengthened. Politicians prepared for the elections to the first Duma. The election laws, drawn up by Count Sergei Witte, allowed for indirect election of representatives. The majority of the representatives would be assigned to rural landowners and peasants, for the czar believed that these people would be less likely to press for political reform than would townspeople and industrial workers.

The czar and his ministers tried to put strict limits on the power of the Duma. For example, the czar made the state council an upper legislative chamber—parallel to Great Britain's House of Lords or to the U.S. Senate—that would have powers equal to those of the Duma. Half the members of this upper chamber were appointed directly by the czar, and the other half were elected by traditionally conservative groups such as the clergy, provincial assemblies, the aristocrats, and managers of businesses.

The week before the Duma's first meeting, the government issued a set of "fundamental laws" specifying what powers the Duma would *not* have. No bills were to become law until they were passed by both houses and signed by the czar. The czar's ministers were responsible to him alone, not to the Duma. The Duma did not have complete control of the budget: If the two houses approved different budget figures, the czar could accept either set, and if no budget was passed by the Duma, the government could continue to use the one adopted the previous year. The czar would keep absolute control over foreign policy, appointments, censorship, the armed forces, the police, and the summoning and dismissal of the Duma. When the Duma was not in session, the czar could rule by decree.

The result of the election laws, the creation of an upper house composed of the state council, and the "fundamental laws" was that the political reforms promised by the October Manifesto could not possibly come about.

## Election and Convening

The two leading political parties, which could both be described as moderate, were the Constitutional Democratic Party, or Cadets, and the Union of October 17, or Octobrists. Both of these parties had come into existence only the previous year, but they competed vigorously against each other in the election campaigns. For the most part, the radical Social Revolutionaries and the Social Democrats boycotted the election.

The Cadets, led by a respected Russian historian named Paul Miliukov, urged that a parliamentary government with a constitutional monarchy be established. The Duma should be fully involved in writing a new constitution, the Cadets said, and there should be a program of land reform to buy huge rural estates from the owners and put land under the control of the common people. The Octobrists, whose leader was Aleksandr Ivanovich Guchkov, had a more conservative platform; they especially disagreed with the Cadets' land-reform proposal.

When the votes were counted, the Cadets had won 180 of the Duma's 520 seats, while the Octobrists won only 12. In all, some forty political groups made up the first Duma.

With a speech from the throne, Nicholas II formally convened the Duma's first meeting on May 10, 1906, in the Winter Palace in Saint Petersburg. From the beginning, it was obvious that the government had no intention of allowing the Duma to exercise any real power. The czar's officials had been alarmed to see that conservative groups had failed to elect a single representative to the Duma. Still, the czar expected the Duma to be fairly conservative in its dealings. He was quite surprised when, shortly after its convocation, the Duma presented an "address to the throne" in which it demanded a universal right to vote, direct elections, abolition of the upper chamber, parliamentary government, and land reform.

Witte resigned as premier, and Ivan Goremykin took his place. On May 26, Goremykin addressed the Duma and rejected all its demands. Undaunted, the Duma continued to call for extensive political reforms over the next two months. During this time, Nicholas considered a proposal to bring Cadets into the government in order to silence the Duma's criticism. Yet Miliukov was reluctant to accept such a compromise, as was Pëtr Stolypin, the czar's minister of the interior. Finally, Nicholas simply dissolved the Duma on July 21. On that same day Stolypin was appointed to succeed Goremykin as premier. He was a hard-liner who became known for his conservative policies.

## Consequences

About two hundred Duma deputies refused to accept the closing down of the Duma. Crossing the border into the Grand Duchy of Finland, they gathered in the town of Viborg and signed the "Viborg Manifesto," written by Paul Miliukov. In this appeal, the deputies said that the government's closing of the Duma was illegal. They also insisted that the government could not collect taxes or draft men for military service without the Duma's consent.

The men who signed this appeal were sentenced to three months in prison. As a result, they were considered criminals and were not eligible to run for reelection to the next Duma, which was scheduled to meet early in 1907. Having been deprived of some of its best political leaders, the drive for a constitutional form of government in Russia lost much of its force in the years before the outbreak of the Revolution of 1917.

*Edward P. Keleher*

**139**

# Germany Launches First U-Boat

> *The development of the German U-boat fleet proved crucial to the course of World War I.*

**What:** Weapons technology
**When:** August 4, 1906
**Where:** Danzig, Germany
**Who:**
RAYMOND LORENZO D'EQUEVILLEY-
   MONTJUSTIN (1860-1931), a Spanish
   engineer
ALFRED VON TIRPITZ (1849-1930), a
   German admiral
SEBASTIAN WILHELM VALENTIN BAUER (1822-
   1875), a German submarine pioneer
JOHN PHILIP HOLLAND (1840-1914), the
   father of the modern submarine
WILLIAM II (1859-1941), the kaiser of
   Germany from 1888 to 1918
GUSTAV KRUPP (1870-1950), a German
   industrialist

## Submarine History

When the *Unterseeboot-eins* (underwater boat number 1) was launched at Danzig on August 4, 1906, few people suspected how important the event would prove to be. The English, the French, the Russians, and the Americans already had submarine fleets, so the German U-boat did not seem much of a threat. When the German fleet went into action during World War I, however, it changed the course of the war.

Submarines were not new in 1906. Cornelis Drebbel, a Dutch inventor, tested the first recorded successful submarine in the Thames River in England in 1620, and underwater craft were used during the American Revolutionary War and the American Civil War. Germany had shown particular interest in submarines: In 1850, Sebastian Wilhelm Bauer built a submarine for government use that incorporated all the basic features of later military submarines. Although the vessel sank during a test dive, Bauer became instrumental in the development of the English

and Russian submarine programs. Later, research by John P. Holland led to the invention of electric engines, horizontal rudders to facilitate diving, and water ballasts.

## German Imperialism

Since Germany's navy had mainly been used for defending its small coast, it did not need to be very large. This began to change in the 1880's, when Germany gained overseas colonies. The leaders of German industries, needing raw materials and new markets for their goods, began to push the German government to build a high-seas fleet that could protect the country's trading ships on any of the world's seas. In 1888, when William II became kaiser of Germany, he was eager to cooperate with these industrialists, thinking that the German empire might become as great as England's.

By 1905, the German navy had grown large enough to be dangerous to England. At the time, the English admiralty made sure that its fleet was at least as large as the fleets of the next three largest naval powers in Europe combined. Because of the growth of the German navy, therefore, the English government decided increase its own navy. In turn, when the English navy began to get larger, Grand Admiral Alfred von Tirpitz decided in 1905 to start working on a submarine fleet that could attack as well as defend.

Because of the work of Gustav Krupp, one of the leaders of German industry, Germany's submarine fleet took little time to build. A Spanish engineer named Raymond Lorenzo d'Equevilley-Montjustin had come to the Krupp firm in 1901 with plans for a double-hulled submarine that would be capable of long-range attacks. The Krupp firm had approached Tirpitz for money, but Tirpitz had refused, thinking that submarines would never be able to attack a surface vessel successfully. In 1902, therefore, the Krupp firm began to develop d'Equevilley-Montjustin's

*In this 1917 photograph, a midget submarine pulls up beside a German U-boat.*

design with its own money. Krupp completed the prototype of d'Equevilley-Montjustin's vessel on June 8, 1903; financial problems, however, forced the builders to leave out many features of the original design.

Krupp's boat was 13 meters long, with a range of 40 nautical kilometers at a surface speed of four knots. Driven by a 65-horsepower electric motor, the boat could travel at 5.5 knots submerged. Krupp invited the kaiser to watch the submarine's test in 1904. Both William II and his sons showed great interest in the vessel, but Tirpitz was still against spending what little money the navy had for further developing a vessel that had not been proved in combat.

When the Russo-Japanese War began in 1904, Krupp convinced the Russian government to buy not only the first submarine he had built but also three more boats of the same design (which

came to be called the Karp class). While building the submarines for the Russian navy, Krupp's engineers greatly improved on the original model. The improved submersibles finally won Tirpitz's approval. In 1905, the admiral agreed to give 1.5 million marks for submarine development and to buy several of the vessels for the German navy.

Tirpitz's decision resulted in the launching of the *U-1* on August 4, 1906. The *U-1* measured 42 meters and displaced 234 tons with a crew of twenty men and officers. Armed with a 46-centimeter bow torpedo tube with three self-propelled torpedoes and an 88-millimeter deck gun, the boat had maximum speeds of 10.8 knots while on the surface and 8.7 knots while submerged. With a cruising range of 2,414 kilometers, the *U-1* was definitely capable of attack. The German navy had an important new weapon.

**141**

## Consequences

The launching of *U-1* did not seem earthshaking in 1906, even to the leaders of the other navies of the world. When World War I began, Germany was well behind the other great powers both in numbers of submarines and in submarine technology. (France had 123 submarines, England had 72, Russia had 41, the United States had 34, and Germany had only 26.) Nevertheless, the German submarine fleet quickly became one of the most important and powerful weapons in the war.

On September 22, 1914, *U-9*, commanded by Lieutenant Captain Otto Weddigen, sank three English armored cruisers in a battle that lasted less than one hour. The battle forced the English admiralty not only to admit that its surface warships were open to attack but also to face the fact that English commercial vessels might be attacked. Since England relied heavily on ship trade for food and other essentials, defeat was certain if German submarines could stop merchant ships from reaching England.

On October 14, 1914, Lieutenant Captain Feldkirchner, commanding the *U-17*, sank an English steamer. Despite the agreement reached between the great powers at Geneva in 1907, which outlawed surprise attacks on civilian ships, the German high command blockaded the British Isles and began to sink commercial ships.

On May 7, 1915, a German U-boat sank the English passenger liner *Lusitania*, killing 1,198 civilians, 124 of whom were Americans. Before the *Lusitania* sailed, the German embassy in the United States had bought a full-page advertisement in *The New York Times* to warn Americans that submarines might attack the liner because it was carrying cargo for the war into a war zone. Although there were reports that the U-boat had launched a second torpedo at the already sinking ship while the passengers were trying to escape, the second explosion actually came when the ammunition that the *Lusitania* was carrying blew up. Because of fierce American protests over the sinking of the *Lusitania*, the Germans stopped attacking civilian ships.

By 1917, though, Germany was desperate. General Erich Ludendorff and Field Marshal Paul von Hindenburg decided to risk American entry into the war and began the blockade again, hoping to knock out England before the United States could intervene. In March, 1917, after three U.S. merchant vessels were sunk, President Woodrow Wilson asked the U.S. Congress for a declaration of war against Germany. The entry of the United States into the war ensured Germany's ultimate defeat.

The bitter feelings between Germany and the United States were a major reason for the U.S. entry into World War II, which again brought defeat for Germany. Thus, the seemingly harmless launching of *U-1* in 1906 had a major impact on the outcomes of the two great global conflicts of the twentieth century.

*Paul Madden*

# Battleship *Dreadnought* Revolutionizes Naval Architecture

*The design of a newer, heavier, and faster battleship allowed Great Britain to retain its naval superiority in the early twentieth century.*

**What:** Military technology
**When:** December, 1906
**Where:** Portsmouth, England
**Who:**
Sir John Arbuthnot Fisher (1841-1920), first sea lord at the British Admiralty from 1904 to 1910
Louis Alexander of Battenberg (1854-1921), British prince and director of naval intelligence from 1902 to 1905
Sir Philip Watts (1846-1926), director of the construction of the Royal Navy from 1902 to 1912

## New Armaments

In 1900, battleships were armed with guns of different calibers, and normally their ranges did not extend beyond three thousand yards. British warships of the *King Edward VII* class, designed in 1901, carried four 12-inch guns, ten 6-inch guns, four 9.2-inch guns, and a battery of small guns for defense against torpedo boats. These ships weighed 16,350 tons and could steam at 18.5 knots.

New developments in naval armaments raised new problems in the design of battleships. First, the accuracy and range of torpedoes were more than doubled, from about three thousand yards to seven thousand yards or more. As a result, opposing formations of battleships would need to keep a greater distance from each other, so that they would have a better chance of avoiding these very dangerous weapons.

Second, telescopic sights and new electrical fire-control equipment allowed accurate gunfire to be delivered at a range of eight thousand or more yards. At such a great distance, though, it was difficult, perhaps even impossible, to tell the difference between the splashes of shells from different calibers of guns—and so it was hard to correct their aim.

With improved weapons to fire farther and more accurately, a new type of battleship was needed.

## Design and Construction

In 1900 Admiral Sir John Arbuthnot Fisher, who was then commander-in-chief of the Royal Navy's Mediterranean station, came up with the idea of a ship that would carry only big guns. With no medium-sized guns, there would be no need to figure out which splashes were caused by which shells. The ship would carry a larger number of heavy guns than older ships did, along with some light, quick-fire artillery to be used against torpedo craft.

When Fisher became First Sea Lord at the Admiralty in October, 1904, he had the chance to try out his new ideas. With technical help from W. H. Gard, chief naval constructor at Portsmouth Dockyard, he had several sketch plans drawn up and submitted to a special Committee on Design. The committee included Rear Admiral Prince Louis Alexander of Battenberg, who as director of Naval Intelligence had information on battleship designs that were being developed in other countries. Other committee members were Sir Philip Watts, the Royal Navy's director of construction; six expert officers of high rank; and six naval architects who were civilians. The committee chose one design from those that had been submitted, and they appointed J. H. Narbeth, a naval constructor, to make the final working drawings. In March, 1905, his plans were accepted.

The final design was for a ship of 17,900 tons

**143**

armed with ten 12-inch guns mounted in twin turrets, three on the center line, and one on each side forward. Eight guns would point out from each side of the ship, and six could point ahead or astern. No medium artillery would be included, but there would be an antidestroyer battery of twenty-seven rapid-firing 12-pound guns. Armor eleven inches thick would protect the body of the ship, and its powerful turbine engines—the first ever fitted in a battleship—would bring it to a speed of twenty-one knots. This ship would be able to overwhelm any existing battleship with superior fire, and its speed guaranteed that it could run down any weaker ship or escape from a large group of enemy ships. The new vessel was to be named HMS *Dreadnought*.

The turbine engines were an extremely important part of the *Dreadnought*'s design. Previous battleships had been powered by reciprocating steam engines. In the *Dreadnought*, turbines weighing no more than the older reciprocating steam engines would provide more horsepower, would give more speed, would use up less fuel at high speeds, and, having fewer moving parts, would be easier to maintain.

The materials needed for the ship's construction were collected quickly. The 12-inch guns and mountings that had been intended for two other ships then being built, the *Lord Nelson* and the *Agamemnon*, were taken for use in the *Dreadnought*. Every effort was made to complete the new battleship rapidly. In great secrecy, the

Naval Historical Center

*The powerful battleship* Dreadnought *featured turbine engines.*

*Dreadnought*'s keel was laid at Portsmouth on October 2, 1905.

With great fanfare, the ship was launched on February 10, 1906. By October 3, it was structurally complete—only a year and a day from its beginning. At once it was headed out to sea for tests. Some who were involved in the *Dreadnought*'s design and construction feared that the shock of firing eight guns at once might damage the ship's hull. That problem did not arise, however, and the ship performed well in its trials. After a few minor adjustments, the *Dreadnought* was formally placed in service in December, 1906.

## Consequences

The successful completion of this new type of battleship had important effects for the naval construction of all the great powers. The first announcement of the *Dreadnought*'s successful trials was the same as saying that all the battleships of the world were now obsolete. As a result, most countries stopped constructing battleships while they assigned their designers to make plans for more modern ships.

Germany did not produce its version of a new battleship until July, 1907. By that time Great Britain already had the *Dreadnought* in commission, and since its design was clearly successful, three similar ships of the *Bellerophon* class were constructed in 1907.

Until this time, Great Britain's fleet of battleships had been twice as large as Germany's. In creating a ship that made the rest of its fleet obsolete, the British were in one sense giving Germany the chance to catch up. Yet if the British navy had not acted when it did, and Germany, Japan, and the United States had suddenly decided to build their own superior battleships, Great Britain would have found itself in the more difficult position of needing to catch up.

As Germany and Great Britain began their battleship-building race, the British pulled ahead. Between 1910 and 1912 the two nations held conferences to try to set limits on annual construction programs, but the negotiations came to nothing. In the end, the Germans accepted the fact that the British would always have a larger navy.

*Steven J. Ramold*

# Fessenden Transmits Music and Voice by Radio

*Reginald Aubrey Fessenden revolutionized radio broadcasting by transmitting music and voice for the first time.*

**What:** Communications
**When:** December 24, 1906
**Where:** Brant Rock, Massachusetts
**Who:**
REGINALD AUBREY FESSENDEN (1866-1932), an American radio pioneer
GUGLIELMO MARCONI (1874-1937), an Italian physicist and inventor

## True Radio

The first major experimenter in the United States to work with wireless radio was Reginald Aubrey Fessenden. This transplanted Canadian was a skilled, self-made scientist, but unlike American inventor Thomas Alva Edison, he lacked the business skills to gain the full credit and wealth that such pathbreaking work might have merited. Guglielmo Marconi, in contrast, is most often remembered as the person who invented wireless (as opposed to telegraphic) radio.

There was a great difference between the contributions of Marconi and Fessenden. Marconi limited himself to experiments with radio telegraphy; that is, he sought to send through the air messages that were currently being sent by wire—signals consisting of dots and dashes. Fessenden sought to perfect radio telephony, or voice communication by wireless transmission. Fessenden thus pioneered the essential precursor of modern radio broadcasting. At the beginning of the twentieth century, Fessenden spent much time and energy publicizing his experiments, thus promoting interest in the new science of radio broadcasting.

Fessenden began his career as an inventor while working for the U.S. Weather Bureau. He set out to invent a radio system by which to broadcast weather forecasts to users on land and at sea. Fessenden believed that his technique of using continuous waves in the radio frequency range (rather than interrupted waves Marconi had used to produce the dots and dashes of Morse code) would provide the power necessary to carry Morse telegraph code yet be effective enough to handle voice communication. He would turn out to be correct. He conducted experiments as early as 1900 at Rock Point, Maryland, about 80 kilometers south of Washington, D.C., and registered his first patent in the area of radio research in 1902.

## Fame and Glory

In 1900, Fessenden asked the General Electric Corporation to produce a high-speed generator of alternating current—or alternator—to use as the basis of his radio transmitter. This proved to be the first major request for a wireless radio apparatus that could project voices and music. It took the engineers three years to design and deliver the alternator. Meanwhile, Fessenden worked on an improved radio receiver. To fund his experiments, Fessenden aroused the interest of financial backers, who put up one million dollars to create the National Electric Signalling Company in 1902.

Fessenden, along with a small group of hand-picked scientists, worked at Brant Rock on the Massachusetts coast south of Boston. Working outside the corporate system, Fessenden sought fame and glory based on his own work, rather than on something owned by a corporate patron.

Fessenden's moment of glory came on December 24, 1906, with the first announced broadcast of his radio telephone. Using an ordinary telephone microphone and his special alternator to generate the necessary radio energy, Fessenden alerted ships up and down the Atlantic coast with his wireless telegraph and arranged for newspaper reporters to listen in from New York City. Fessenden made himself the center of the show. He played the violin, sang, and read from the Bible. Anticipating what would become standard practice fifty years later, Fessenden also transmitted the sounds of a phonograph recording. He ended his first broadcast by wishing those listening "a Merry Christmas." A similar, equally well-publicized demonstration came on December 31.

Although Fessenden was skilled at drawing attention to his invention and must be credited, among others, as one of the engineering founders of the principles of radio, he was far less skilled at making money with his experiments, and thus his long-term impact was limited. The National Electric Signalling Company had a fine beginning and for a time was a supplier of equipment to the United Fruit Company. The financial panic of 1907, however, wiped out an opportunity to sell the Fessenden patents—at a vast profit—to a corporate giant, the American Telephone and Telegraph Corporation.

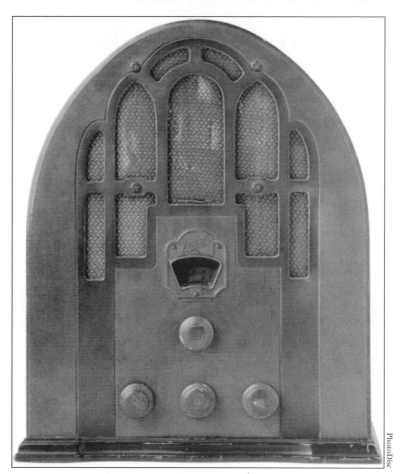

*Reginald Aubrey Fessenden's experiments enabled later inventors to create tabletop radios such as the one pictured here.*

## Consequences

Had there been more receiving equipment available and in place, a massive audience could have heard Fessenden's first broadcast. He had the correct idea, even to the point of playing a crude phonograph record. Yet Fessenden, Marconi, and their rivals were unable to establish a regular series of broadcasts. Their "stations" were experimental and promotional.

It took the stresses of World War I to encourage broader use of wireless radio based on Fessenden's experiments. Suddenly, communicating from ship to ship or from a ship to shore became a frequent matter of life or death. Generating publicity was no longer necessary. Governments fought over crucial patent rights. The Radio Corporation of America (RCA) pooled vital knowledge. Ultimately, RCA came to acquire the Fessenden patents. Radio broadcasting commenced, and the radio industry, with its multiple uses for mass communication, was off and running.

*Douglas Gomery*

**147**

# Lumière Brothers Invent Color Photography

*The Lumière brothers introduced the autochrome plate, the first commercially successful process in which a single exposure in a regular camera produced a color image.*

**What:** Photography
**When:** 1907
**Where:** Lyons, France
**Who:**
Louis Lumière (1864-1948), a French inventor and scientist
Auguste Lumière (1862-1954), an inventor, physician, physicist, chemist, and botanist
Alphonse Seyewetz, a skilled scientist and assistant of the Lumière brothers

## Adding Color

In 1882, Antoine Lumière, painter, pioneer photographer, and father of Auguste and Louis, founded a factory to manufacture photographic gelatin dry-plates. After the Lumière brothers took over the factory's management, they expanded production to include roll film and printing papers in 1887 and also carried out joint research that led to fundamental discoveries and improvements in photographic development and other aspects of photographic chemistry.

While recording and reproducing the actual colors of a subject was not possible at the time of photography's inception (about 1822), the first practical photographic process, the daguerreotype, was able to render both striking detail and good tonal quality. Thus, the desire to produce full-color images, or some approximation to realistic color, occupied the minds of many photographers and inventors, including Louis and Auguste Lumière, throughout the nineteenth century.

As researchers set out to reproduce the colors of nature, the first process that met with any practical success was based on the additive color theory expounded by the Scottish physicist James Clerk Maxwell in 1861. He believed that any color can be created by adding together red, green, and blue light in certain proportions. Maxwell, in his experiments, had taken three negatives through screens or filters of these additive primary colors. He then took slides made from these negatives and projected the slides through the same filters onto a screen so that their images were superimposed. As a result, he found that it was possible to reproduce the exact colors as well as the form of an object.

Unfortunately, since colors could not be printed in their tonal relationships on paper before the end of the nineteenth century, Maxwell's experiment was unsuccessful. Although Frederick E. Ives of Philadelphia, in 1892, optically united three transparencies so that they could be viewed in proper alignment by looking through a peephole, viewing the transparencies was still not as simple as looking at a black-and-white photograph.

## The Autochrome Plate

The first practical method of making a single photograph that could be viewed without any apparatus was devised by John Joly of Dublin in 1893. Instead of taking three separate pictures through three colored filters, he took one negative through one filter minutely checkered with microscopic areas colored red, green, and blue. The filter and the plate were exactly the same size and were placed in contact with each other in the camera. After the plate was developed, a transparency was made, and the filter was permanently attached to it. The black-and-white areas of the picture allowed more or less light to shine through the filters; if viewed from a proper distance, the colored lights blended to form the various colors of nature.

In sum, the potential principles of additive color and other methods and their potential applications in photography had been discovered and even experimentally demonstrated by 1880.

Yet a practical process of color photography utilizing these principles could not be produced until a truly panchromatic emulsion was available, since making a color print required being able to record the primary colors of the light cast by the subject.

Louis and Auguste Lumière, along with their research associate Alphonse Seyewetz, succeeded in creating a single-plate process based on this method in 1903. It was introduced commercially as the autochrome plate in 1907 and was soon in use throughout the world. This process is one of many that takes advantage of the limited resolving power of the eye. Grains or dots too small to be recognized as separate units are accepted in their entirety and, to the sense of vision, appear as tones and continuous color.

## Consequences

While the autochrome plate remained one of the most popular color processes until the 1930's, soon this process was superseded by subtractive color processes. Leopold Mannes and Leopold Godowsky, both musicians and amateur photographic researchers who eventually joined forces with Eastman Kodak research scientists, did the most to perfect the Lumière brothers' advances in making color photography practical. Their collaboration led to the introduction in 1935 of Kodachrome, a subtractive process in which a single sheet of film is coated with three layers of emulsion, each sensitive to one primary color. A single exposure produces a color image.

Color photography is now commonplace. The amateur market is enormous, and the snapshot is almost always taken in color. Commercial and publishing markets use color extensively. Even photography as an art form, which has been in black and white for most of its history, has turned increasingly to color.

*Genevieve Slomski*

# Thomson Confirms Existence of Isotopes

*J. J. Thomson's experiments proved that atoms of a pure chemical substance did not necessarily have identical atomic weights.*

**What:** Physics; Chemistry
**When:** 1907-1910
**Where:** Cambridge, England
**Who:**
SIR JOSEPH JOHN THOMSON (1856-1940), a British physicist
FREDERICK SODDY (1877-1956), a British chemist
EUGEN GOLDSTEIN (1850-1930), a German physicist

## The Puzzle of Radioactivity

Early in the nineteenth century, the English chemist John Dalton (1766-1844) proposed what would become the modern theory of the atom. Dalton believed that if he were able to keep on dividing a piece of a pure chemical substance (for example, an element such as aluminum), he would finally arrive at its smallest part, an atom, which would be incapable of further division. All atoms of the same chemical substance were thought to be exactly alike in size, shape, and weight.

Following the discovery of natural radioactivity in the late 1890's, much effort went into the investigation of several families of radioactive elements. Radioactivity involves the steady, spontaneous emission of either electrically charged particles (called alpha and beta particles) or penetrating radiation (gamma rays) by certain unstable elements. Thorium, for example, was observed to decay by means of a series of such emissions, transforming into a whole chain of different but related "daughter elements," each of which decayed and became a chemically different substance, until the stable element lead was produced.

By the early twentieth century, scientists using the techniques of analytic chemistry had become skilled in separating mixtures of elements. In this context, it came as a surprise when chemist Frederick Soddy discovered that some of the atoms produced in the radioactive series, although chemically identical, seemed to have different atomic weights. Thus was born the idea of the "isotope." Although it was Soddy who coined the word in 1913, it was the crucial work of the physicist J. J. Thomson that proved the existence of isotopes and provided a new method of chemical analysis.

## Thomson's Positive Rays

At the time of Thomson's pioneering work, the structure of the atom itself was still unclear. Within a few years, however, the work of Ernest Rutherford (1871-1937) and his coworkers at Manchester University in England would demonstrate the modern concept of the atom: a miniature solar system in which most of the mass was concentrated in a tiny core of positively charged particles, the nucleus, which was surrounded by a group of negative particles called electrons. The atom as a whole was electrically neutral.

The electron itself had been discovered by Thomson about ten years before his work on isotopes. In that earlier work, he passed electricity through a gas contained in a glass tube at very low pressure. A battery was connected to two metal plates, or "electrodes," that were sealed in the walls of the tube. The electrode on the positive battery terminal was called the anode; the other, the cathode. When most of the air was removed from the tube, an eerie blue glow would be sent out by the gas remaining in the tube, and electricity would begin to flow. At still lower pressures, the glow disappeared completely, and a strange greenish light appeared in the glass at the end of the tube opposite the cathode.

Experiments of this kind showed that any obstacle placed between the electrodes would cast a sharp shadow on the glass wall. It seemed that something was streaming from the cathode to the anode. Thomson subsequently concluded that "cathode rays," which consisted of electrons, had caused the effect that he had observed. He was able to measure the amount of electrical charge that the electrons carried for each unit of their weight. This characteristic factor is known as a particle's charge-to-mass ratio. In 1906, Thomson was awarded the Nobel Prize in Physics for this work.

At about the same time, Eugen Goldstein, a German physicist working in Berlin, had shown that if the cathode in the tube were filled with holes, a second kind of ray—this time electrically positive—could also be observed flowing through the tube. These rays, however, were headed in the opposite direction, toward the negative cathode, and as they passed through the

holes, they created brilliant streams of light in the gas.

Although the cathode rays themselves consisted of electrons, the positive rays seemed to be more complex. A standard technique for exploring the nature of a stream of charged particles was to shoot them through the two poles of a magnet. Electrically charged moving particles were known to have their paths bent as they moved through a magnetic field, and the amount of bending could be shown to depend on their charge-to-mass ratio and speeds: The tracks of heavier particles would be bent less than those of lighter ones, and the tracks of faster particles would be bent less than those of slower ones. Charged particles, however, can also be pushed by electric fields. In his painstaking experiments, using a clever combination of the two kinds of fields, magnetic and electric, Thomson found

*Sir Joseph John Thomson (left), shown receiving a medal from Herbert Asquith in 1906.*

that he could actually separate and identify the charged particles that made up the positive rays.

Thomson believed that the electrons making up the cathode ray were forced into the tube by repulsion from the negative cathode. Apparently, the positive particles were then formed in the gas by collision with the cathode rays. As an atom lost an electron, it became a positive particle called an "ion"—an atom minus one of its electrons. By examining the records of the tracks that those positive rays made on impact with a photographic film, Thomson was able to determine the weights of the positive particles. For example, when a tiny amount of the rare gas neon was in the tube, two quite different atomic weights of neon showed up. One had a weight of 20 units; the other, 22 units. These were the two isotopes of the element neon.

## Consequences

Thomson's work was a crucial step in the creation of the modern theory of the atom. Although his own ideas about how positive and negative charges are arranged inside an atom proved incorrect, his work paved the way for Rutherford's dynamic model, which soon became widely accepted.

In the Rutherford model, there are equal numbers of positive particles (protons) in the nucleus, and they are balanced by an equal number of negative electrons orbiting the nucleus. The extremely light electrons determine all of an element's chemical properties. Hence, isotopes of different atomic weights must have nuclei of different weights. According to Rutherford's model, each isotope's nuclei had to have exactly the same number of the heavy, positive protons, so it seemed that a third kind of particle—heavy but electrically neutral—might also be inside the nucleus. Twenty years later, a third particle, called the neutron, was discovered and identified. Thus, isotopes were finally understood as atoms of an element that have a total number of neutrons inside their nuclei that is different from the total number of neutrons in the nuclei of other standard atoms of that element.

In many practical applications, isotopes of certain elements are used for a variety of purposes. Archaeologists, for example, make use of two of carbon's four known isotopes to determine the ages of objects that may be many thousands of years old. Radioactive isotopes are used as chemical tracers within the human body for medical diagnosis. Also, the possibility of the fusion of pairs of the three isotopes of the lightest element, hydrogen, offers hope for the creation of nonpolluting sources of energy.

*Victor W. Chen*

# Einstein Develops General Theory of Relativity

*Albert Einstein incorporated an explanation of gravitational forces into his special theory of relativity.*

**What:** Physics
**When:** 1907-1919
**Where:** Switzerland and Germany
**Who:**
ALBERT EINSTEIN (1879-1955), a German-Swiss-American physicist
DAVID HILBERT (1862-1943), a German mathematician
SIR ARTHUR STANLEY EDDINGTON (1882-1944), an English astronomer, philosopher, and physicist

## A Matter of Some Gravity

Although Sir Isaac Newton's law of gravity was very successful at making predictions, it did not explain how gravity worked. In fact, Newton stated that explaining gravity was not his goal: "Gravity must be caused by an agent acting constantly according to certain laws; but whether this agent should be material or immaterial, I have left to the consideration of my readers." For almost two hundred years, however, Newton's readers cared little about this question. Newton's law can be described as an "as if" law. Two bodies act as if there is a force between them that acts like the force of gravity that Newton proposed. Newton did not address the question of how or why this force operated.

Furthermore, Newton's physics could not explain why, in calculating gravitational attraction between objects, gravitational mass turns out to be exactly equal to inertial mass. (Gravitational mass determines the strength of the gravity acting on an object, and inertial mass determines that object's resistance to any force.) It is because of this equality that all bodies that fall under the influence of gravity alone have the same acceleration, their different masses notwithstanding. It was this strange equality that directed Einstein's thoughts toward a theory of gravity when almost no one else considered it to be an important question.

## It Is All Relative

When Einstein developed his special theory of relativity in 1905, he did it in an environment in which many other scientists were working on the same problem. The questions involved were the burning issues of the time, and other scientists were also coming close to the answers. The situation in which the general theory of relativity was developed was much different. Einstein did receive some help from his friend Marcel Grossman, a mathematician, and there was a later parallel effort to develop the same theory by the famous mathematician David Hilbert, who was inspired to do so by Einstein. Aside from these minor exceptions, however, Einstein's work on general relativity was entirely his own, and it was performed in an atmosphere in which there was little independent interest in the problem.

Einstein began to work on general relativity after he examined the deficiencies of the special theory of relativity. Gravity and relativity were incompatible. In particular, in order to incorporate an understanding of gravity into the special theory of relativity, it was necessary to deny the equality of gravitational mass and inertial mass. Because that equality could be established experimentally to a high degree of accuracy, however, it was impossible to ignore. Furthermore, the triumph of special relativity was that it established that all motion was relative. There was no longer any concept of absolute velocity; only the idea of relative velocity remained. Acceleration, however, was left as an absolute. Einstein thought

**153**

that acceleration should be relative if velocity was relative. The apparent discrepancy, along with the problem of gravity, disturbed Einstein and led him to develop his theory of gravity, which is known as the general theory of relativity.

The crucial step in the development of the general theory of relativity was the publication of the "Principle of Equivalence" in 1907. In this paper, it was proposed that, in any small region of space, one could not distinguish between gravity and acceleration. This meant that a person in a closed room who saw objects fall when they were dropped had no way of knowing whether those objects fell because the room was at rest on the surface of a planet or because the room was in a rocket ship that was accelerating in the direction opposite to the direction of the falling objects. Because gravity and acceleration were equivalent, the equivalence of gravitational mass and inertial mass was thus explained.

In 1911, Einstein used the principle of equivalence to establish that, because light seen by an accelerated observer is bent, gravity must also bend light. Between 1911 and 1916, Einstein worked on developing the complicated mathematics of his complete theory. In doing so, he discovered the correct value for the bending of light. Light follows the shortest distance between two points. When the shortest distance between two points appears to be curved, it means that the area of space that is involved is curved. For example, a straight line drawn on the two-dimensional surface of a globe will, if it is projected onto a flat map, appear to be curved. By using this line of reasoning, Einstein concluded that

the curved path of light near a mass means that the four-dimensional space-time around that mass is curved. Furthermore, in 1917, Einstein used general relativity to show that the total mass of the universe affects the structure or shape of the universe as a whole.

**Consequences**

Newton's theory of gravity was almost—but not quite—perfect, but Einstein's theory, as Einstein himself found in 1915, corrected those imperfections. More important to the acceptance of this theory, however, was the verification of its predictions for the bending of light. The experiment that was needed required, among other things, a total eclipse of the sun so that light from the stars could be checked to see whether it was bent as it passed the sun. This experiment was carried out in Africa in 1919 by the British astronomer Sir Arthur Eddington, and its results verified Einstein's theory.

Since that time, the "geometrification" of space has led to the idea of microscopic "wormholes" connecting one point in space to another. On a larger scale, this geometrification is manifested in searches for "black holes" from which not even light can escape. Such black holes are caused by the extreme warping of space that results when stars collapse. On the largest scale, this new view of space and time provides the basis for the various models that have been proposed to explain the creation and evolution of the universe.

*Carl G. Adler*

# Second Hague Peace Conference Convenes

*Though the 1907 gathering at The Hague was called the "Second International Peace Conference," the main topics of discussion and negotiation turned out to be the rules of warfare.*

**What:** International relations
**When:** June, 1907
**Where:** The Hague, the Netherlands
**Who:**
THEODORE ROOSEVELT (1858-1919), president of the United States from 1901 to 1909
ELIHU ROOT (1845-1937), U.S. secretary of state from 1905 to 1909
NICHOLAS II (1868-1918), czar of Russia from 1894 to 1917
ALEKSANDR PETROVICH IZVOLSKI (1856-1919), Russian minister of foreign affairs
SIR HENRY CAMPBELL-BANNERMAN (1836-1908), prime minister of Great Britain from 1905 to 1908
SIR EDWARD GREY (1862-1933), foreign secretary of Great Britain from 1905 to 1916
WILLIAM II (1859-1941), emperor of Germany from 1888 to 1918
JOSEPH HODGES CHOATE (1832-1917), chief delegate from the United States to the Second Hague Peace Conference

## Threats to World Peace

Since the adjournment of the International Peace Conference, or the First Hague Conference, on July 29, 1899, there had been no indication that the world's great powers—including France, Germany, Great Britain, Japan, Russia, and the United States—were willing to sacrifice what they saw as their own interests in order to create peace. Great Britain had involved itself in the Boer War, the great powers had joined to suppress the Boxer Rebellion in China, and sporadic fighting had occurred in the Philippines.

The country most of the other powers feared, however, was Germany. In 1900, Germany had decided to begin a massive program to build warships. This "Naval Law," if fully carried out, would have meant that within twenty years Great Britain's supremacy at sea would be in serious doubt.

In response, Great Britain began to make alliances with other nations: the Hay-Paunceforte Treaty with the United States in 1901, an alliance with Japan in 1902, and the Entente Cordiale with France in 1904. Also, Arthur James Balfour, prime minister of Great Britain from 1902 to 1905, had taken steps to modernize the British naval and military forces.

This series of events made the world's balance of power seem shakier than ever. In 1905, there was a crisis in Morocco, one of the areas in North Africa where European colonial rivalries were especially intense. William II, Emperor of Germany, was blamed for the crisis. In the same year, Japan had surprised the world with its easy victories over Russia at Port Arthur and Mukden, humiliating Russia even further in defeating the Russian Baltic Fleet in the Tsushima Strait.

## Talk of Peace and War

World leaders who favored peace and disarmament began to ask for a second International Peace Conference at The Hague to deal with many questions that had not been resolved in the 1899 conference. Having helped to negotiate a settlement to the Russo-Japanese War, U.S. president Theodore Roosevelt seemed the obvious sponsor. Yet Nicholas II, Czar of Russia, who wanted to regain influence and prestige after the embarrassment of losing to the Japanese, was the one who issued the invitations.

Russia and Germany made it known from the beginning that they had no intention of discussing disarmament or the limiting of arms.

Library of Congress

*Nicholas II (left), depicted in this formal portrait with George V, king of England, issued invitations to the conference.*

It cannot be surprising, then, that the Second Hague Conference focused more on ways of conducting war than on ways of creating peace. The conference opened on June 15, 1907. Forty-four nations were represented—eighteen more than had been at the 1899 conference. The principal topics discussed were arbitration in international disputes, rules of land warfare, rules of war at sea, and maritime law. Great Britain did bring to the floor a proposal for considering disarmament or a moratorium on arms buildup, but within less than half an hour it had been voted away with the recommendation that it be given "serious study."

In wars at sea, the British strategy had been to form blockades. This practice was challenged at the conference, as was the submarine warfare for which Germany had been preparing. Yet Great Britain remained determined to keep its right to blockade harbors and capture ships, while the Germans were equally determined to use contact mines and submarines to break blockades.

In regard to the settlement of international disputes, delegates debated whether arbitration should be voluntary or required. The Germans insisted that it could not be compulsory, and in the final document—the Convention of Pacific Settlement of International Disputes—their view prevailed.

Though some of the European nations objected, the Americans pushed through a resolution specifying that the Hague conferences should continue. A third conference was to be held in another eight years, the same length of time that separated the first two conferences. By 1915, though, World War I was in full swing.

William II said that if disarmament were brought up at all in the conference, his delegates would leave, while Aleksandr Petrovich Izvolski, Russian minister of foreign affairs, called disarmament "a craze of Jews, socialists and hysterical women."

Though Elihu Root, U.S. secretary of state, thought that disarmament and related questions should be discussed, Roosevelt considered it fairly worthless to do so. The United States was pursuing the building of battleships of the Dreadnought class, and Roosevelt claimed that they would be a greater deterrent to war than any peace society.

## Consequences

In the first decade of the twentieth century, Great Britain, Germany, Russia, France, Japan, and the United States were all competing to prove themselves as world powers. In the race for power, the main methods were forming alliances and building up military forces and weapons.

In this international mood, a peace conference had little chance of making headway toward an actual reduction in arms. The participants in the Second Hague Conference did come to some agreements about what would and would not be permitted in conducting wars, but once World War I broke out, many of these agreements were disregarded in the heat of battle.

Though the terrible devastation of the two world wars caused many people to reconsider disarmament as a worthwhile goal, debates continue about how peace is to be kept. It has long been debated whether the best deterrent to war is a large military buildup, including the development of nuclear weapons—"peace through strength"—or a reduction of arms and a placement of greater concern on meeting important human needs.

*John H. Greising*

# Britain, France, and Russia Form Triple Entente

> *The growing cooperation among Great Britain, Russia, and France, strengthened by the Anglo-Russian Entente signed in 1907, would prove very important at the outbreak of World War I.*

**What:** International relations
**When:** August 31, 1907
**Where:** London and Saint Petersburg
**Who:**
EDWARD VII (1841-1910), king of Great Britain from 1901 to 1910
ALEKSANDR PETROVICH IZVOLSKI (1856-1919), Russian minister of foreign affairs from 1906 to 1910
SIR EDWARD GREY (1862-1933), foreign secretary of Great Britain from 1905 to 1916
SIR CHARLES HARDINGE (1858-1944), British ambassador to Russia from 1904 to 1906 and permanent undersecretary at the British Foreign Office from 1906 to 1910
SIR ARTHUR NICOLSON (1849-1928), British ambassador to Russia from 1906 to 1910

## The Need for Accord

During 1904 and 1905, Great Britain entered into the Entente Cordiale with France, which was already allied with Russia. At the same time, relations between Germany and Great Britain were continuing to deteriorate. The first Moroccan crisis and Germany's determination to build up its navy added to British misgivings.

Great Britain first attempted to form an alliance with Russia in April, 1904, when talks were held between King Edward VII and Aleksandr Petrovich Izvolski, who at that time was Russia's envoy to Denmark.

Meanwhile, the British and French governments were completing the Entente Cordiale. Since February of that year, Russia had been at war with Japan. Japan had been an ally of Great Britain since 1902, but their agreement required one to come to the aid of the other only if that other was involved in war with two countries. Because of this requirement, Great Britain remained free to stay out of the Russo-Japanese War and to build its own alliance with Russia.

Both British foreign secretary Sir Edward Grey and Sir Charles Hardinge, British ambassador to Russia from 1904 to 1906, earnestly wished to settle Great Britain's differences with Russia over Persia and India. In July, 1905, Emperor William II of Germany and Czar Nicholas II of Russia signed the Björkö Treaty, agreeing that they would come to each other's aid in case of attack by another European power. German officials had wanted to extend the two countries' promises to each other through the treaty, but that goal had been opposed by the Russian Foreign Office and by leaders in France, Russia's main ally.

Concerned about the move toward friendship between Germany and Russia, Grey and Hardinge continued to press for an Anglo-Russian agreement. In 1907, they received the help of Sir Arthur Nicolson, who became the new British ambassador to Russia after Hardinge became permanent undersecretary at the British Foreign Office. Nicolson worked hard to write an agreement that the Russians could accept.

Since May, 1906, Aleksandr Izvolski had been minister of foreign affairs in Russia. He had already been advocating closer ties between Russia and Great Britain, for four reasons. First, he thought such ties would allow Russia to mend its relations with Japan, which in 1905 had renewed its alliance with Great Britain. Second, he thought that an Anglo-Russian entente would strengthen Russia's alliance with France. Third, the Russians had already joined the British in expressing a desire to resolve their differences regarding Persia, India, and other parts of Asia.

Finally, if Russia and Great Britain could create ties, there would be no need for an alliance with Germany. With British and French protection, Russia could challenge German and Austrian efforts to extend influence in the Middle East, and Russian ships could eventually gain access to the Dardanelles, a strait opening to the Aegean Sea.

## The Agreement Is Made

Negotiations began in June, 1906, but dragged on until August, 1907. It took time to break down the suspicions between the two countries. Nicholas II, an autocratic czar, had little in common with British Liberalism, and the British Liberal press raised an outcry against attacks on Jews in Russia and criticized the closing of the Duma, the Russian parliament.

Izvolski insisted that the Dardanelles should be open to Russian warships; he got Grey to promise only that in the future would Great Britain not oppose Russia in this regard. Izvolski accepted the British negotiators' demand that Persia be divided into British and Russian spheres of influence, though Russia had earlier wished to dominate all of Persia in order to have access to the Persian Gulf. The British government also insisted that Russia simultaneously make reconciliation with Japan; on July 30, 1907, Russia and Japan signed a treaty agreeing to respect each other's rights in the Far East. A month later, Izvolski and Nicolson signed the convention that established the Anglo-Russian Entente.

The entente had to do only with Asia—specifically Afghanistan, Tibet, and Persia. With the Russian promise that each should respect the territorial integrity of Tibet (which was ruled by China) and of Afghanistan, Great Britain was assured that these two states would stand in the way of any future Russian advance toward India, one of Great Britain's important colonies.

Even more important was the agreement to recognize the "independence" and "integrity" of Persia while dividing it up into three spheres of influence. Russia received the northern zone, which was the largest but did not include the Persian Gulf. The gulf was declared to be a neutral zone. The British received a desert wasteland in the south that contained roads leading to India.

The agreement contained no commitment of the two nations to fight for each other in case of war. Still, the Anglo-Russian agreement completed a network of treaties that bound together Great Britain, Russia, and France. Newspapers in these countries began to call this network the "Triple Entente."

## Consequences

The agreement did not create immediate cooperation and understanding on many issues. Great Britain's commitments to France and Russia were limited, and the agreement had to do only with Asia. Yet it did do away with some of the conflict between the British and the Russians.

The Anglo-Russian Entente was important as a step toward cooperation among Great Britain, France, and Russia—a cooperation that became more important in the next few years. Crises in Morocco and the Balkans drew these three powers together in opposition to Germany and Austria-Hungary. The members of the Triple Entente began to do some of their military planning together, anticipating a serious conflict with Germany. Those preparations served them well when they entered World War I as allies in 1914.

*Edward P. Keleher*

# Haber Converts Atmospheric Nitrogen into Ammonia

*Fritz Haber developed the first successful method for converting nitrogen from the atmosphere and combining it with hydrogen to synthesize ammonia, a compound used as a fertilizer.*

**What:** Chemistry
**When:** 1908
**Where:** Germany
**Who:**
FRITZ HABER (1868-1934), a German chemist who won the 1918 Nobel Prize in Chemistry

## The Need for Nitrogen

The nitrogen content of the soil, essential to plant growth, is maintained normally by the deposition and decay of old vegetation and by nitrates in rainfall. If, however, the soil is used extensively for agricultural purposes, more intensive methods must be used to maintain soil nutrients such as nitrogen. One such method is crop rotation, in which successive divisions of a farm are planted in rotation with clover, corn, or wheat, for example, or allowed to lie fallow for a year or so. The clover is able to absorb nitrogen from the air and deposit it in the soil through its roots. As population has increased, however, farming has become more intensive, and the use of artificial fertilizers—some containing nitrogen—has become almost universal.

Nitrogen-bearing compounds, such as potassium nitrate and ammonium chloride, have been used for many years as artificial fertilizers. Much of the nitrate used, mainly potassium nitrate, came from Chilean saltpeter, of which a yearly amount of half a million tons was imported at the beginning of the twentieth century into Europe and the United States for use in agriculture. Ammonia was produced by dry distillation of bituminous coal and other low-grade fuel materials. Originally, coke ovens discharged this valuable material into the atmosphere, but more economical methods were found later to collect and condense these ammonia-bearing vapors.

At the beginning of the twentieth century, Germany had practically no source of fertilizer-grade nitrogen; almost all of its supply came from the deserts of northern Chile. As demand for nitrates increased, it became apparent that the supply from these vast deposits would not be enough. Other sources needed to be found, and the almost unlimited supply of nitrogen in the atmosphere (80 percent nitrogen) was an obvious source.

## Temperature and Pressure

When Fritz Haber and coworkers began his experiments on ammonia production in 1904, Haber decided to repeat the experiments of the British chemist Sir William Ramsay and Sydney Young, who in 1884 had studied the decomposition of ammonia at about 800 degrees Celsius. They had found that a certain amount of ammonia was always left undecomposed. In other words, the reaction between ammonia and its constituent elements—nitrogen and hydrogen—had reached a state of equilibrium.

Haber decided to determine the point at which this equilibrium took place at temperatures near 1,000 degrees Celsius. He tried several approaches, reacting pure hydrogen with pure nitrogen, and starting with pure ammonia gas and using iron filings as a catalyst. (Catalytic agents speed up a reaction without affecting it otherwise).

Having determined the point of equilibrium, he next tried different catalysts and found nickel to be as effective as iron, and calcium and manganese even better. At 1,000 degrees Celsius, the rate of reaction was enough to produce practical amounts of ammonia continuously.

**160**

Further work by Haber showed that increasing the pressure also increased the percentage of ammonia at equilibrium. For example, at 300 degrees Celsius, the percentage of ammonia at equilibrium at 1 atmosphere pressure was very small, but at 200 atmospheres, the percentage of ammonia at equilibrium was far greater. A pilot plant was constructed and was successful enough to impress a chemical company, Badische Anilin-und Soda-Fabrik (BASF). BASF agreed to study Haber's process and to investigate different catalysts on a large scale. Soon thereafter, the process became a commercial success.

### Consequences

With the beginning of World War I, nitrates were needed more urgently for use in explosives than in agriculture. After the fall of Antwerp, 50,000 tons of Chilean saltpeter were discovered in the harbor and fell into German hands. Because the ammonia from Haber's process could be converted readily into nitrates, it became an important war resource. Haber's other contribution to the German war effort was his development of poison gas, which was used for the chlorine gas attack on Allied troops at Ypres in 1915. He also directed research on gas masks and other protective devices.

At the end of the war, the 1918 Nobel Prize in Chemistry was awarded to Haber for his development of the process for making synthetic ammonia. Because the war was still fresh in everyone's memory, it became one of the most controversial Nobel awards ever made. A headline in *The New York Times* for January 26, 1920, stated: "French Attack Swedes for Nobel Prize Award: Chemistry Honor Given to Dr. Haber, Inventor of German Asphyxiating Gas." In a letter to the *Times* on January 28, 1920, the Swed-

*Fritz Haber.*

The Nobel Foundation

ish legation in Washington, D.C., defended the award.

Haber left Germany in 1933 under duress from the anti-Semitic policies of the Nazi authorities. He was invited to accept a position with Cambridge University, England, and died on a trip to Basel, Switzerland, a few months later, a great man whose spirit had been crushed by the actions of an evil regime.

*Joseph Albert Schufle*

**161**

# Hardy and Weinberg Model Population Genetics

> *Godfrey Harold Hardy and Wilhelm Weinberg independently presented the first model for evaluating changes in gene frequency within a population and gave birth to the field of population genetics.*

**What:** Biology; Genetics
**When:** 1908
**Where:** Stuttgart, Germany, and Cambridge, England
**Who:**
WILHELM WEINBERG (1862-1937), a German physician, geneticist, and medical statistician
GODFREY HAROLD HARDY (1877-1947), a professor of mathematics at Trinity College and Oxford University, and a leading mathematician
GREGOR JOHANN MENDEL (1822-1884), an Austrian botanist
CHARLES DARWIN (1809-1882), an English naturalist

## Darwin and Mendel

Within a few decades following the publication of *On the Origin of Species* (1859) by Charles Darwin, the theory of natural selection had gained considerable approval in the scientific community and had revolutionized the way biologists viewed the natural world. This work provided a comprehensive explanation for both the origin and the maintenance of the seemingly endless variation in nature.

Despite almost immediate acceptance, the theory was incomplete in that it failed to explain how individual characteristics are inherited and how the process of inheritance could translate into the kinds of changes in populations and species that originally were predicted in Darwin's theory.

The first of these problems was overcome in 1900, when the work of Gregor Johann Mendel was rediscovered independently by several researchers. In his series of breeding experiments on garden peas, Mendel had demonstrated that many inherited traits are determined by factors known now as genes. *Genes* are the intracellular units by which hereditary characteristics are passed from one organism to another. From the frequency of traits in his populations of pea plants, Mendel reasoned that an organism receives one gene from each parent for all such heritable traits. In addition, he argued that alleles (alternate forms of the same gene) separate independently and randomly from one another when gametes (egg and sperm) form but combine again during fertilization.

Almost immediately following the rediscovery of Mendel's laws of inheritance, several scientists began to realize the implications for the study of population genetics and evolutionary change. The first observations were noted by the American zoologist W. E. Castle in 1903. More complete analyses were presented independently in 1908 by the English mathematician Godfrey Harold Hardy and the German physician Wilhelm Weinberg. These later works came to be known collectively as the Hardy-Weinberg law and eventually became the foundation for the study of population genetics.

## The Conditions for Nonevolution

The Hardy-Weinberg law states that, given simple patterns of Mendelian inheritance, the frequency of alleles in a population will remain constant from generation to generation, if certain ideal conditions are met. In other words, if these conditions hold true, allelic frequencies will not change, the genetic structure of the population will remain constant over time, and evolutionary change will not occur.

These ideal conditions are as follows. First, the population must be a large, randomly breeding population. In other words, all individuals in the population must have equal reproductive success. If this condition is not met, and certain individuals experience greater reproductive success than others, or if nonrandom breeding occurs as a result of small population size, then certain genes will be overrepresented in the next generation and the population's gene frequencies will change. Second, the population must be closed; that is, there must be no immigration or emigration of individuals in or out of the population. Third, there must be no spontaneous changes in alleles (mutations). Finally, all alleles must share equal probability of transmission to the next generation. For example, if some individuals possess alleles or combinations of alleles that, under certain environmental conditions, enhance their chances of survival and subsequent reproduction, then their genes will be represented more than those of others in the next generation, gene and allelic frequencies will change, and the population will adapt to environmental conditions. This is the essence of Darwin's theory of natural selection and the primary mechanism by which evolutionary changes proceed.

Given these conditions, the Hardy-Weinberg law asserts that changes in gene frequency in a population (evolution) will not occur. When any one or more of these conditions are violated, however, gene frequencies will be altered and evolution will take place. Thus, by demonstrating the conditions necessary for evolution *not* to occur, Hardy and Weinberg were able to illustrate those factors that actually contribute to evolutionary change.

## Consequences

The Hardy-Weinberg law was a critical breakthrough in evolutionary biology that effectively linked Mendel's laws of inheritance with Darwin's theory of natural selection. It demonstrated clearly how cellular mechanisms of inheritance can translate into the microevolutionary changes that Darwin had predicted.

The synthesis of Mendel's and Darwin's work resulted in renewed interest in evolutionary biology and soon gave birth to the new field of population genetics. This field was advanced greatly during the 1920's, and it continues to be one of the major fields of biological research.

In addition to its impact on basic research, the Hardy-Weinberg law has had several practical applications. Perhaps the most important of these is its use as a conceptual teaching model. The Hardy-Weinberg model is employed in nearly every college-level biology text as a starting point for discussions on evolution, adaptation, and population genetics. A second important application derived from the model concerns the manner and degree to which harmful alleles manifest themselves within a population. The Hardy-Weinberg model shows how lethal alleles, such as those that code for fatal genetic diseases, can be maintained in a population at low frequencies.

*Michael A. Steele*

# Hughes Revolutionizes Oil-Well Drilling

> *Howard R. Hughes and Walter B. Sharp developed a rotary cone drill bit that enabled oil-well drillers to penetrate hard rock formations.*

**What:** Engineering
**When:** 1908
**Where:** Houston, Texas
**Who:**
HOWARD R. HUGHES (1869-1924), an American lawyer, drilling engineer, and inventor
WALTER B. SHARP (1860-1912), an American drilling engineer, inventor, and partner to Hughes

## Digging for Oil

The modern rotary drill rig is basically unchanged in its essential components from its earlier versions of the 1900's. A drill bit is attached to a line of hollow drill pipe. The latter passes through a hole on a rotary table, which acts essentially as a horizontal gear wheel and is driven by an engine. As the rotary table turns, so do the pipe and drill bit.

During drilling operations, mud-laden water is pumped under high pressure down the sides of the drill pipe, and jets out with great force through the small holes in the rotary drill bit against the bottom of the borehole. This fluid then returns outside the drill pipe to the surface, carrying with it rock material cuttings from below. Circulated rock cuttings and fluids are regularly examined at the surface to determine the precise type and age of rock formation and for signs of oil and gas.

A key part of the total rotary drilling system is the drill bit, which has sharp cutting edges that make direct contact with the geologic formations to be drilled. The first bits used in rotary drilling were paddlelike "fishtail" bits, fairly successful for softer formations, and tubular coring bits for harder surfaces. In 1893, M. C. Baker and C. E. Baker brought a rotary water-well drill rig to Corsicana, Texas, for modification to deeper oil drilling. This rig led to the discovery of the large Corsicana-Powell oil field in Navarro County, Texas. This success also motivated its operators, the American Well and Prospecting Company, to begin the first large-scale manufacture of rotary drilling rigs for commercial sale.

In the earliest rotary drilling for oil, short fishtail bits were the tool of choice, insofar as they were at that time the best at being able to bore through a wide range of geologic strata without needing frequent replacement. Even so, in the course of any given oil well, many bits were required typically in coastal drilling in the Gulf of Mexico. Especially when encountering locally harder rock units such as limestone, dolomite, or gravel beds, fishtail bits would typically either curl backward or break off in the hole, requiring the time-consuming work of pulling out all drill pipe and "fishing" to retrieve fragments and clear the hole.

Because of the frequent bit wear and damage, numerous small blacksmith shops established themselves near drill rigs, dressing or sharpening bits with a hand forge and hammer. Each bit-forging shop had its own particular way of shaping bits, producing a wide variety of designs. Nonstandard bit designs were frequently modified further as experiments to meet the specific requests of local drillers encountering specific drilling difficulties in given rock layers.

## Speeding the Process

In 1907 and 1908, patents were obtained in New Jersey and Texas, for steel, cone-shaped drill bits incorporating a roller-type coring device with many serrated teeth. Later in 1908, both patents were bought by lawyer Howard R. Hughes.

Although comparatively weak rocks such as sands, clays, and soft shales could be drilled rapidly (at rates exceeding 30 meters per hour), in harder shales, lime-dolostones, and gravels, drill rates of 1 meter per hour or less were not uncommon. Conventional drill bits of the time had average operating lives of three to twelve hours. Economic drilling mandated increases in both bit life and drilling rate. Directly motivated by his petroleum prospecting interests, Hughes and his partner, Walter B. Sharp, undertook what were probably the first recorded systematic studies of drill bit performance while matched against specific rock layers.

Although many improvements in detail and materials have been made to the Hughes cone bit since its inception in 1908, its basic design is still used in rotary drilling. One of Hughes's major innovations was the much larger size of the cutters, symmetrically distributed as a large number of small individual teeth on the outer face of two or more cantilevered bearing pins. In addition, "hard facing" was applied to drill bit teeth to increase usable life. Hard facing is a metallurgical process basically consisting of wedding a thin layer of a hard metal or alloy of special composition to a metal surface to increase its resistance to abrasion and heat. A less noticeable but equally essential innovation, not included in other drill bit patents, was an ingeniously designed gauge surface that provided strong uniform support for all the drill teeth. The force-fed oil lubrication was another new feature included in Hughes's patent and prototypes, reducing the power necessary to rotate the bit by 50 percent over that of prior mud or water lubricant designs.

## Consequences

In 1925, the first superhard facing was used on cone drill bits. In addition, the first so-called self-cleaning rock bits appeared from Hughes, with significant advances in roller bearings and bit tooth shape translating into increased drilling efficiency. The much larger teeth were more adaptable to drilling in a wider variety of geological formations than earlier models. In 1928, tungsten carbide was introduced as an additional bit facing hardener by Hughes metallurgists. This, together with other improvements, resulted in the Hughes ACME tooth form, which has been in almost continuous use since 1926.

Many other drilling support technologies, such as drilling mud, mud circulation pumps, blowout detectors and preventers, and pipe properties and connectors have enabled rotary drilling rigs to reach new depths (exceeding 5 kilometers in 1990). The successful experiments by Hughes in 1908 were critical initiators of these developments.

*Gerardo G. Tango*

# Spangler Makes First Electric Vacuum Cleaner

> *The first portable domestic vacuum cleaner successfully adapted to electricity, James Murray Spangler's machine was an important element in the electrification of domestic appliances in the early twentieth century.*

**What:** Engineering
**When:** 1908
**Where:** Canton, Ohio
**Who:**
MELVILLE R. BISSELL (1843-1889), the inventor and marketer of the Bissell carpet sweeper in 1876
H. CECIL BOOTH (1871-1955), a British civil engineer
WILLIAM HENRY HOOVER (1849-1932), an American industrialist
JAMES MURRAY SPANGLER (1848-1915), an American inventor

### From Brooms to Bissells

During most of the nineteenth century, the floors of homes were cleaned primarily with brooms. Carpets were periodically dragged out of the home by the boys and men of the family, stretched over rope lines or fences, and given a thorough beating to remove dust and dirt. In the second half of the century, carpet sweepers, perhaps inspired by the success of street-sweeping machines, began to appear. Although there were many models, nearly all were based upon the idea of a revolving brush within an outer casing that moved on rollers or wheels when pushed by a long handle.

Melville Bissell's sweeper, patented in 1876, featured a knob for adjusting the brushes to the surface. The Bissell Carpet Company, also formed in 1876, became the most successful maker of carpet sweepers and dominated the market well into the twentieth century. Electric vacuum cleaners were not feasible until homes were wired for electricity and the small electric motor was invented.

Thomas Edison's success with an incandescent lighting system in the 1880's and Nikola Tesla's invention of a small electric motor, which was used in 1889 to drive a Westinghouse Electric Corporation fan, opened the way for the application of electricity to household technologies.

### Cleaning with Electricity

In 1901, H. Cecil Booth, a British civil engineer, observed a London demonstration of an American carpet cleaner that blew compressed air at the fabric. Booth was convinced that the process should be reversed so that dirt would be sucked out of the carpet. In developing this idea, Booth invented the first successful suction vacuum sweeper.

Booth's machines, which were powered by gasoline or electricity, worked without brushes. Dust was extracted by means of a suction action through flexible tubes with slot-shaped nozzles. Some machines were permanently installed in buildings that had wall sockets for the tubes in every room. Booth's British Vacuum Cleaner Company also employed horse-drawn mobile units from which white-uniformed men unrolled long tubes that they passed into buildings through windows and doors. His company's commercial triumph came when it cleaned Westminster Abbey for the coronation of Edward VII in 1902. Booth's company also manufactured a 1904 domestic model that had a direct-current electric motor and a vacuum pump mounted on a wheeled carriage. Dust was sucked into the nozzle of a long tube and deposited into a metal container. Booth's vacuum cleaner used electricity from overhead light sockets.

The portable electric vacuum cleaner was invented in 1907 in the United States by James

**166**

Murray Spangler. When Spangler was a janitor in a department store in Canton, Ohio, his asthmatic condition was worsened by the dust he raised with a large Bissell carpet sweeper. Spangler's modifications of the Bissell sweeper led to his own invention. On June 2, 1908, he received a patent for his Electric Suction Sweeper. The device consisted of a cylindrical brush in the front of the machine, a vertical-shaft electric motor above a fan in the main body, and a pillow case attached to a broom handle behind the main body. The brush dislodged the dirt, which was sucked into the pillow case by the movement of air caused by a fan powered by the electric motor. Although Spangler's initial attempt to manufacture and sell his machines failed, Spangler had, luckily for him, sold one of his machines to a cousin, Susan Troxel Hoover, the wife of William Henry Hoover.

The Hoover family was involved in the production of leather goods, with an emphasis on horse saddles and harnesses. William Henry Hoover, president of the Hoover Company, recognizing that the adoption of the automobile was having a serious impact on the family business, was open to investigating another area of production. In addition, Mrs. Hoover liked the Spangler machine that she had been using for a couple of months, and she encouraged her husband to enter into an agreement with Spangler. An agreement made on August 5, 1908, allowed Spangler, as production manager, to manufacture his machine with a small work force in a section of Hoover's plant. As sales of vacuum cleaners increased, what began as a sideline for the Hoover Company became the company's main line of production.

Few American homes were wired for electricity when Spangler and Hoover joined forces; not until 1920 did 35 percent of American homes have electric power. In addition to this inauspicious fact, the first Spangler-Hoover machine, the Model O, carried the relatively high price of seventy-five dollars. Yet a full-page ad for the Model O in the December, 1908, issue of the *Saturday Evening Post* brought a deluge of requests.

American women had heard of the excellent performance of commercial vacuum cleaners, and they hoped that the Hoover domestic model would do as well in the home.

**Consequences**

As more and more homes in the United States and abroad became wired for electric lighting, a clean and accessible power source became available for household technologies. Whereas electric lighting was needed only in the evening, the electrification of household technologies made it necessary to use electricity during the day. The electrification of domestic technologies therefore matched the needs of the utility companies, which sought to maximize the use of their facilities. They became key promoters of electric appliances. In the first decades of the twentieth century, many household technologies became electrified. In addition to fans and vacuum cleaners, clothes-washing machines, irons, toasters, dish-washing machines, refrigerators, and kitchen ranges were being powered by electricity.

The application of electricity to household technologies came as large numbers of women entered the work force. During and after World War I, women found new employment opportunities in industrial manufacturing, department stores, and offices. The employment of women outside the home continued to increase throughout the twentieth century. Electrical appliances provided the means by which families could maintain the same standards of living in the home while both parents worked outside the home.

It is significant that Bissell was motivated by an allergy to dust and Spangler by an asthmatic condition. The employment of the carpet sweeper, and especially the electric vacuum cleaner, not only made house cleaning more efficient and less physical but also led to a healthier home environment. Whereas sweeping with a broom tended only to move dust to a different location, the carpet sweeper and the electric vacuum cleaner removed the dirt from the house.

*Thomas W. Judd*

**167**

# Steinmetz Warns of Future Air Pollution

*Charles Proteus Steinmetz warned of air pollution from the burning of coal and of water pollution from sewage disposal into rivers, summarizing early recognition of environmental impacts from population growth and urbanization.*

**What:** Earth science; Environment
**When:** 1908
**Where:** New York, New York
**Who:**
CHARLES PROTEUS STEINMETZ (1865-1923), an American electrical engineer and professor
WILLIAM THOMPSON SEDGWICK (1855-1921), a professor and lecturer
ALLEN HAZEN (1869-1930), a chemist and sanitary engineer

## A Prophetic Message

In a 1908 lecture delivered at the New York Electrical Trade School, Charles Proteus Steinmetz presented a prophetic message on future impacts of the development of electricity. Steinmetz challenged students to reduce the cost of electricity by finding more efficient methods of distributing electrical demand evenly over the course of a day and throughout the year. Electrical consumption would continue to grow, and it was up to the electrical engineer to make electricity economical whether it was created by steam power from coal or by water power. This need for greater efficiency of electrical use, he warned, would soon shift from economic reasons to reasons of necessity as the nation faced declining supplies of coal and posed a greater impact on the free-running streams for hydroelectric power.

Steinmetz cautioned that as coal reserves run out, there would be increasing pressure to develop the nation's watercourses for electrical generation. "[T]here will be no more rapid creeks and rivers," he said, "streams which furnish electric power will be slow-moving pools, connected with one another by power stations." Reserves of hard, anthracitic (low-sulfur) coal al-

ready were in short supply, and energy produced by the burning of soft, bituminous (high-sulfur) coal created serious air pollution problems. "Probably even before the soft coal is used up," Steinmetz predicted, "we will have awakened to the viciousness of poisoning nature and ourselves with smoke and coal gas."

Smoke pollution from the burning of high-sulfur coal had become a nuisance for most industrial cities by the beginning of the twentieth century. A major problem for the smoke abatement campaign that ensued as part of the Progressive Era reforms was the prevailing belief that equated smoke with economic growth and prosperity. The Department of Public Utilities in Boston, for example, reported that prior to 1910, industries depicting factories on their letterheads "invariably represented the stack with a black plume of smoke trailing away from it to typify activity and prosperity. . . . [I]f a stack was not belching out great volumes of dense smoke, it signified that the plant was shut down." Cities such as Milwaukee, Chicago, and Pittsburgh were commonly plagued by the "smoke evil," which blocked the sun, blackened the lungs, covered buildings with soot, and dirtied laundry that had been hung out to dry.

## Drinking Wastewater

Rapid growth in population and industrialization, and the consequent urbanization created by this growth, had placed heavy demands on the nation's streams for both water consumption and waste disposal. As a result, residents of many cities soon found themselves drinking the sewage of their upstream neighbors. In the late nineteenth century, the growing trend of exporting pollution to downstream communities was greatly accelerated by the introduction of municipal sewerage systems.

By 1908, leading public health officials were convinced that untreated sewage should not be discharged into rivers used for drinking water. Not only was there a risk of waterborne disease but also sewage disposal in waterways limited their use for recreation and industry. Sanitary engineers, however, supported the disposal of sewage into rivers, arguing that drinking water drawn from these sources could be purified by the employment of new filtration technology, which had proven effective in disease control. They said it was much cheaper to purify the water taken from rivers than to purify the sewage prior to disposal in those rivers.

### Consequences

In general, dirty air was viewed as one of the costs of industrial prosperity. This perspective prevailed with respect to stream quality as well. Rivers became the common receptors of raw sewage, which received treatment prior to disposal only when deemed a nuisance. Population growth, along with increased urbanization and industrialization, combined to place heavy loads of contamination into the nation's waterways. Sanitary engineers and public officials chose to take advantage of the capacity of large volumes of river or lake water to assimilate and dilute wastes. It cost less to purify drinking water drawn from polluted sources than to treat sewage prior to dumping it into the river.

The precedent of waste dilution established in the early part of the twentieth century allowed the routine disposal of toxic compounds such as mercury and petroleum into waterways. By the 1960's, Detroit's industrial area was dumping hundreds of barrels of oil per day into the Detroit River, which then flowed into Lake Erie. The Cuyahoga River in Cleveland carried so much oil and other flammable wastes that it would sometimes catch fire. On June 22, 1969, two railroad bridges over the Cuyahoga River were destroyed by river fires. Eventually, public pressure became intense enough to provoke legislative action. Lake Erie and the Cuyahoga River became national environmental icons, nagging a renewed public consciousness that had been aroused in 1962 with the publication of Rachel Carson's *Silent Spring.*

In 1972, Canadian prime minister Pierre Elliott Trudeau and U.S. president Richard M. Nixon signed a water quality agreement that pledged to protect and restore the Great Lakes. This agreement led to the allocation of billions of dollars of federal money to municipalities for construction of sewage treatment plants and to farmers for agricultural improvements designed to curb runoff of nutrients and insecticides. The Clean Air Acts of 1965 and 1970 provided citizens with some relief from air pollution. In the 1980's and 1990's, contamination of air and water continued to provoke debate over the standards of purity that society must achieve.

*Robert Lovely*

National Archives

*Charles Proteus Steinmetz.*

# Geiger and Rutherford Develop Geiger Counter

*Hans Geiger and Ernest Rutherford developed the first electronic device able to detect atomic particles.*

**What:** Physics
**When:** February 11, 1908
**Where:** Manchester, England
**Who:**
HANS GEIGER (1882-1945), a German physicist
ERNEST RUTHERFORD (1871-1937), a British physicist
SIR JOHN SEALY EDWARD TOWNSEND (1868-1957), an Irish physicist
SIR WILLIAM CROOKES (1832-1919), an English physicist
WILHELM CONRAD RÖNTGEN (1845-1923), a German physicist
ANTOINE-HENRI BECQUEREL (1852-1908), a French physicist

## Discovering Natural Radiation

When radioactivity was discovered and first studied, the work was done with rather simple devices. In the 1870's, Sir William Crookes learned how to create a very good vacuum in a glass tube. He placed electrodes in each end of the tube and studied the passage of electricity through the tube. This simple device became known as the "Crookes tube." In 1895, Wilhelm Conrad Röntgen was experimenting with a Crookes tube. It was known that when electricity went through a Crookes tube, one end of the glass tube might glow. Certain mineral salts placed near the tube would also glow. In order to observe carefully the glowing salts, Röntgen had darkened the room and covered most of the Crookes tube with dark paper. Suddenly, a flash of light caught his eye. It came from a mineral sample placed some distance from the tube and shielded by the dark paper; yet when the tube was switched off, the mineral sample went dark. Ex-

perimenting further, Röntgen became convinced that some ray from the Crookes tube had penetrated the mineral and caused it to glow. Since light rays were blocked by the black paper, he called the mystery ray an "X ray," with "X" standing for unknown.

Antoine-Henri Becquerel heard of the discovery of X rays, and in February, 1886, set out to discover if glowing minerals themselves emitted X rays. Some minerals, called "phosphorescent," begin to glow when activated by sunlight. Becquerel's experiment involved wrapping photographic film in black paper and setting various phosphorescent minerals on top and leaving them in the sun. He soon learned that phosphorescent minerals containing uranium would expose the film.

A series of cloudy days, however, brought a great surprise. Anxious to continue his experiments, Becquerel decided to develop film that had not been exposed to sunlight. He was astonished to discover that the film was deeply exposed. Some emanations must be coming from the uranium, he realized, and they had nothing to do with sunlight. Thus, natural radioactivity was discovered by accident with a simple piece of photographic film.

## Rutherford and Geiger

Ernest Rutherford joined the world of international physics at about the same time that radioactivity was discovered. Studying the "Becquerel rays" emitted by uranium, Rutherford eventually distinguished three different types of radiation, which he named "alpha," "beta," and "gamma" after the first three letters of the Greek alphabet. He showed that alpha particles, the least penetrating of the three, are the nuclei of helium atoms (a group of two neutrons and a proton tightly bound together). It was later shown that beta particles are electrons. Gamma rays, which are far more penetrating than either alpha or beta particles, were shown to be similar to X rays, but with higher energies.

Rutherford became director of the associated research laboratory at Manchester University in 1907. Hans Geiger became an assistant. At this time, Rutherford was trying to prove that alpha particles carry a double positive charge. The best way to do this was to measure the electric charge that a stream of alpha particles would bring to a target. By dividing that charge by the total number of alpha particles that fell on the target, one could calculate the charge of a single alpha particle. The problem lay in counting the particles and in proving that every particle had been counted.

Basing their design upon work done by Sir John Sealy Edward Townsend, a former colleague of Rutherford, Geiger and Rutherford constructed an electronic counter. It consisted of a long brass tube, sealed at both ends, from which most of the air had been pumped. A thin wire, insulated from the brass, was suspended down the middle of the tube. This wire was connected to batteries producing about thirteen hundred volts and to an electrometer, a device that could measure the voltage of the wire. This voltage could be increased until a spark jumped between the wire and the tube. If the voltage was turned down a little, the tube was ready to operate. An alpha particle entering the tube would ionize (knock some electrons away from) at least a few atoms. These electrons would be accelerated by the high voltage and, in turn, would ionize more atoms, freeing more electrons. This process would continue until an avalanche of electrons struck the central wire and the electrometer registered the voltage change. Since the tube was nearly ready to arc because of the high voltage, every alpha particle, even if it had very little energy, would initiate a discharge. The most complex of the early radiation detection

devices—the forerunner of the Geiger counter—had just been developed. The two physicists reported their findings in February, 1908.

## Consequences

Their first measurements showed that one gram of radium emitted 34 thousand million alpha particles per second. Soon, the number was refined to 32.8 thousand million per second. Next, Geiger and Rutherford measured the amount of charge emitted by radium each second. Dividing this number by the previous number gave them the charge on a single alpha particle. Just as Rutherford had anticipated, the charge was double that of a hydrogen ion (a proton). This proved to be the most accurate determination of the fundamental charge until the American physicist Robert Andrews Millikan conducted his classic oil-drop experiment in 1911.

Another fundamental result came from a careful measurement of the volume of helium emitted by radium each second. Using that value, other properties of gases, and the number of helium nuclei emitted each second, they were able to calculate Avogadro's number more directly and accurately than had previously been possible. (Avogadro's number enables one to calculate the number of atoms in a given amount of material.)

The true Geiger counter evolved when Geiger replaced the central wire of the tube with a needle whose point lay just inside a thin entrance window. This counter was much more sensitive to alpha and beta particles and also to gamma rays. By 1928, with the assistance of Walther Müller, Geiger made his counter much more efficient, responsive, durable, and portable. Today, there are probably few radiation facilities in the world that do not have at least one Geiger counter or one of its compact modern relatives.

*Charles W. Rogers*

**171**

# *Muller v. Oregon* Upholds Sociological Jurisprudence

> *In this important case, the Supreme Court agreed that the constitutionality of laws should be judged not only by legal precedent but also by their effect on people's well-being.*

**What:** Law; Labor
**When:** February 24, 1908
**Where:** Washington, D.C.
**Who:**
LOUIS DEMBITZ BRANDEIS (1856-1941), an
  American lawyer
FLORENCE KELLEY (1859-1932), general
  secretary of the National Consumers'
  League, and a social reformer
JOSEPHINE GOLDMARK (1877-1950), a
  leader of the National Consumers'
  League

## Defeat of Reform Laws

At the beginning of the twentieth century, many—perhaps most—lawyers and judges in the United States believed in "natural" economic laws and held that property rights were more important than human rights. According to their thinking, judges in the United States were entitled to review legislation to see whether it agreed with or violated state or federal constitutions. To attempt to control the use of property through laws, these people said, was unconstitutional and went against the natural order of things.

Labor leaders and social reformers working to improve the welfare of poor working people had begun to see the courts, then, as the greatest threat to progress. When states had passed laws attempting to improve people's living and working conditions, federal courts had often struck them down. In addition, the Supreme Court had accepted the idea that corporations were persons under the Fourteenth Amendment. No state, it said, could deprive a corporation (person) of life, liberty, or property without due process of law.

In 1905, the Supreme Court set specific limitations on how far a state could go in making laws about hours and other working conditions for its industrial laborers. In *Lochner v. New York*, the Supreme Court ruled that laws such as a New York law to limit the working hours of bakers to ten hours per day and sixty hours per week were "mere meddlesome interferences with the rights of the individual." A state could not interfere with the freedom of employer and employee to make a labor contract unless there were obvious important reasons such as health.

Social reformers became alarmed that state restrictions on the working hours of women might also be in danger. Thus it was that Florence Kelley, chief factory inspector of Illinois and general secretary of the National Consumers' League, joined Josephine Goldmark, another leader in the National Consumers' League, to hire Louis Dembitz Brandeis to defend an Oregon law that limited a workday to ten hours for women workers in industry.

## The Arguments

Brandeis, known as "the people's attorney," had tried for years to show that American law in his day did not match the new economic and social facts of life in America. Lawyers and judges who cared only for legal precedents and natural law knew very little about industrial conditions in the twentieth century. Brandeis said, "A lawyer who has not studied economics and sociology is very apt to become a public enemy." He held that the law was a living organism that could be changed and adjusted to fit people's real lives in American cities and factories. In his brief in *Muller v. Oregon*, Brandeis saw an opportunity to prove the validity of what has been called "sociological jurisprudence"—taking economic and social considerations into account, as well as legal precedents, in deciding the constitutionality of a law.

Brandeis's arguments from legal precedent took up only two pages of his brief (the document that laid out his argument). In more than one hundred pages, he attempted to show that Oregon had adopted its ten-hour law for women in order to guard the public health, safety, and welfare. "Long hours of labor are dangerous for women," Brandeis said, more than for men, because of physiological differences. "Overwork . . . is more dangerous to the health of women than of men, and entails upon them more lasting injury." When working women became very fatigued, he went on, there was often a "general deterioration of health"—anemia, difficulties in childbearing, and industrial accidents.

Furthermore, he said, overwork was demoralizing. A breakdown in the health and morals of women "inevitably lowers the entire community physically, mentally, and morally." There would be a rise in infant mortality, he predicted, if women were forced to work long hours. On the other hand, he argued that reasonable working hours actually raised "the tone of the entire community."

Brandeis brought together his economic and social evidence in a very convincing presentation. In response, the Court decided to uphold Oregon's ten-hour law for women.

## Consequences

Several years later, attorney and law professor Felix Frankfurter wrote, "The *Muller* case is epoch-making, not because of its decision, but because of the authoritative recognition by the Supreme Court that the way in which Mr. Brandeis presented the case . . . laid down a new technique for counsel." When constitutional questions were being argued, Frankfurter said, courts were now obligated to insist on considering social and economic effects before deciding the issue.

Brandeis himself continued to learn about the realities of life among the American working class, and he became an expert in labor economics. In 1916, President Woodrow Wilson nominated him to serve on the Supreme Court. Though the appointment was bitterly fought by conservatives, Brandeis eventually took his seat on the Court, where he was able to help shape the direction of American law for twenty-three years. His principles of sociological jurisprudence have been a basic part of legal and constitutional practice ever since.

*William M. Tuttle*

*Florence Kelley.*

# Hale Discovers Magnetic Fields in Sunspots

> *George Ellery Hale discoverd magnetic fields in sunspots, giving astronomers valuable information about how sunspots are formed.*

**What:** Astronomy
**When:** June 26, 1908
**Where:** Pasadena, California
**Who:**
GEORGE ELLERY HALE (1868-1938), an American astronomer
PIETER ZEEMAN (1865-1943), a Dutch physicist

## Split Spectra

In the seventeenth century, the Italian astronomer Galileo used a small telescope to observe dark spots on the surface of the sun. Since then, astronomers have invented new methods of observing the sun and have used these new methods to learn more about sunspots and other features of the sun.

In 1908, Hale observed the sun with a "spectroheliograph," a special instrument he had developed. This instrument worked by filtering the light of the sun so that light of different wavelengths could be viewed separately. This was useful for studying the processes going on in the sun and for identifying the various chemical elements contained within it. When he observed the sun and focused on a wavelength coming from hydrogen atoms, he noticed huge swirls of hydrogen gas resembling a tornado. These "vortices" seemed to be associated with the formation of sunspots. Hale considered what the consequences of this whirling motion might be. Research by the English physicist Joseph John Thomson had shown that hot bodies emit electrons, and because the sun is very hot, Hale thought that perhaps the sun was emitting electrons that, if caught up in this whirling motion, might create a magnetic field in the sunspots.

Hale had a way to check this hypothesis. A star's spectrum (a visual representation of all the kinds of light that it emits) can reveal much about the temperature of the star's gases, the velocity of its rotation, and other information. In 1896, Pieter Zeeman had discovered that if a source of light is placed in a magnetic field, the lines in the spectrum of the light source will be split into two or more parts by the magnetic field. When Hale learned of this effect (the Zeeman effect), he conducted laboratory work of his own to observe what the effect looked like in certain test cases. Once he had this information, he was ready to compare any observed splitting behavior in solar spectral lines with the laboratory results; if a magnetic field was associated with sunspots, one would expect to find a close match between the splitting of the solar line and the splitting of the laboratory line.

## A Special Telescope

The sun emits light at many wavelengths, each of which is associated with a different color. Light produced by the sun is perceived as white because the colors of these many wavelengths are blended and viewed together; a prism or finely spaced grating must be used to spread the light back into its separate colored components. Hale expended much care and effort in building the 18-meter tower telescope on Mount Wilson. This tower had mirrors on its top to catch the sun's light, which it reflected through a telescope with a 30-centimeter lens and down to a spectrograph about 9 meters underground. This method minimized the temperature changes that can distort and muddy the images obtained from telescopes, thus producing images that were unusually steady and sharp.

Hale spent much of the summer of 1908 running up and down the ladder of the tower in search of the Zeeman effect. On June 26, 1908, Hale observed a doubling of spectral lines. He immediately compared his observations of the sun with his laboratory observations and found that they were related. For the first time, an extra-terrestrial magnetic field had been detected and related to laboratory observations. Hale continued to make observations for two more weeks, and by July 6, he was confident that what he had found was indeed the Zeeman effect, indicating that there were magnetic fields in sunspots. Zeeman considered Hale's results and agreed that Hale's hypothesis was the best explanation for the observed phenomena.

## Consequences

Astronomers work by understanding processes that can be observed on Earth and applying them in the heavens. With Hale's discovery, astronomers realized that they could apply their knowledge of terrestrial magnetic and electric phenomena to the distant stars. Zeeman had found—in addition to the fact that spectral lines split in the presence of a magnetic field—that the distance between components of the split line is directly proportional to the strength of the magnetic field that causes the split. Hale measured the separations in lines split in the laboratory by a magnetic field of known strength, measured the separations in split lines in the spectrum of a sunspot, and derived the strength of the sunspot's magnetic field. He found impressively large magnetic fields. Also, Hale was able to study how the strength of the field varied at different places in a sunspot; he discovered that the magnetic field is strongest at the center and weakest toward the edges. This had implications for what the structure of the sunspot might be like.

Hale then turned his attention to determining whether the sun had an overall magnetic field. Although Hale worked at this question periodically for the rest of his life, it was never answered. Later astronomers not only have determined that the sun has a magnetic field overall but also have learned how this field changes over the sun's twenty-two year sunspot cycle, discovered how the field is affected by the sun's rotation, and determined the field's role in sunspot formation. Understanding the sun's magnetic properties has been crucial in understanding solar activity.

Another question Hale considered was whether the magnetic fields associated with sunspots could

*George Ellery Hale.*

be strong enough to cause magnetic storms on Earth. Hale was not in a position to answer this question, since it required records on both solar and terrestrial magnetic events over a period of time. The question indicates why the study of the sun is important to astronomers. Sunspots are often associated with energetic events on the sun's surface, which can send charged particles and energetic radiation out into space to interact with Earth's atmosphere. An interaction of this type can cause such benign phenomena as the "aurora borealis," or northern lights; it can also interfere with communications on Earth. In a long-term space colony, the radiation from solar flares could be hazardous to the colonists. Because sunspots and flares affect humans, astronomers are interested in studying the magnetic processes that cause them.

*Mary Hrovat*

**175**

# Kamerlingh Onnes Liquefies Helium

*Heike Kamerlingh Onnes transformed helium gas into liquid helium, initiating the study of matter at temperatures approaching the lowest achievable temperature, absolute zero.*

**What:** Physics; Chemistry
**When:** July 9, 1908
**Where:** Leiden, the Netherlands
**Who:**
HEIKE KAMERLINGH ONNES (1853-1926), a Dutch physicist
SIR JAMES DEWAR (1842-1923), a Scottish chemist and physicist

## The Search for Liquid Air

Perhaps the most familiar example of liquefaction is rain, which is caused by the condensation of water vapor in the air. In the late eighteenth century, Antoine-Laurent Lavoisier predicted that other constituents of air would also liquefy if they became cold enough. Lacking effective cooling techniques, scientists wondered whether all gases could be liquefied. Early researchers tried to liquefy gases by compression, forcing molecules closer together. The Dutch scientist Martinus van Marum liquefied ammonia by compression, but attempts by others to liquefy air at high pressures failed. Studies of gases in the nineteenth century suggested the reason for this failure: A gas will liquefy only if its temperature and pressure are below characteristic critical values. These conditions were not then obtainable for the pressurized air.

One way to cool a gas is to force it to expand quickly. Pursing one's lips and blowing on one's palm illustrates this effect, which is the basis of household refrigerators and air conditioners. In 1877, Louis-Paul Cailletet liquefied oxygen and nitrogen by using more extreme expansion. This produced temperatures below −120 degrees Celsius, at which point the liquids that had been formed quickly evaporated.

In 1883, Polish physicists Zygmunt Florenty von Wróblewski and Karol S. Olszewski improved oxygen liquefaction using a modified Cailletet-type apparatus, in which gas was expanded through a valve into a tube with a closed end that was immersed in liquid ethylene. Reducing the pressure above the ethylene with a vacuum pump caused the ethylene to boil rapidly, cooling to below −130 degrees Celsius. This technique kept the oxygen in a liquefied state.

## Toward the Ultimate Liquefaction

Several advances paved the way toward achieving lower temperatures. The first was the use of the Joule-Thomson effect, in which a gas cools by expanding through small openings in a porous material. Compressed gas is sent through such an opening, cooling and partially liquefying in the process. The liquid settles in a flask, and the expanded gas is returned to the original container. Along the way, the expanded gas is cooled by a heat exchanger fluid that makes thermal contact with the cold liquid-gas mixture. The process is repeated again and again, enhancing the cooling effect and making it possible to produce large quantities of liquid. Devices of this type were patented independently by William Hampson in England and Carl Paul Gottfried von Linde in Germany in 1895.

In the 1890's, Sir James Dewar used this kind of apparatus to liquefy hydrogen, which has a critical temperature of −240 degrees Celsius. He first cooled the hydrogen gas to −205 degrees Celsius by putting it in thermal contact with liquefied air under reduced pressure. In order to store significant quantities of the liquid hydrogen, he invented a double-walled glass container with exceptional insulating qualities.

Dewar evacuated and sealed the space between the walls to minimize heat transfer to the liquid hydrogen. His remarkable container, which has remained useful for a century, is called a "cryostat thermos" (*cryo* is the Greek word for cold) or, simply, a "dewar." The perfection of this container literally changed the study of matter at low temperatures. Dewar continued his work with hydrogen, subjecting it to a pressure less than 1 percent that of normal atmospheric pressure. At that pressure, hydrogen boiled rapidly and, ultimately, solidified.

This development set the stage for the liquefaction of helium, the only gas with a critical temperature lower than that of hydrogen. For ten years after Dewar's breakthrough, scientists in Poland, Holland, and England tried unsuc-

cessfully to liquefy helium. During that time, however, Heike Kamerlingh Onnes built a large research facility at Leiden, The Netherlands, that was equipped to produce large amounts of liquid hydrogen and liquid air. He used his exceptional laboratory facilities, capable staff, and personal experimental skills to liquefy helium in 1908.

Kamerlingh Onnes's successful helium experiment began at 5:45 A.M. on July 9, 1908, with the first seven hours devoted to the production of 75 liters of liquid air and 20 liters of liquid hydrogen. These were used to precool the apparatus and the helium gas within it. The circulation of helium through a Joule-Thomson process began at about 4:30 P.M., and the successful production of liquid helium was confirmed about three hours later. In his fourteen-hour experiment, Kamerlingh Onnes achieved the ultimate liquefaction, bringing the helium to a temperature of −268 degrees Celsius.

Kamerlingh Onnes proceeded to the next logical step, which was to boil the helium gas under reduced pressure in an attempt to solidify it. Using a strong vacuum pump, he reduced the pressure above the helium, but solidification did not occur. Later, it was discovered that helium differs from all other known materials in that it solidifies only under a pressure of about twenty-five times that of normal atmospheric pressure.

California Institute of Technology

*Heike Kamerlingh Onnes (left) and physicist Johannes Diderik van der Waals.*

## Consequences

The ability to achieve temperatures below −268 degrees Celsius brought the discovery of remarkable physical phenomena that could not be explained by existing physical theories and that led to the development of quantum theory.

In 1908, Kamerlingh Onnes found that the electrical resistance of mercury dropped sharply at −268 degrees Celsius. This was the first detection of superconductivity, a discovery for which Kamerlingh Onnes was awarded the Nobel Prize in Physics in 1913. Superconductivity has made possible sophisticated imaging techniques in medicine (making it possible to view the interior of a body), various developments in high-energy physics, and the construction of high-speed trains that levitate magnetically above the tracks that guide them.

Liquid helium exhibits the property of superfluidity, or a sharp drop in resistance to fluid flow, at temperatures below −271 degrees Celsius. In 1972, helium 3, the less common isotope of helium, was also found to exhibit superfluidity.

The liquefaction of helium ended the quest to liquefy all gases. The accomplishment also began the study of the properties of materials near the low-temperature limit of matter, absolute zero (−273.15 degrees Celsius).

*Harvey S. Leff*

# Austria Annexes Bosnia and Herzegovina

*When Austria-Hungary annexed two Serbo-Croatian countries that it had administered since 1878 under the terms of the Congress of Berlin, other European powers reacted with anger but, in the end, allowed the annexation to stand.*

**What:** International relations
**When:** October 7, 1908
**Where:** The Balkans
**Who:**

COUNT ALOIS LEXA VON AEHRENTHAL (1854-1912), foreign minister of Austria-Hungary from 1906 to 1912

ALEKSANDR PETROVICH IZVOLSKI (1856-1919), Russian minister of foreign affairs from 1906 to 1910

PËTR ARKADEVICH STOLYPIN (1862-1911), premier of Russia from 1906 to 1911

SIR EDWARD GREY (1862-1933), foreign secretary of Great Britain from 1905 to 1916

BERNHARD VON BÜLOW (1849-1929), prince and chancellor of the German Empire from 1900 to 1909

## Secret Negotiations

Bosnia and Herzegovina, Balkan states which were predominantly Serbo-Croatian, had been under the rule of Austria since 1878, according to the terms of the Congress of Berlin. Serbia, a neighboring nation, had been ruled since 1903 by Peter I, of the Karageorgevich dynasty, which was traditionally hostile to Austria. With the Serbian Radical Party in power, Serbia aimed to gather all the southern Slav countries under Habsburg control into a greater Serbian or Yugoslav state. Alois Lexa von Aehrenthal, who had been appointed foreign minister of Austria-Hungary in 1906, was well aware of Serbia's ambitions. He believed that Austria-Hungary should annex Bosnia and Herzegovina so as to frustrate Serbian efforts to dominate the region.

With the knowledge of Czar Nicholas II but not the government, the Russian minister of foreign affairs, Aleksandr Petrovich Izvolski, secretly let Aehrenthal know that Russia would be willing to support Austria-Hungary's annexation of Bosnia and Herzegovina. There was, however, one condition: Austria-Hungary, also known as the Dual Monarchy, must approve the opening of the Dardanelles Strait to Russian warships. Aehrenthal received this proposal positively, and the two statesmen then awaited the right moment to seal their bargain.

That moment came in the summer of 1908, with the outbreak of the Young Turk Revolution. The goal of this uprising was to revive the Ottoman Empire, especially in the Balkans. While the Turks were preoccupied with civil war, Aehrenthal and Izvolski met in Buchlau, Moravia, on September 15 and orally repeated their earlier agreement to support each other's aims in the Balkans. Unfortunately, they made no written account of their decisions, nor did they set a date for the annexation. As a result, within three weeks a serious misunderstanding led to a crisis in Europe.

## The Annexation

Izvolski left the conference with the impression that nothing would be done immediately (or so he later claimed). He set out to visit various European capitals to try to obtain the Great Powers' agreement to Russia's upcoming access to the Dardanelles.

Meanwhile, Aehrenthal was preparing for the annexation. He told Bulgaria that it should proclaim its independence, which it did on October 5. Two days later, Aehrenthal announced that Austria had annexed Bosnia and Herzegovina.

The protest that arose was so great that it brought Austria to the brink of war with the other European nations. Serbia and Montenegro were outraged, viewing the annexation as a deliberate

Library of Congress

*The marketplace in Sarajevo, Bosnia, Austria-Hungary, around 1910.*

changed much. Sir Edward Grey, the British foreign secretary, did reject Izvolski's plea for the opening of the Dardanelles to Russian warships.

Relations among the European states remained in a crisis for almost six months. With the reluctant support of Russia, Serbia prepared for war with Austria. Izvolski insisted that the dispute be brought before an international conference, while Aehrenthal firmly refused.

Great Britain and France did not give strong support to Russia, but Germany's decision to give full support to its ally Austria-Hungary finally brought the crisis to an end. On March 21, 1909, Prince Bernhard von Bülow, chancellor of the German Empire, practically ordered Izvolski to back down and accept the annexation. Izvolski did so, and on March 31, 1909, Serbia very reluctantly followed suit.

step to make sure that a great southern Slav state would not be formed. The Young Turks were angry at Austria's violation of Ottoman sovereignty.

In Russia, where there was much sympathy with the southern Slavs, the annexation and Izvolski's role in helping to make it possible were loudly condemned. Pětr Arkadevich Stolypin, premier of Russia, ordered Izvolski to retract his support for the annexation. Izvolski did so by denying any involvement and by calling on Great Britain and France to come to the aid of Serbia and the Turks.

That help, however, did not come. France and Great Britain were not pleased with the annexation, but they recognized the fact that Austria-Hungary had been in control of Bosnia and Herzegovina for thirty years already, so the status quo in the western Balkans would not really be

## Consequences

Though peace was restored to the Balkans, it was a shaky peace that lasted for only a few years. Relations between Austria and Serbia got progressively worse, until Archduke Francis Ferdinand of Austria was assassinated in 1914, by a Serbian nationalist.

Meanwhile, the uneasy truce that Austria and Russia had maintained in the Balkans from 1878 to 1908, was now shattered beyond repair. After 1908, Russian diplomats worked hard to create a league of Balkan states as a barrier against any Austrian ambitions to expand the Austro-Hungarian empire further. A Balkan League was indeed founded in 1912; yet instead of uniting against Austria, it used its strength to disrupt the Ottoman Empire still further.

*Edward P. Keleher*

# Belgium Annexes Congo Free State

---

*Because Africans had been mistreated under the reign of King Leopold II, the Belgian government took control of his private domain, the Congo Free State, and reformed it.*

---

**What:** Human rights; Political reform
**When:** November 1, 1908
**Where:** Brussels, Belgium, and the Congo
**Who:**
LEOPOLD II (1835-1909), king of Belgium from 1865 to 1909, and founder of the Congo Independent State
EDMOND DENE MOREL (1873-1924), a British journalist
JULES RENKIN (1862-1934), Belgian minister for colonies from 1908 to 1918

## Colonial Abuses

The Congo Independent State (also called the Congo Free State) was different from any other European colony in Africa. It was considered a sovereign country and was ruled by King Leopold II of Belgium, but the constitution made no link between Belgium and the Congo State until 1908. Mostly, the Congo was run as Leopold's private business.

From the time he gained control of the Congo in the 1870's, Leopold announced that he wanted to end the slave trade and bring commerce, prosperity, and "civilization" to the Africans. The Congo State did work to stop the slave trade, but otherwise Leopold did not show much concern for the rights of Africans.

The Congo Independent State ruled by military force. All land considered "unoccupied" was taken over by the government. Local officials were allowed to collect taxes as they wished, and many of them required payments of ivory, groundnuts, and wild rubber. Other taxes had to be paid in labor: cutting wood for river steamers, transporting officials in canoes, or serving as porters on expeditions. Women had to provide cas-

sava bread for the state's workers and soldiers, and residents also had to bring meat and fish to the government stations. Naturally, villages that were close to government stations carried the heaviest tax burden.

Those who had invested money in the Congo, including Leopold, tried to get quick profits. Beginning in the early 1890's, Leopold allowed private firms to take control of vast territories. The state held half of the stock in many of these companies, which included the Société Anversoise du Commerce du Congo (known as Anversoise), the Anglo-Belgian India Rubber and Exploring Company (called Abir), and the Compagnie du Kasai.

The worst abuses of Africans occurred in the rubber industry. Local officials set quotas of wild rubber that each village was required to produce; these officials were given a commission on the amount gathered. Soon missionaries began reporting that to punish the villagers for not meeting their quotas, the officials and their assistants were taking hostages (especially women), whipping people, and sometimes amputating their hands.

## Protest Rises

In the outside world, these reports led to an outcry against the Congo State. Leopold tried to prevent the publication of a report by British consul Roger Casement, but it was published in 1904. In response, British journalist Edmond Dene Morel founded the Congo Reform Association (CRA). Morel's campaign soon spread to other countries, especially the United States.

Finally, Leopold formed a commission to look into Casement's accusations. This commission decided that there was no evidence that European officials had ordered mutilation or murder; they claimed that the amputations were the result of a "native custom." Nevertheless, the

**181**

commission's report, released in 1905, did back up most of Casement's findings.

According to the commission, there were abuses in the tax system, the companies in the Congo were not well supervised, and unauthorized military expeditions were being sent out by the companies. Leopold set up another commission to recommend reforms.

Meanwhile, the push for reform was gaining steam overseas. Sir Edward Grey, who had already criticized the Congo State, became foreign secretary of Great Britain. By 1906, the Hearst newspapers in the United States had taken up the campaign against Leopold. A reform movement also appeared in Belgium itself, as the Socialist Party gained strength in the parliament.

By 1906, Belgium was being pressured hard by other nations. Leopold introduced reforms in that year, but Morel and the CRA said that they were not enough. In November, Grey told a group of reformers that he hoped Belgium would annex the Congo. In December, the Belgian parliament debated the issue and voted to consider annexation.

Negotiations with Leopold were long and difficult, but a treaty was finally written to transfer the Congo to Belgium, along with a new colonial charter. The Belgian Chambers made these documents into law in August and September, 1908. The annexation came into effect on November 1, 1908, although it was not announced in the Congo until November 16.

## Consequences

Morel was not satisfied with the annexation, for King Leopold still had certain powers in the new Belgian Congo. The companies that had invested in the Congo were allowed to remain, although their powers of administration were reduced. The new colonial law did not bring an end to forced labor, and most of Leopold's officials continued in their jobs. So Morel and the CRA convinced the British and American governments not to recognize the annexation immediately.

Because of the efforts of Morel and of Jules Renkin, the new Belgian minister of colonies, there were soon reforms in the Belgian Congo. Renkin presented his reform program in October, 1909. Free trade would be introduced over a period of three years, and consumer goods would be imported for Africans. Taxes would now be paid in currency.

In December, 1909, Leopold died. His successor, Albert I, wanted to encourage humane policies in Africa and had even visited the colony himself—something Leopold had never done. In 1912, Renkin was given the power to remove officials who were corrupt; that same year, the old governor, Baron Théophile Wahis, was replaced. In June, 1913, the CRA dissolved itself, having decided that its job was done. The United States and Great Britain granted recognition to the Belgian Congo.

Gradually, the Belgian Congo became similar to other European colonies in Africa. The Europeans discriminated against the Africans, who still had few rights under the law and no voice in government. For the most part, they were not subjected to open cruelty. The government gave subsidies to mission schools for Africans, but discouraged them from getting a more advanced education.

Under this "paternalistic" system, adult Africans were treated as children. When independence came suddenly to the Belgian Congo in 1960, the frustrations of years of European rule were expressed in a long, bloody civil war.

*T. K. Welliver*

# Boule Reconstructs Neanderthal Man Skeleton

*Marcellin Boule's reconstruction of a near-complete Neanderthal skeleton called into question the possibility of finding the "missing link" between higher apes and humans.*

**What:** Anthropology
**When:** December, 1908
**Where:** Paris, France
**Who:**
MARCELLIN BOULE (1861-1942), a French paleontologist
RUDOLF VIRCHOW (1821-1902), a German professor of anatomy
A. BOUYSSONIE, J. BOUYSSONIE, and L. BARDON, the French priests who discovered the Neanderthal skeleton

## The "Missing Link"

Late in the fall of 1908, the French paleontologist Marcellin Boule reconstructed a nearly complete skeleton of *Homo neanderthalensis* in his Paris laboratory. The fossil remains had been brought to him by three priests, who had found the bones in a cave at La Chapelle-aux-Saints in the Corrèze in France. By determining the age of the layer of earth in which the fossils were found, as well as the age of the artifacts found nearby, Boule could tell that the bones were from the Mousterian period (between 100,000 and 40,000 B.C.E.). This dating supported Boule's theory that *Homo neanderthalensis* was the "missing link" in the evolution of humans from the higher apes. This theory would be debated for nearly half a century.

Boule had very few firm guidelines to follow when reconstructing the Neanderthal skeleton. Nevertheless, a book that he published with fellow paleontologist Henri Vallois (*L'Homme fossile*) makes it clear that Boule believed that any later discoveries would support his theory. After he had set forth his ideas, several new discoveries were made: at Le Moustier and La Ferrassie, both in Dordogne, in 1909; at La Quina in the French Charente district in 1911; and in Palestine and on the Italian peninsula in the 1920's and 1930's.

## Human or Ape?

When Boule finished piecing together the Chapelle-aux-Saints fossil, he noticed several things that seemed to indicate links between apes and humans. These included a bent-over skeletal posture, a very large jaw, a prominent brow ridge, a sloping forehead, and little or no chin. Finally, the curvature of the leg bones, as well as a certain "pigeon toe" effect in the feet, suggested basic simian (apelike) features that have disappeared in *Homo sapiens* (modern human beings).

Boule knew that considerable support existed for different theories about *Homo neanderthalensis*. In the mid-nineteenth century, the German anatomist Rudolf Virchow had argued that Neanderthal remains should be assigned a place along with fossils of *Homo sapiens*, as a "cousin" who lived at the same time as modern humans. It was wrong, according to this school of thought, to limit the main period of *Homo neanderthalensis* to the Mousterian age simply to support the theory of evolution proposed by the English naturalist Charles Darwin. Neanderthals were not necessarily early humans who had been replaced by modern humans. Instead, they may simply have been unable to adapt to the same conditions as those that confronted modern humans.

Evolutionists, including Boule, wanted very much to find a pre-*sapiens* link between apes and *Homo sapiens*. They were willing to overlook the ideas of Virchow and others in order to find one. Boule's reconstruction of the Chapelle-aux-Saints skeleton seemed to meet this need.

**183**

## Consequences

There were two main theories about where *Homo neanderthalensis* belonged in the theory of

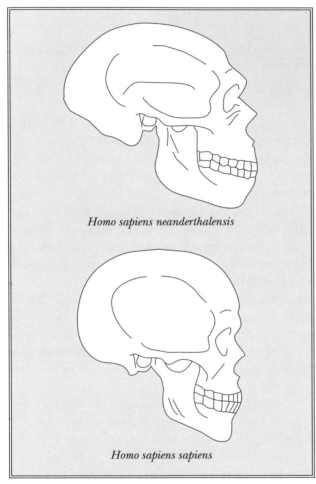

*Homo sapiens neanderthalensis*

*Homo sapiens sapiens*

*At top, the Neanderthal skull shows a smaller brain case, protruding brow ridges, and a massive jawbone in comparison with the modern human skull below.*

evolution. Boule suggested that *Homo sapiens* had replaced *Homo neanderthalensis*. The other theory was that the two species existed side by side until the superior adaptive qualities of *Homo sapiens* allowed them to survive while *Homo neanderthalensis* declined in numbers and eventually disappeared.

Well before Boule's time, many scientists were unwilling to accept *Homo neanderthalensis* as an evolutionary "cousin" of *Homo sapiens*. One of these was the geologist Thomas Huxley, who believed that *Homo neanderthalensis* was more like an ape than a human.

Virchow, on the other hand, refused to see *Homo neanderthalensis* as belonging to a species separate from *Homo sapiens*. Virchow believed that some of the Neanderthal's apelike features could be explained by "ailments"—specifically, the effects of rickets from malnutrition—suffered by the basically human individuals.

With the passage of time and the discovery of additional Neanderthal specimens, scientists began to grow skeptical of Boule's theory. In 1957, when W. L. Strauss and A. J. E. Cave took a close look at the skeleton Boule had reconstructed, they found that Virchow might be right in suggesting that disease had affected some of the Neanderthal's features. They also questioned the accuracy of Boule's work, especially in the foot area, where angles may have been exaggerated to make the foot look more like that of an ape. Strauss and Cave's work helped to lift *Homo sapiens neanderthalensis* to the status of a branch of *Homo sapiens*, rather than a "missing link."

*Byron D. Cannon*

# Johannsen Coins *Gene*, *Genotype*, and *Phenotype*

> *Wilhelm Ludwig Johannsen helped found the modern science of genetics with his experimental support for and creation of several important concepts.*

**What:** Biology; Genetics
**When:** 1909
**Where:** Copenhagen, Denmark
**Who:**
WILHELM LUDWIG JOHANNSEN (1857-1927), a Danish plant physiologist and chemist
GREGOR JOHANN MENDEL (1822-1884), an Austrian monk who pioneered the study of genetics
CHARLES DARWIN (1809-1882), an English naturalist
THOMAS HUNT MORGAN (1866-1945), an American embryologist and geneticist and winner of the 1933 Nobel Prize in Physiology or Medicine

## Heredity Dispute

Natural selection, one of the primary mechanisms of evolutionary change, was first described in 1859 in *On the Origin of Species* by Charles Darwin. He described a process in which certain individual organisms are better able to survive and reproduce offspring than other organisms because of inherited differences. Over time, the inherited properties of a population gradually and continuously change, and the changes sometimes lead to a new population of organisms. Darwin also tried to account for the origins of inherited differences, which he discussed in *Variation of Animals and Plants Under Domestication* (1868). Darwin believed that environment sometimes could directly influence the inherited characteristics that were passed to the offspring (soft inheritance).

Not even some of Darwin's strongest supporters, such as the biologists Thomas Henry Huxley and August Weissmann or the biostatistician Francis Galton, accepted all of his theory. Weissmann, for example, favored the idea of natural selection. He believed, however, that environment had nothing to do with the kinds of observable characteristics that an offspring might inherit from its parents. Galton argued that the inherited characteristics were passed from generation to generation with little change.

Moreover, both Galton and Huxley questioned Darwin's emphasis on small, heritable differences and continuous change among members of a population. They argued instead for the influence of mutations as the agents of change and believed that the evolution that resulted proceeded through sudden, spontaneous jumps in time. (A mutation is defined as a change occurring within the genetic makeup of a cell that is passed along to the cell's offspring.)

## Bridging the Gap

Wilhelm Johannsen was a Danish plant physiologist and chemist. His experiments in plant heredity would offer persuasive evidence to support the mutation theory suggested by the Dutch botanist Hugo de Vries. As a result, Johannsen would play a significant role in helping to bridge the gap between those who favored Darwin's views and those who favored Galton's.

Johannsen undertook many years of experiments in order to test his ideas. He eventually introduced three important concepts, which he called the "phenotype," the "genotype," and the "gene." The phenotype describes what the organism actually is according to its physical appearance, biological processes, and so forth. Throughout its lifetime, the phenotype experiences changes continuously. Modern science, for example, has confirmed that an individual's phenotype differs during childhood, adulthood, and old age. The genotype is the cell's genetic makeup, the total number of genes that an offspring receives from its parents.

Johannsen also searched for a word to describe broadly the units of inheritance at the heart of his theory. De Vries had called these units "factors" or "pangens" after Darwin's own descriptions. Johannsen borrowed from both men and coined the term "gene" as a unit of calculation or accounting. He summarized the results of his learning in a book published in 1909.

## Consequences

The geneticist Leslie Clarence Dunn notes that Johannsen's 1909 text was the first and most influential textbook of genetics on the European continent. In it, Johannsen introduced the new science and defined its key concepts. Although there were a few critics of his pure-line work, most geneticists in 1910 accepted his theory and rejected any significant role for natural selection in the evolutionary process.

The science Johannsen helped found, genetics, is thriving. Since his discoveries, an ever-increasing number of behavioral, anatomical, and physiological traits have been shown to have an inherited component. Knowing how specific traits (including genetic diseases such as cystic fibrosis) are transmitted permits genetic counselors to advise parents about the risk of passing the condition on to their children. Research on the molecular biology of the gene and chromosome (the "genotype") is a fast-growing area and includes, for example, manipulation of the genetic material of food plants and animals in order to increase production or resistance to disease; identification of potent mutagens, particularly those affecting deoxyribonucleic acid (DNA), in the sperm or egg; and mapping of a specific gene's locations on a particular chromosome.

*Joan C. Stevenson*

# Taft Conducts Dollar Diplomacy

*In a cooperative effort between American government and business, President William Howard Taft began a policy of using business investments to try to bring stability to struggling nations.*

**What:** Economics; International relations
**When:** 1909-1913
**Where:** Washington, D.C.
**Who:**
WILLIAM HOWARD TAFT (1857-1930), president of the United States from 1909 to 1913
PHILANDER CHASE KNOX (1835-1921), U.S. secretary of state from 1909 to 1913

### Dollars Instead of Bullets

The policy that has become known as "dollar diplomacy" was begun by President William Howard Taft and his secretary of state, Philander C. Knox. A lawyer from Pennsylvania, Knox was sympathetic to big business but was also concerned to build support for the United States within other nations and their governments. The best way to help both American business and American government, Taft and Knox decided, was to bring American money and business know-how to countries whose support for the United States was important and that needed help to improve their economy and standard of living.

This kind of American intervention would be peaceful—sending dollars instead of bullets—and American companies and foreign populations would both benefit. In 1910, Knox said, "The problem of good government is inextricably interwoven with that of economic prosperity and sound finance; financial stability contributes perhaps more than any other one factor to political stability." If Americans brought money and jobs to other countries, Knox believed, the governments of those countries would become more stable and would be better allies of the United States.

### Testing the Theory

The Taft administration used dollar diplomacy in two areas, the Caribbean and China. In the Caribbean, Taft and Knox followed the program begun by Theodore Roosevelt's administration in the Dominican Republic. That country had been politically unstable, and American officials had feared that foreign nations would take advantage of the situation and try to take control there. So it was that in 1905, Roosevelt had made an agreement with leaders of the Dominican Republic. The Dominican government would receive a loan from American banks to pay off its debts. In exchange for the loan, the American president would be allowed to appoint the head of the Dominican customs service—and in the Dominican Republic, as in all Caribbean states, the customs service was the main source of income for the government. The arrangement seemed to work perfectly. After the Americans took over the collection of customs money, the Dominican Republic enjoyed a time of peace, and the financial situation of its government improved. This state of affairs lasted through most of the Taft presidency.

Taft and Knox tried to use the same principles in dealing with Nicaragua. After supporting the overthrow of the powerful dictator José Santos Zelaya in 1909, Taft sent Thomas C. Dawson to Nicaragua to help the new government restore order. Dawson persuaded the Nicaraguans to install an American collector of customs and, in return, obtained a loan for Nicaragua from New York banks. Though the U.S. Senate refused several times to ratify this agreement, Taft appointed a collector of customs by executive order, and the New York bankers made several loans to Nicaragua. As extra security for the loans, the American banks took a controlling interest in the Nicaraguan National Bank and the Nicaraguan state railways.

**187**

In spite of these efforts to make Nicaragua financially stable, in 1912 the country's majority political party began a revolt against the American-supported President Adolfo Díaz. The Taft administration responded by sending warships and marines to keep Díaz in power. Unfortunately, bullets along with dollars were needed to keep Nicaragua in the kind of order that the U.S. government wanted to see.

The situation in China was quite different from that of the Caribbean. In China there was more active competition for political and economic influence, and the United States had less influence there than did several other great powers. Under Roosevelt, the American policy in the Far East had depended upon good relations with Japan: Japanese cooperation was considered essential to protect the Philippines as an American possession. In dealing with China, then, dollar diplomacy represented a major change in U.S. policy.

Taft and Knox tried to increase American influence in China mainly by funneling American money into that country. Their approach never changed. They demanded that American banking groups be included equally with banks from other countries in backing every foreign loan floated by China. Yet the New York bankers Taft had relied upon in Nicaragua did not have as much interest in investing in China, and they also did not have enough money at their disposal to make the large investments the Taft administration was seeking. To raise the money, the American banks had to rely on loans from the money markets of Great Britain and France. Thus the "American" investment in China was not really American.

Because American funds for investing in China were hard to come by, and because other world powers were unhappy with U.S. attempts to compete there, in 1912, the Taft administration decided not to continue its aggressive financial involvement in China. Instead, the United States returned to the more moderate "open door" approach of earlier administrations.

## Consequences

The attempt to use dollar diplomacy in Nicaragua showed that Taft and Knox's thinking was too simplistic. Political stability there did not depend only on paying debts and keeping up with budgets. Other important factors in Nicaraguan political life were political rivalry, struggles for prestige and power, the large gap in standard of living between a wealthy few and many poor, and resentment against American interference. All these factors worked against Taft's efforts from the very beginning.

Even the Dominican Republic, considered the "model" of dollar diplomacy, actually had succeeded not because of American policy but because of a skillful president, Roman Caceras. When Caceras was assassinated, a new wave of unrest arose in the Dominican Republic, and it was ended only in 1916, when the United States Marines occupied the capital city and other important areas of the country.

Dollar diplomacy, then, was not nearly as successful as some have made it out to be. It did bring the United States forward as one of the world's imperialistic powers in the years before World War I. Dollar diplomacy was part of the long history of U.S. interventions south of its borders—interventions that created great tensions between the United States and Latin American countries in the second half of the twentieth century.

*Karl A. Roider*

*William Howard Taft.*

# Millikan Conducts Oil-Drop Experiment

*Robert Andrews Millikan measured electrical charges on tiny oil drops and determined that the electron is the fundamental unit of electricity.*

**What:** Physics
**When:** January-August, 1909
**Where:** Chicago, Illinois
**Who:**
ROBERT ANDREWS MILLIKAN (1868-1953), an American physicist, professor, research director, chief executive officer, and winner of the 1923 Nobel Prize in Physics
HARVEY FLETCHER (1884-1981), an American physicist and research director

## Measuring Electric Charge

The first measurement of the electric charge carried by small water droplets was made in 1897 at Cambridge, England. The method timed the rate of fall of an ionized cloud of water vapor inside a closed chamber. The experiment was improved in 1903 by using a beam of X rays to produce the cloud between horizontal plates charged by a battery. The rate of descent of the top surface of the cloud between the plates was measured with an electric field that was switched on and off. The procedure, although an improvement, suffered from instabilities and irregularities on the top of the cloud. The cloud surface was difficult to delineate and resulted in measurements that fluctuated as much as 100 percent.

In 1909, a young graduate student, Harvey Fletcher at the University of Chicago (then College of Chicago), went to Robert Andrews Millikan to receive suggestions for work on a doctoral thesis in physics. Millikan suggested improving upon the measurement of electronic charge previously performed at Cambridge, England. Millikan's initial plan was to use an electric field not only strong enough to increase the speed of fall of the upper surface of the ionized cloud but also powerful enough to keep the top of the cloud surface stationary when the electric field was reversed. This would allow the rate of evaporation to be easily observed and compensated for in the computations. This technical improvement would permit the researcher for the first time to make measurements on isolated droplets and eliminate the experimental uncertainties and assumptions involved in using the cloud method.

Millikan's improvement included the construction of a 10,000-volt small cell storage battery with enough strength to hold the top surface of the cloud suspended long enough to measure the rate of evaporation of the droplets. When the electric field was turned on, however, the result was a complete surprise to Millikan. The top of the cloud surface instantaneously dissipated, and, since the experimental result assumed a rate of fall for the ionized cloud, Millikan saw this result as a complete failure. Repeated tests showed that whenever the cloud was dispersed, a few droplets would remain. By nature, however, these droplets had the proper charge-to-mass ratio to allow the downward force of gravity or weight of the droplet to be balanced by the upward pull of the electric field on the droplet's charge. This procedure became known as the "balanced drop method."

With practice, Millikan found that he could reduce evaporation by turning off the field shortly before certain droplets in the field of view changed motion from slow downward to upward. This made it possible to time the motion for a longer period. From Stokes's law, he found the weight of the droplet. Also, by knowing the strength of the electric field, Millikan calculated the electric charge necessary to balance its weight. The experimenters soon realized that the droplets always carried multiples of whole

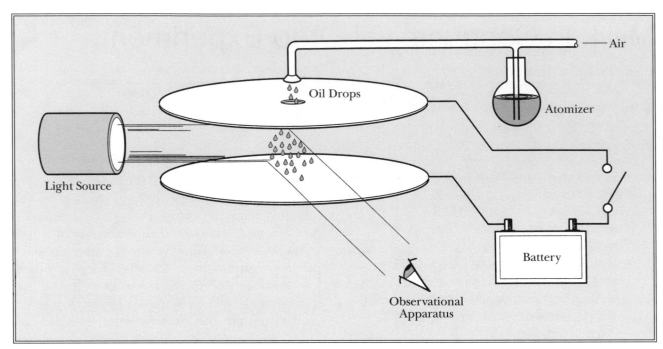

*Millikan's oil-drop experiment.*

number (1, 2, 3, 4, and so on) charges—never fractional amounts.

### "Little Starlets"

The actual experimental arrangement used by Millikan and Fletcher consisted of a small box with a volume of 2 or 3 centimeters fastened to the end of a microscope. A tube was placed from the box to an expansion chamber secured by an adjustable petcock valve that allowed a rapid expansion of air to form a water vapor cloud in the box. Surrounding the box on both ends were two brass conducting plates about 20 centimeters in diameter and 4 millimeters thick. A small hole was bored into the top plate to allow the oil mist from an atomizer to enter the region between the plates, which were separated by approximately 2 centimeters. A small arc light with two condensing lenses created a bright, narrow beam that was, in turn, permitted to pass between the plates.

An instrument called a "cathetometer" was placed on the microscope so that the microscope could be raised or lowered to the proper angle for best illumination (which from practice turned out to be about 120 degrees). The plate separation made it possible to apply a potential difference and produce an electric field. The apparatus was operated by turning on the light; focusing the microscope, which was placed about 1 meter from the plates; and then spraying oil over the top plate, while switching on the battery. When viewed through the microscope, the oil droplets looked like "little starlets" that had the colors of the rainbow.

When the electric field was first switched on, one would notice that the droplets would move at different speeds; some moved slowly upward, while the others moved rapidly downward. Superimposed on the droplets' downward fall was a small random back-and-forth motion (known as "Brownian movement") caused by the collision of the tiny droplets with thermally agitated air molecules within the chamber. When the electric field was reversed by changing the polarity of the battery, the same droplets that were moving downward moved upward, and vice versa. The experimenters deduced that the nature of this motion indicated that some of the droplets were negatively charged, while the others carried a positive charge.

### Consequences

No fractional amounts of the basic charge of oil droplets were ever observed—only whole number increments. This implied that the unit charge obtained could not be subdivided into smaller charges and was independent of the droplet size. These exact values showed that the electronic charge was not merely a statistical mean, as previous experimenters had believed. The experiment was, in fact, direct evidence for the existence of the electron as a finite-sized particle carrying a fundamental charge. It also made it possible to examine the attractive or repulsive properties of isolated electrons and to determine that electrical phenomena in solutions and gases are caused by electrical units that have fundamentally the same charge.

The experiment was an improvement over previous measurements in that Millikan was able to control precisely the strength of the electrical field while varying the droplet size. He also demonstrated that a completely discharged oil droplet fell at the same rate as an uncharged droplet with the electric field on. This indicated that something fundamental, which he chose to call "electricity," could be placed on or removed from the droplet only in exact amounts.

*Michael L. Broyles*

# National Association for the Advancement of Colored People Is Formed

*The NAACP was founded as a national organization that would promote the rights and civil liberties of African Americans.*

**What:** Social reform; Political reform
**When:** February 12, 1909
**Where:** New York City
**Who:**
WILLIAM EDWARD BURGHARDT DU BOIS (1868-1963), a professor at Atlanta University
BOOKER TALIAFERRO WASHINGTON (1856-1915), the founder and president of Tuskegee Institute
WILLIAM MONROE TROTTER (1872-1934), a radical African American leader
WILLIAM ENGLISH WALLING (1877-1936), a journalist and labor organizer

## Rights Under Attack

By the beginning of the twentieth century, many of the civil rights achieved by African Americans during the Reconstruction period were under severe attack. Supported by Supreme Court rulings in their favor, Southern states passed laws that had the effect of taking away the vote from most African American citizens. Other laws promoted segregation—preventing African Americans from using and participating in public institutions on an equal basis with whites.

In the North, white people were generally not as open in practicing racial discrimination, but it was still taken for granted. Tensions between the two races sometimes erupted into violence—lynchings or riots in the cities. In these confrontations, African Americans received little help from the press, the courts, and the law-enforcement agencies.

Within the African American community, there were various opinions about how to respond to the loss of rights. The most prominent spokesman for African Americans, educator Booker T. Washington, had for some time been urging people of color to work on getting skills and education so they could gain better jobs in factories and businesses. For the time being, he said, these goals should take priority over the push for civil and political rights.

Professor W. E. B. Du Bois, at Atlanta University, disagreed with Washington. In 1905, Du Bois founded the Niagara Movement, an organization of well-educated African Americans who called for public protest against the loss of rights. They did not think it was right to ignore discrimination and merely try to get good jobs. Du Bois's point of view was the one that inspired a group of people in 1909 to start an organization dedicated to fighting racial discrimination in all areas of American life.

## The NAACP Is Born

In August, 1908, there was a bloody race riot in Springfield, Illinois. White mobs destroyed much of the black section of Springfield and lynched two African Americans. More than fifty others were left dead or injured, and about two thousand African American residents left the city.

Shocked that the hometown of Abraham Lincoln could be the site of such violence, a group of white liberals decided that something should be done. William English Walling, a Kentucky journalist and labor organizer, wrote several articles in *The Independent*, condemning the Springfield riot and calling for "a powerful body of citizens" to come to the aid of African Americans.

Early in 1909, Walling met with Mary White Ovington, the socialist descendant of an abolitionist family, and Henry Moskovitz, a New York social worker, to discuss ways of gaining support for his idea. They invited Oswald Garrison

Villard (grandson of William Lloyd Garrison, a famous abolitionist) to join them, and the group soon grew to more than fifteen, including two prominent African American clergymen, Bishop Alexander Walters and the Reverend William Henry Brooks.

This planning committee decided to hold a "Conference on the Status of the Negro" on May 31 and June 1 in New York City. About three hundred men and women, including many white liberals, attended the two-day meeting. There they established a permanent organization and listened to scientific evidence to disprove the idea that black people were genetically inferior

to whites. Du Bois presented a speech in which he argued that the problems of African Americans were as much political as economic—that is, it was not simply that African Americans were poor but also that they were being treated unfairly under the current laws and government.

Booker T. Washington had been invited to the conference but had been told that the new organization would be an aggressive one. He decided not to attend. Yet his absence did not mean that there was complete agreement on what was to be done. There were heated arguments before the selection of a Committee of Forty on Permanent Organization, and before resolutions were passed demanding equal rights for African Americans and protection against violence. Villard's proposals were opposed strongly by William Monroe Trotter, editor of the *Boston Guardian*, and by J. Milton Waldron, president of the National Negro Political League. Both of these leaders favored more radical positions than those adopted by the majority, and in the end, neither of them was chosen for the Committee of Forty.

During the next year, Villard and others struggled to raise funds and plan for a second conference. Though the white press generally ignored them, and though there were open disputes with Washington, the committee succeeded in making a plan for organizing the new group so that it could move forward in its task.

At the second annual conference, the group, now named the National Association for the Advancement of Colored People, appointed Du Bois as director of publicity and research. In making this choice the organization clearly identified itself as progressive. Du Bois had been frustrated with his own Niagara Movement and saw his new post as a chance to become more effective as a spokesman for change. Within six months, he had launched the NAACP magazine, *Crisis*, which soon became an important opinion maker on issues of race.

*W. E. B. Du Bois.*

**193**

## Consequences

For the first time since Reconstruction, there was a major drive to end discrimination against African Americans in the United States. The way the NAACP was formed reflected many of the strengths and weaknesses that continued to mark the organization in its first few years. There were a large number of white leaders, and the organization depended on financial support from white liberals. Its program emphasized political and civil rights and aimed for change by means of laws and court decisions.

There were those within the African American community who criticized the NAACP for these characteristics. In the early years, most of the protests came from those who believed that the new organization was too aggressive. Even at the beginning, however, there were also some, such as Trotter, who insisted that the NAACP did not go far enough in pressing for change.

*John C. Gardner*

# Mexican Revolution Creates Constitutional Government

> *The Mexican Revolution succeeded in gaining broader representation in government for the people.*

**What:** Civil war; Military conflict; Political reform

**When:** 1910's

**Where:** Mexico

**Who:**

PORFIRIO DÍAZ (1830-1915), president of Mexico from 1876 to 1880 and 1884 to 1911

FRANCISCO INDALECIO MADERO (1873-1913), a leader in the Mexican Revolution, and president of Mexico from 1911 to 1913

VICTORIANO DE LA HUERTA (1854-1916), provisional president of Mexico from 1913 to 1914

VENUSTIANO CARRANZA (1859-1920), a leader of the Constitutionalist Movement, and president of Mexico from 1914 to 1920

ÁLVARO OBREGÓN (1880-1928), a major figure in the Constitutionalist Movement, and president of Mexico from 1920 to 1924

FRANCISCO (PANCHO) VILLA (1877-1923) and

EMILIANO ZAPATA (1877?-1919), popular revolutionary leaders

## The Díaz Regime

In 1876 Mexico had come under the rule of Porfirio Díaz, whose motto had been the restoration of constitutional government. Although he did operate under a constitution, he dominated the country as a dictator. He allowed large groups of Mexicans to lose important political and economic rights.

Díaz invited investors from foreign countries to bring their wealth to Mexico. Wealthy U.S. investors were happy to accept the invitation at a time when railroads, mining (especially of petroleum), and land were all being developed in Mexico. Petroleum production increased more than a thousandfold between 1901 and 1911, under mostly American control. This industry was almost completely free from taxes during Díaz's regime. Foreign individuals and companies also invested in land, sometimes establishing large haciendas (ranches) that spread across millions of acres.

## Reform and Revolution

More and more, Díaz was criticized within Mexico for dominating the country and failing to protect the people's political rights. A challenge to his presidency came in the election of 1910, when Francisco Madero ran against Díaz. When Madero lost, he called the election fraudulent and declared himself in revolt. His protest encouraged others to rise up against Díaz. In 1911, the regime fell from power, and Díaz went into exile; Madero was elected the new president.

Yet Madero kept many of the Díaz officials in power, and his middle-of-the-road policies did not satisfy either the *porfiristas* (those loyal to Díaz) or those who wanted sweeping change in Mexico. There were various rebellions, and finally, General Victoriano Huerta led a successful overthrow of Madero in February, 1913. Madero and his vice president were killed.

Huerta, a *porfirista*, came under immediate attack from liberals and revolutionaries across Mexico. In the south there was an uprising led by Emiliano Zapata, while in the north another band of revolutionaries gathered around Pancho Villa. At one time, more than two hundred "revolutionary" groups claimed to be the legitimate government of Mexico.

Library of Congress

*General Venustiano Carranza (bearded) sits and relaxes in the field with his followers.*

Zapata issued his Plan of Ayala, calling for reform in agriculture and the ownership of land. His supporters were mostly Indians and peasants. Villa, based in the northern state of Chihuahua, was a very independent leader, but he did fight alongside the Constitutionalists (or *carrancistas*), led by Venustiano Carranza. Carranza issued the Plan of Guadalupe, which called for the restoration of constitutional government in Mexico.

These men's groups, along with many others, succeeded in overthrowing Huerta in August, 1914. Within a short time, however, the leaders of the various revolutionary factions were fighting among themselves.

Carranza and the Constitutionalists won in the end for two reasons. One was the unusual talent of Carranza's ally Álvaro Obregón, a military leader who defeated Villa in April, 1915, and became known as the Hero of Celaya. Obregón was a very persuasive man who was able to attract the Mexican labor movement and other important

groups to support the Constitutionalists. The second reason for Carranza's success was that the Woodrow Wilson administration gave recognition to his government in October, 1915.

Though Carranza never united all Mexico under his government, he did have several important successes. Most important, in the fall of 1916 he authorized the calling of a convention to write a new constitution for Mexico. This constitution made use of some of the more revolutionary ideas, such as land reform, labor reform, reform of the petroleum and other mining industries, and restrictions on the powers of the Roman Catholic clergy. Clearly, it not only offered political freedoms but also launched major changes in Mexico's economic system and the living conditions of its people.

Carranza himself was not committed to the more radical changes promised by the constitution. Many Mexicans began to criticize him for not fulfilling promises he had made since 1913, and he was blamed for the continuing civil strife, crime, and corruption in the country. Villa and Zapata were still leading armed uprisings against his government. In 1919, however, Carranza "solved" one problem by allowing some *carrancistas* to assassinate Zapata.

The constitution did not allow presidents to have more than one term in office, so Carranza chose Ignacio Bonillas, Mexico's ambassador to the United States, to succeed him in 1920. This decision was a serious mistake. The logical candidate for the presidency was Obregón, a national hero who was spoken of as "a genius in war and peace." Obregón had announced his own candidacy in June, 1919. Trying to thwart him, Carranza ordered federal troops to move into Sonora, which was the base of Obregón's power. As a result, a new revolutionary movement appeared on the scene, led by Obregón and a few other Sonorans.

Their political program, the Plan of Agua Prieta, was proclaimed on April 23, 1920, and rebuked Carranza for trying to impose a president on Mexico by force. He was declared unfit for office, and the revolutionaries moved to seize power and appoint a provisional president. Carranza fled the capital and was killed on May 21, 1920. Adolfo de la Huerta became provisional president until November 30, 1920, and suc-

ceeded in making an agreement with Villa so that he would stop his attacks on the government.

Meanwhile, Obregón was elected president; he took office on December 1, 1920. His election marked the end of the Mexican Revolution and brought calm to the country. His Institutional Revolutionary Party (PRI) remained in power without interruption until 2000.

## Consequences

The Mexican Revolution—which began with Madero's attempt to succeed Díaz as Mexico's president—launched a reform movement that led to the social, economic, and political reorganization of the country. Though the revolutionaries did not succeed in making all the changes they had hoped to make, there was some evidence of real and lasting change. The government of Mexico had been given to the common people. Wealthy people of Spanish descent no longer held exclusive power; Mexicans of *mestizo* (mixed Indian and Spanish) background were now well represented. Land was eventually redistributed, especially under the government of Lázaro Cárdenas, who was president from 1934 to 1940.

*Daniel D. DiPiazza*

# Crystal Sets Lead to Modern Radios

*The invention of crystal sets, the first radio receivers, led to the development of the modern radio.*

**What:** Communications
**When:** 1910
**Where:** United States
**Who:**
H. H. DUNWOODY (1842-1933), an American inventor
SIR JOHN A. FLEMING (1849-1945), a British scientist-inventor
HEINRICH HERTZ (1857-1894), a German physicist
GUGLIELMO MARCONI (1874-1937), an Italian engineer-inventor
JAMES CLERK MAXWELL (1831-1879), a Scottish physicist
GREENLEAF W. PICKARD (1877-1956), an American inventor

### From Morse Code to Music

In the 1860's, James Maxwell Clerk demonstrated that electricity and light had electromagnetic and wave properties. The conceptualization of electromagnetic waves led Maxwell to propose that such waves, made by an electrical discharge, would eventually be sent long distances through space and used for communication purposes. Then, near the end of the nineteenth century, the technology that produced and transmitted the needed Hertzian (or radio) waves was devised by Heinrich Rudolph Hertz, Guglielmo Marconi (inventor of the wireless telegraph), and many others. The resultant radio broadcasts, however, were limited to the dots and dashes of the Morse code.

Then, in 1901, H. H. Dunwoody and Greenleaf W. Pickard invented the crystal set. Crystal sets were the first radio receivers that made it possible to hear music and the many other types of now-familiar radio programs. In addition, the simple construction of the crystal set enabled countless amateur radio enthusiasts to build "wireless receivers" (the name for early radios) and to modify them. Although, except as curiosities, crystal sets were long ago replaced by more effective radios, they are where it all began.

### Crystals, Diodes, Transistors, and Chips

Radio broadcasting works by means of electromagnetic radio waves, which are low-energy cousins of light waves. All electromagnetic waves have characteristic vibration frequencies and wavelengths. This article will deal mostly with long radio waves of frequencies from 550 to 1,600 kilocycles (kilohertz), which can be seen on amplitude-modulation (AM) radio dials. Frequency-modulation (FM), shortwave, and microwave radio transmission use higher-energy radio frequencies.

The broadcasting of radio programs begins with the conversion of sound to electrical impulses by means of microphones. Then, radio transmitters turn the electrical impulses into radio waves that are broadcast together with higher-energy carrier waves. The combined waves travel at the speed of light to listeners. Listeners hear radio programs by using radio receivers that pick up broadcast waves through antenna wires and reverse the steps used in broadcasting. This is done by converting those waves to electrical impulses and then into sound waves. The two main types of radio broadcasting are AM and FM, which allow the selection (modulation) of the power (amplitude) or energy (frequency) of the broadcast waves.

The crystal set radio receiver of Dunwoody and Pickard had many shortcomings. These led to the major modifications that produced modern radios. Crystal sets, however, began the radio industry and fostered its development. Today, it is possible to purchase somewhat modified forms of crystal sets, as curiosity items. All crystal sets, original or modern versions, are crude AM radio receivers that are composed of four components:

an antenna wire, a crystal detector, a tuning circuit, and a headphone or loudspeaker.

Antenna wires (aerials) pick up radio waves broadcast by external sources. Originally simple wires—modern aerials—are made to work better by means of insulation and grounding. The crystal detector of a crystal set is a mineral crystal that allows radio waves to be selected (tuned). The original detectors were crystals of a lead-sulfur mineral, galena. Later, other minerals (such as silicon and carborundum) were also found to work. The tuning circuit is composed of 80 to 100 turns of insulated wire, wound on a 0.33-inch support. Some surprising supports used in homemade tuning circuits include cardboard toilet-paper-roll centers and Quaker Oats cereal boxes. When realism is desired in collector crystal sets, the coil is usually connected to a wire probe selector called a "cat's whisker." In some

such crystal sets, a condenser (capacitor) and additional components are used to extend the range of tunable signals. Headphones convert chosen radio signals to sound waves that are heard by only one listener. If desired, loudspeakers can be used to enable a roomful of listeners to hear chosen programs.

An interesting characteristic of the crystal set is the fact that its operation does not require an external power supply. Offsetting this are its short reception range and a great difficulty in tuning or maintaining tuned-in radio signals. The short range of these radio receivers led to, among other things, the use of power supplies (house current or batteries) in more sophisticated radios. Modern solutions to tuning problems include using manufactured diode vacuum tubes to replace crystal detectors, which are a kind of natural diode. The first manufactured di-

*The crystal set in this 1912 display is probably the largest ever made.*

Hulton Archive

**199**

odes, used in later crystal sets and other radios, were invented by John Ambrose Fleming, a colleague of Marconi's. Other modifications of crystal sets that led to sophisticated modern radios include more powerful aerials, better circuits, and vacuum tubes. Then came miniaturization, which was made possible by the use of transistors and silicon chips.

## Consequences

The impact of the invention of crystal sets is almost incalculable, since they began the modern radio industry. These early radio receivers enabled countless radio enthusiasts to build radios, to receive radio messages, and to become interested in developing radio communication systems. Crystal sets can be viewed as having spawned all the variant modern radios. These in-

clude boom boxes and other portable radios; navigational radios used in ships and supersonic jet airplanes; and the shortwave, microwave, and satellite networks used in the various aspects of modern communication.

The later miniaturization of radios and the development of sophisticated radio system components (for example, transistors and silicon chips) set the stage for both television and computers. Certainly, if one tried to assess the ultimate impact of crystal sets by simply counting the number of modern radios in the United States, one would find that few Americans more than ten years old own fewer than two radios. Typically, one of these is run by house electric current and the other is a portable set that is carried almost everywhere.

*Sanford S. Singer*

# Electric Washing Machine Is Introduced

American manufacturers used electricity to run hand-operated washing tubs and wringers, making the job of washing clothes much easier.

**What:** Engineering
**When:** 1910
**Where:** United States
**Who:**

O. B. WOODROW, a bank clerk who claimed to be the first to adapt electricity to a remodeled hand-operated washing machine

ALVA J. FISHER (1862-1947), the founder of the Hurley Machine Company, who designed the Thor electric washing machine, claiming that it was the first successful electric washer

HOWARD SNYDER, the mechanical genius of the Maytag Company

## Hand Washing

Until the development of the electric washing machine in the twentieth century, washing clothes was a tiring and time-consuming process. With the development of the washboard, dirt was loosened by rubbing. Clothes and tubs had to be carried to the water, or the water had to be carried to the tubs and clothes. After washing and rinsing, clothes were hand-wrung, hang-dried, and ironed with heavy, heated irons. In nineteenth century America, the laundering process became more arduous with the greater use of cotton fabrics. In addition, the invention of the sewing machine resulted in the mass production of inexpensive ready-to-wear cotton clothing. With more clothing, there was more washing.

One solution was hand-operated washing machines. The first American patent for a hand-operated washing machine was issued in 1805. By 1857, more than 140 patents had been issued; by 1880, between 4,000 and 5,000 patents had been granted. While most of these machines were never produced, they show how much the public wanted to find a mechanical means of washing clothes. Nearly all the early types prior to the Civil War (1861-1865) were modeled after the washboard.

Washing machines based upon the rubbing principle had two limitations: They washed only one item at a time, and the rubbing was hard on clothes. The major conceptual breakthrough was to move away from rubbing and to design machines that would clean by forcing water through a number of clothes at the same time.

An early suction machine used a plunger to force water through clothes. Later electric machines would have between two and four suction cups, similar to plungers, attached to arms that went up and down and rotated on a vertical shaft. Another hand-operated washing machine was used to rock a tub on a frame back-and-forth. An electric motor was later substituted for the hand lever that rocked the tub. A third hand-operated washing machine was the dolly type. The dolly, which looked like an upside-down three-legged milking stool, was attached to the inside of the tub cover and was turned by a two-handled lever on top of the enclosed tub.

## Machine Washing

The hand-operated machines that would later dominate the market as electric machines were the horizontal rotary cylinder and the underwater agitator types. In 1851, James King patented a machine of the first type that utilized two concentric half-full cylinders. Water in the outer cylinder was heated by a fire beneath it; a hand crank turned the perforated inner cylinder that contained clothing and soap. The inner-ribbed design of the rotating cylinder raised the clothes as the cylinder turned. Once the clothes reached the top of the cylinder, they dropped back down into the soapy water.

The first underwater agitator-type machine, the second type, was patented in 1869. In this machine, four blades at the bottom of the tub were attached to a central vertical shaft that was turned by a hand crank on the outside. The agitation created by the blades washed the clothes by driving water through the fabric. It was not until 1922, when Howard Snyder of the Maytag Company developed an underwater agitator with reversible motion, that this type of machine was able to compete with the other machines. Without reversible action, clothes would soon wrap around the blades and not be washed.

Claims for inventing the first electric washing machine came from O. B. Woodrow, who founded the Automatic Electric Washer Company, and Alva J. Fisher, who developed the Thor electric washing machine for the Hurley Machine Corporation. Both Woodrow and Fisher made their innovations in 1907 by adapting electric power to modified hand-operated, dolly-type machines. Since only 8 percent of American homes were wired for electricity in 1907, the early machines were advertised as adaptable to electric or gasoline power but could be hand-operated if the power source failed. Soon, electric power was being applied to the rotary cylinder, oscillating, and suction-type machines. In 1910, a number of companies introduced washing machines with attached wringers that could be operated by electricity. The introduction of automatic washers in 1937 meant that washing machines could change phases without the action of the operator.

## Consequences

By 1907 (the year electricity was adapted to washing machines), electric power was already being used to operate fans, ranges, coffee percolators, flatirons, and sewing machines. By 1920, nearly 35 percent of American residences had been wired for electricity; by 1941, nearly 80 percent had been were wired. The majority of American homes had washing machines by 1941; by 1958, this had risen to an estimated 90 percent.

The growth of electric appliances, especially washing machines, is directly related to the decline in the number of domestic servants in the United States. The development of the electric washing machine was, in part, a response to a decline in servants, especially laundresses. Also, rather than easing the work of laundresses with technology, American families replaced their laundresses with washing machines.

Commercial laundries were also affected by the growth of electric washing machines. At the end of the nineteenth century, they were in every major city and were used widely. Observers noted that just as spinning, weaving, and baking had once been done in the home but now were done in commercial establishments, laundry work had now begun its move out of the home. After World War II (1939-1945), however, although commercial laundries continued to grow, their business was centered more and more on institutional laundry, rather than residential laundry, which they had lost to the home washing machine.

Some scholars have argued that, on one hand, the return of laundry to the home resulted from marketing strategies that developed the image of the American woman as one who is home operating her appliances. On the other hand, it was probably because the electric washing machine made the task much easier that American women, still primarily responsible for the family laundry, were able to pursue careers outside the home.

*Thomas W. Judd*

# Hale Telescope Begins Operation on Mount Wilson

> *Hale built several telescopes that served as tools for observation, and he developed new techniques for building more powerful telescopes.*

**What:** Astronomy
**When:** 1910
**Where:** Mount Wilson, California
**Who:**
GEORGE ELLERY HALE (1868-1938), an American astronomer, inventor, and builder of observatories at Mount Wilson and Mount Palomar, California

## The Ideal Location

In the late nineteenth century, observational astronomy had not advanced far beyond the Italian astronomer Galileo's first telescope in either power or design. Most powerful telescopes were still small, refracting designs. These telescopes were very similar to those used by ship captains; they contained a fixed eyepiece lens and an adjustable lens that could be positioned to focus the light of an image onto the eyepiece. Larger, more powerful reflecting telescopes used a concave mirror to collect light from an object and focus it onto a series of lenses or an adjustable eyepiece. All these telescopes were only capable of observing; they could not be combined with other, more advanced instruments to analyze the images they captured.

George Ellery Hale, after graduating from the Massachusetts Institute of Technology (MIT), moved to Chicago. At MIT, Hale had had access to the 38-centimeter refracting telescope at Harvard University and the experience of its astronomers. In Chicago, Hale became involved in the building of a 102-centimeter reflecting telescope at Lake Geneva, Wisconsin. In 1903, Hale left Chicago to establish an observatory at Mount Wilson, California.

Because of its high elevation, Mount Wilson was an ideal location for an observatory. It was usually above the surrounding clouds. Also, the large amount of trees and foliage growing on the mountainsides helped to absorb the radiant heat that can distort the rays of light coming into the telescope during solar observations.

## Many Telescopes

The first project Hale undertook was the Snow Solar Telescope, which began observations near the end of 1903. Hale immediately encountered his first major design problem. The Snow Telescope was a horizontally designed scope that rested upon the ground. The heat waves that were reflected off the earth became so intense that the mirrors of the telescope would expand and distort during the middle part of the day; therefore, observing was limited to early morning and evening. Hale thought that elevating the telescope would bring it out of the intense waves of heat and provide a cleaner image. To test his hunch, he climbed a tall evergreen with a small telescope in hand and compared the image he saw from the top of the tree with the one he had observed at the foot of the tree. Convinced that he was correct in his theory, Hale began construction of an elevated telescope.

The final version of the telescope consisted of a rotating dome atop an 18-meter tower that housed a mirror for gathering light. This light was reflected down a shaft to the base of the tower, where it was focused by a series of lenses and mirrors. Hale's design was a great success, allowing solar observers to track the sun throughout the day. In addition, the reduction in the waves of heat that reached the gathering mirror provided the observers with more detailed images. The success of this tower led to the construction of a more powerful 46-meter tower.

Hale added a spectrograph to the towers to analyze the data collected by the telescopes. The spectrograph is capable of breaking light into its component parts, which scientists can then analyze to determine the components of the light-producing bodies and the chemical reactions that are taking place within them. The spectrograph allowed scientists to discover the composition of the sun. Since spectrographs need to maintain a constant temperature, Hale housed them in wells dug beneath each tower.

Solar observation, however, was not the only role of the Mount Wilson Observatory. The success of the 102-centimeter telescope at Lake Geneva inspired Hale to build a 153-centimeter telescope at Mount Wilson. In 1906, Los Angeles businessman John D. Hooker provided the funds for a 254-centimeter telescope, the largest of its day. During a visit to Mount Wilson in 1910, the American industrialist and humanitarian Andrew Carnegie announced that he would provide half a million dollars to mount and house the telescope.

The glass used for the 254-centimeter mirror was from France. Hale and his colleagues were shocked and disappointed when they inspected it, not because the glass was made from green wine bottles but because it was full of bubbles. Hale was afraid that the bubbles would cause the glass to warp or crack when the temperature changed. The group did not have many options, however, so they decided to take a chance and proceeded with the grinding and coating. The bubbles turned out to cause no harm, and the construction of what came to be known as the Hooker telescope was completed successfully in 1917.

## Consequences

Hale's innovative designs for solar telescopes, coupled with the introduction of the spectrograph attached to the telescopes, provided the foundation for intense solar research. Scientists were now able to observe solar flares, solar storms, and other solar disturbances and to understand the internal chemical changes that occurred with and caused these disturbances.

The 254-centimeter telescope provided scientists with a view 200,000 times greater than that of the human eye. The American astronomer Edwin Powell Hubble used this view to great advantage. At the time the 254-centimeter telescope was completed, there was controversy among astronomers about the nature of the Milky Way galaxy: What were the clouds that could be seen in the Milky Way? Using the 254-centimeter telescope, Hubble discovered that these clouds were actually distant galaxies and that there were hundreds of galaxies that had not previously been seen. Furthermore, Hubble discovered that these galaxies were traveling away from the Milky Way and that the galaxies farther away were traveling at a greater speed than those closer to the Milky Way. This was the first hard evidence for what would become known as the "big bang theory."

Hale soon discovered that the information observed through the 254-centimeter telescope provoked questions that could only be solved by a deeper look into space. Because of this success, Hale began to build a 508-centimeter telescope in Palomar, California.

*Mary Hrovat*

# Rous Discovers Cancer-Causing Viruses

*Peyton Rous discovered that chicken liver sarcoma (connective tissue cancer) is caused by a virus.*

**What:** Medicine; Health
**When:** 1910
**Where:** New York, New York
**Who:**
PEYTON ROUS (1879-1970), an American pathologist who shared the 1966 Nobel Prize in Physiology or Medicine
DAVID BALTIMORE (1938- ), an American virologist who shared the 1975 Nobel Prize in Physiology or Medicine
HOWARD M. TEMIN (1934-1994), an American virologist who shared the 1975 Nobel Prize in Physiology or Medicine

## Smaller than Life

In the early twentieth century, viruses were shown to be noncellular in structure, consisting only of nucleic acid—that is, deoxyribonucleic acid (DNA) and ribonucleic acid (RNA)—wrapped within a protective protein covering. Viruses are immobile and inactive outside cells. They can function only within a host cell, and then only to reproduce and destroy the host cell. They are intracellular parasites, always invading cells, robbing cellular resources, reproducing, and destroying. Because of the noncellular structure and unusual nature of viruses, there is considerable debate over their classification as a lifeform.

Once a virus is carried by air or fluid to the cells of a given host species, it may be only by chance that it physically contacts a cell. Once physical contact is made, a rapid series of chemical reactions between the virus protein covering and the cell membrane triggers the injection of the viral DNA or RNA into the cell. Once it is inside the host cell, the viral nucleic acid can follow two possible infection routes, depending upon

cellular conditions and certain enzymes encoded by the viral nucleic acid: the lysogenic cycle and the lytic cycle. In the lysogenic cycle, the viral nucleic acid encodes a repressor enzyme that prevents viral reproduction, followed by the viral DNA inserting itself into the host cell DNA and lying dormant indefinitely. During cellular stress, the dormant virus can enter the lytic cycle. In the lytic cycle, the viral nucleic acid commandeers the cell's resources, which are directed to synthesize up to several thousand new viruses, each of which can infect new cells.

## Viral Chicken Cancer

In 1909, Peyton Rous began research in pathology at the Rockefeller Institute for Medicinal Research (now Rockefeller University) in New York City. Rous was interested in the physiology of cancer within mammals and birds. He discovered a type of connective tissue cancer in chickens, later called "Rous sarcoma," which causes gross hypertrophy (enlargement) of certain organs, particularly the liver and gallbladder. Rous sarcoma eventually is fatal.

In his experiments, Rous grafted sarcoma tumor cells from diseased hens to healthy hens; the healthy hens contracted the disease. He then cultivated hen tumor cells, extracted a fluid not containing cells, and injected this fluid into healthy hens. Again, the healthy hens contracted the disease. His results pointed toward one possible conclusion: Some noncellular component of the tumor extract was capable of producing cancer in healthy hens. The most plausible explanation was a virus. Further experiments yielded identical results.

Rous hypothesized that a Rous sarcoma virus caused this chicken sarcoma. Nevertheless, his work was derided by his peers, who unsuccessfully repeated his experiments with other species. The failure of many to accept his conclusion reflected a considerable lack of understanding

The Nobel Foundation

*Peyton Rous.*

## Consequences

Rous's discovery paved the way for a better understanding of the origin of viruses. Viruses most likely evolved from cells because viruses are noncellular, because they must reproduce inside cells, and because they have the same genetic code as living cells. It is possible that, more than one billion years ago, a small group of genes capable only of reproduction and of manufacturing a protective protein covering escaped from a cell and temporarily existed outside cells in an inactive, dormant state. Viruses could be intercellular messengers whose functions went awry.

The Rous sarcoma virus was the first of more than twenty-four oncogenic viruses discovered during the twentieth century. Several of these viruses can, in addition to causing cancer, also cause various other diseases. For example, the Epstein-Barr virus can cause a rare type of lymph node cancer called "Burkitt's lymphoma." This virus also causes infectious mononucleosis and may be responsible for certain cases of chronic fatigue. Similarly, the hepatitis B virus can cause liver cancer. Hepatitis is also a noncancerous liver disease that afflicts about two hundred million people worldwide every year.

In the 1960's, molecular virologists Howard M. Temin and David Baltimore demonstrated that RNA retroviruses, such as Rous sarcoma virus, could encode DNA from RNA using a special viral enzyme called "reverse transcriptase." This phenomenon went against established scientific dogma, which maintained that DNA encodes RNA. The list of such RNA retroviruses includes the notorious human immunodeficiency virus (HIV), the causative agent of acquired immune deficiency syndrome (AIDS) in humans. While HIV causes AIDS, it does not cause cancer; instead, it destroys an individual's immune system cells such that a person's body is unable to defend itself from secondary infections, such as pneumonia, and spontaneous cancers.

*David Wason Hollar, Jr.*

of both viruses and cancer by the medical and scientific community of that time. Despite the negative reactions, Rous continued his studies of liver and gallbladder physiology.

With greater understanding of viruses during subsequent decades of the twentieth century, Rous's viral theory of cancer began to be recognized. From his studies of Rous sarcoma virus, his theory maintained that some cancers could be caused by viruses. The discovery of more tumor-causing viruses during the 1950's resulted in Rous sharing the 1966 Nobel Prize in Physiology or Medicine.

# Electric Refrigerators Revolutionize Home Food Preservation

> *Using the ideas of Marcel Audiffren, Fred Wolf developed the first electric refrigerator. Christian Steenstrup invented a hermetically sealed refrigerator that improved production and lowered costs.*

**What:** Food science
**When:** 1910-1939
**Where:** Detroit, Michigan, and Schenectady, New York
**Who:**
MARCEL AUDIFFREN, a French monk
CHRISTIAN STEENSTRUP (1873-1955), an American engineer
FRED WOLF, an American engineer

## Ice Preserves America's Food

Before the development of refrigeration in the United States, a relatively warm climate made it difficult to preserve food. Meat spoiled within a day and milk could spoil within an hour after milking. In early America, ice was stored below ground in icehouses that had roofs at ground level. George Washington had a large icehouse at his Mount Vernon estate. By 1876, America was consuming more than 2 million tons of ice each year, which required 4,000 horses and 10,000 men to deliver.

Several related inventions were needed before mechanical refrigeration was developed. James Watt invented the condenser, an important refrigeration system component, in 1769. In 1805, Oliver Evans presented the idea of continuous circulation of a refrigerant in a closed cycle. In this closed cooling cycle, a liquid refrigerant evaporates to a gas at low temperature, absorbing heat from its environment and thereby producing "cold," which is circulated around an enclosed cabinet. To maintain this cooling cycle, the refrigerant gas must be returned to liquid form through condensation by compression. The first closed-cycle vapor-compression refrig-erator, which was patented by Jacob Perkins in 1834, used ether as a refrigerant.

Iceboxes were used in homes before refrigerators were developed. Ice was cut from lakes and rivers in the northern United States or produced by ice machines in the southern United States. An ice machine using air was patented by John Gorrie at New Orleans in 1851. Ferdinand Carre introduced the first successful commercial ice machine, which used ammonia as a refrigerant, in 1862, but it was too large for home use and produced only a pound of ice per hour. Ice machinery became very dependable after 1890 but was plagued by low efficiency. Very warm summers in 1890 and 1891 cut natural ice production dramatically and increased demand for mechanical ice production. Ice consumption continued to increase after 1890; by 1914, 21 million tons of ice were used annually. The high prices charged for ice and the extremely low efficiency of home iceboxes gradually led the public to demand a substitute for ice refrigeration.

## Refrigeration for the Home

Domestic refrigeration required a compact unit with a built-in electric motor that did not require supervision or maintenance. Marcel Audiffren, a French monk, conceived the idea of an electric refrigerator for home use around 1910. The first electric refrigerator, which was invented by Fred Wolf in 1913, was called the Domelre, which stood for domestic electric refrigerator. This machine used condensation equipment that was housed in the home's basement. In 1915, Alfred Mellowes built the first refrigerator to contain all of its components; this machine was known as Guardian's Frigerator.

General Motors acquired Guardian in 1918 and began to mass produce refrigerators. Guardian was renamed Frigidaire in 1919. In 1918, the Kelvinator Company, run by Edmund Copeland, built the first refrigerator with automatic controls, the most important of which was the thermostatic switch. Despite these advances, by 1920 only a few thousand homes had refrigerators, which cost about $1,000 each.

The General Electric Company (GE) purchased the rights to the General Motors refrigerator, which was based on an improved design submitted by one of its engineers, Christian Steenstrup. Steenstrup's innovative design included a motor and reciprocating compressor that were hermetically sealed with the refrigerant. This unit, known as the GE Monitor Top, was first produced in 1927. A patent on this machine was filed in 1926 and granted to Steenstrup in 1930. Steenstrup became chief engineer of GE's electric refrigeration department and accumulated thirty-nine additional patents in refrigeration over the following years. By 1936, he had more than one hundred patents to his credit in refrigeration and other areas.

Further refinement of the refrigerator evolved with the development of Freon, a nonexplosive, nontoxic, and noncorrosive refrigerant discovered by Thomas Midgley, Jr., in 1928. Freon used lower pressures than ammonia did, which meant that lighter materials and lower temperatures could be used in refrigeration.

During the years following the introduction of the Monitor Top, the cost of refrigerators dropped from $1,000 in 1918 to $400 in 1926, and then to $170 in 1935. Sales of units increased from 200,000 in 1926 to 1.5 million in 1935.

Initially, refrigerators were sold separately from their cabinets, which commonly were used wooden iceboxes. Frigidaire began making its own cabinets in 1923, and by 1930, refrigerators that combined machinery and cabinet were sold.

Throughout the 1930's, refrigerators were well-insulated, hermetically sealed steel units that used evaporator coils to cool the food compartment. The refrigeration system was transferred from on top of to below the food storage area, which made it possible to raise the food storage area to a more convenient level. Special lightbulbs that produced radiation to kill taste- and odor-bearing bacteria were used in refrigerators. Other developments included sliding shelves, shelves in doors, rounded and styled cabinet corners, ice cube trays, and even a built-in radio.

The freezing capacity of early refrigerators was inadequate. Only a package or two of food could be kept cool at a time, ice cubes melted, and only a minimal amount of food could be kept frozen. The two-temperature refrigerator consisting of one compartment providing normal cooling and a separate compartment for freezing was developed by GE in 1939. Evaporator coils for cooling were placed within the refrigerator walls, providing more cooling capacity and more space for food storage. Frigidaire introduced a Cold Wall compartment, while White-Westinghouse introduced a Colder Cold system. After World War II, GE introduced the refrigerator-freezer combination.

**Consequences**

Audiffren, Wolf, Steenstrup, and others combined the earlier inventions of Watt, Perkins, and Carre with the development of electric motors to produce the electric refrigerator. The development of domestic electric refrigeration had a tremendous effect on the quality of home life. Reliable, affordable refrigeration allowed consumers a wider selection of food and increased flexibility in their daily consumption. The domestic refrigerator with increased freezer capacity spawned the growth of the frozen food industry. Without the electric refrigerator, households would still depend on unreliable supplies of ice.

*Garrett L. Van Wicklen*

# Antisyphilis Drug Salvarsan Is Introduced

> *Paul Ehrlich's Salvarsan was the first successful chemotherapeutic for the treatment of syphilis, ushering in a new age of medicine.*

**What:** Medicine; Health
**When:** April, 1910
**Where:** Frankfurt am Main, Hesse, Germany
**Who:**
PAUL EHRLICH (1854-1915), a German research physician and chemist
WILHELM VON WALDEYER (1836-1921), a German anatomist
FRIEDRICH VON FRERICHS (1819-1885), a German physician and professor
SAHACHIRO HATA (1872-1938), a Japanese physician and bacteriologist
FRITZ SCHAUDINN (1871-1906), a German zoologist

## The Great Pox

The ravages of syphilis on humankind are seldom discussed openly. A disease that struck all varieties of people and was transmitted by direct and usually sexual contact, syphilis was both feared and reviled. Many segments of society across all national boundaries were secure in their belief that syphilis was divine punishment of the wicked for their evil ways.

It was not until 1903 that bacteriologists Élie Metchnikoff and Pierre-Paul-Émile Roux demonstrated the transmittal of syphilis to apes, ending the long-held belief that syphilis was exclusively a human disease. The disease destroyed families, careers, and lives, driving its infected victims mad, destroying the brain, or destroying the cardiovascular system. It was methodical and slow, but in every case, it killed with singular precision. There was no hope of a safe and effective cure prior to the discovery of Salvarsan.

Prior to 1910, conventional treatment consisted principally of mercury or, later, potassium iodide. Mercury, however, administered in large doses, led to severe ulcerations of the tongue, jaws, and palate. Swelling of the gums and loosening of the teeth resulted. Dribbling saliva and the attending fetid odor also occurred. These side effects of mercury treatment were so severe that many preferred to suffer the disease to the end rather than undergo the standard cure. About 1906, Metchnikoff and Roux demonstrated that mercurial ointments, applied very early, at the first appearance of the primary lesion, were effective.

Once the spirochete-type bacteria invaded the bloodstream and tissues, the infected person experienced symptoms of varying nature and degree—high fever, intense headaches, and excruciating pain. The patient's skin often erupted in pustular lesions similar in appearance to smallpox. It was the distinguishing feature of these pustular lesions that gave syphilis its other name: the "Great Pox." Death brought the only relief then available.

## Poison Dyes

Paul Ehrlich became fascinated by the reactions of dyes with biological cells and tissues while a student at the University of Strasbourg under Wilhelm von Waldeyer. It was von Waldeyer who sparked Ehrlich's interest in the chemical viewpoint of medicine. Thus, as a student, Ehrlich spent hours at this laboratory experimenting with different dyes on various tissues. In 1878, he published a book that detailed the discriminate staining of cells and cellular components by various dyes.

Ehrlich joined Friedrich von Frerichs at the Charité Hospital in Berlin, where Frerichs allowed Ehrlich to do as much research as he wanted. Ehrlich began studying atoxyl in 1908, the year he won jointly with Metchnikoff the Nobel Prize in Physiology or Medicine for his work on immunity. Atoxyl was effective against trypanosome—a parasite responsible for a variety of infections, notably sleeping sickness—but also imposed serious side effects upon the patient, not the least of which was a tendency to cause blindness. It was Ehrlich's study of atoxyl, and several hundred derivatives sought as alternatives to atoxyl in trypanosome treatment, that led to the development of derivative 606 (Salvarsan). Although compound 606 was the first chemotherapeutic to be used effectively against syphilis, it was discontinued as an atoxyl alternative and shelved as useless for five years.

The discovery and development of compound 606 was enhanced by two critical events. First, the Germans Fritz Schaudinn and Erich Hoffmann discovered that syphilis is a bacterially caused disease. The causative microorganism is a spirochete so frail and gossameric in substance that it is nearly impossible to detect by casual microscopic examination; Schaudinn chanced upon it one day in March, 1905.

This discovery led, in turn, to German bacteriologist August von Wassermann's development of the now famous test for syphilis: the Wassermann test. Second, a Japanese bacteriologist, Sahachiro Hata, came to Frankfurt in 1909 to study syphilis with Ehrlich. Hata had studied syphilis in rabbits in Japan. Hata's assignment was to test every atoxyl derivative ever developed under Ehrlich for its efficacy in syphilis treatment. After hundreds of tests and clinical trials, Ehrlich and Hata announced Salvarsan as a "magic bullet" that could cure syphilis, at the April, 1910, Congress of Internal Medicine in Wiesbaden, Germany.

The announcement was electrifying. The remedy was immediately and widely sought, but it was not without its problems. A few deaths resulted from its use, and it was not safe for treatment of the gravely ill. Some of the difficulties inherent in Salvarsan were overcome by the development of neosalvarsan in 1912 and sodium salvarsan in 1913. Although Ehrlich achieved much, he fell short of his own assigned goal, a chemotherapeutic that would cure in one injection.

## Consequences

The significance of the development of Salvarsan as an antisyphilitic chemotherapeutic agent cannot be overstated. Syphilis at that time was as frightening and horrifying as leprosy and was a virtual sentence of slow, torturous death. Salvarsan was such a significant development that Ehrlich was recommended for a 1912 and 1913 Nobel Prize for his work in chemotherapy.

It was several decades before any further significant advances in "wonder drugs" occurred, namely, the discovery of prontosil in 1932 and its first clinical use in 1935. On the heels of prontosil—a sulfa drug—came other sulfa drugs. The sulfa drugs would remain supreme in the fight against bacterial infection until the antibiotics, the first being penicillin, were discovered in 1928; however, they were not clinically recognized until World War II (1939-1945). With the discovery of streptomycin in 1943 and Aureomycin in 1944, the assault against bacteria was finally on a sound basis. Medicine possessed an arsenal with which to combat the pathogenic microbes that for centuries before had visited misery and death upon humankind.

*Eric R. Taylor*

*Paul Ehrlich (left).*

# Union of South Africa Is Established

> *Two British colonies and two former Afrikaner republics in southern Africa came together to form the Union of South Africa, a self-governing independent nation that would be ruled by whites for eight decades.*

**What:** Political independence; Political reform
**When:** May 31, 1910
**Where:** South Africa
**Who:**

SIR ALFRED MILNER (1854-1925), British high commissioner for South Africa

HORATIO HERBERT KITCHENER (1850-1916), British commander in chief in the Boer War and negotiator for Great Britain after the war

JOSEPH CHAMBERLAIN (1836-1914), British secretary of state for the colonies

LOUIS BOTHA (1862-1919), prime minister of the Transvaal (1907-1910) and first prime minister of the Union of South Africa (1910-1919); cofounder of *Het Volk*

JAN CHRISTIAN SMUTS (1870-1950), state attorney for the Cape Colony and cofounder of *Het Volk*

## Afrikaner versus English

In the nineteenth century, the two most serious problems within South Africa were political unity among the whites (South Africans of English and Dutch descent) and what the whites called "native policy." The end of the South African, or Boer, War of 1899-1902 (what the Boers, or Dutch settlers, called the Second War of Freedom and the Africans called the White Man's War) left Great Britain with a few problems to resolve. Though the British had won the war, Africans (blacks and people of mixed race) outnumbered whites within South Africa, and the defeated Afrikaners (Boers) outnumbered whites of English origin. Leaders of both political parties in Great Britain agreed that established white communities in the British Empire should run their own affairs. How could British supremacy be established, then, within South Africa?

Great Britain's high commissioner for South Africa, Sir Alfred Milner, had a solution: The region should become "a self-governing white community, supported by well-treated and justly governed black labour from Cape Town to Zambesi." To make sure that white South Africans remained loyal to the British Empire, Milner proposed that large numbers of British people be encouraged to move to South Africa and share in the wealth of its natural resources (which included extensive gold fields). Meanwhile, the Afrikaners could be persuaded to loosen their ties to Dutch culture.

General Horatio Herbert Kitchener, who had been British commander in chief in the Boer War, and Joseph Chamberlain, secretary of state for the colonies, wanted to treat both the Afrikaners and the Africans more liberally. In the drafting of the 1902 Treaty of Vereeniging, Milner's views won out. Yet making his ideas a reality was not easy. There was an economic recession after the Boer War, and the great wave of British immigration did not occur. Furthermore, the terms of the treaty and the way British diplomacy was carried out encouraged the Afrikaners to be even more nationalistic—and more hostile to the British—than before. As a result, the British government changed its policy to create equality between settlers of British and Dutch origin.

## Forming the Union

Two Afrikaner leaders came to the fore during and just after the Boer War: Louis Botha and Jan Smuts. Botha was a farmer, while Smuts was an intellectual lawyer. Together, however, they

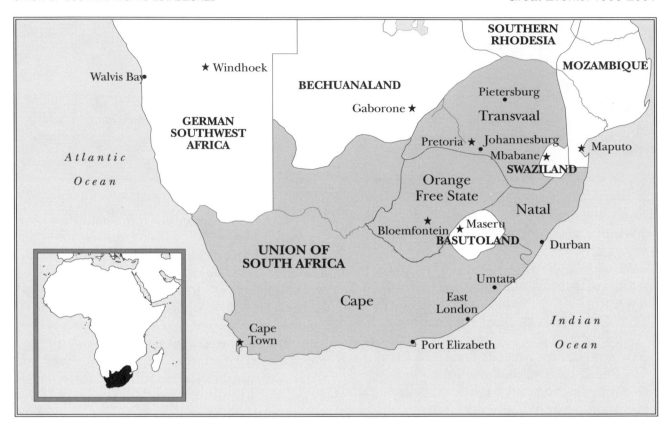

founded a policy of *Het Volk* (the people), which stood for Afrikaner self-government along with four levels of conciliation within South Africa. First, Afrikaners were to be reconciled to one another, and second to citizens of British descent. Third, regional disputes would be solved by joining the British colonies of Cape and Natal with the Boer colonies of Transvaal and the Orange Free State. The fourth level, external conciliation, aimed for the formation of a South African nation within a liberalized British Empire.

By 1908, Botha and Smuts had almost convinced leaders of the South African colonies that conciliation would be politically wise and morally right. Yet for the majority of South Africa's people, *Het Volk* was a white solution to a white problem. The most liberal whites in the most liberal colony (the Cape) wanted South Africa to be run along the lines of the Cape's own system—giving voting rights without any stated racial restrictions, but under other restrictions that would ensure that the vast majority of those who qualified would be white men. Smuts responded by saying that all white parties should aim "to do justice to the Natives and to take all wise and pru-

dent measures for their civilization and improvement. But I don't believe in politics for them."

For many in the white minority the idea of union had become a top priority by 1907, for economic reasons and for self-preservation; whites wanted to build strength among themselves in order to resist rebellions among the Zulus and other African groups. Whites who were loyal to the British Empire wanted union to help lessen conflicts among the South African colonies; a unified and peaceful South Africa, they thought, might attract large numbers of British immigrants and fulfill Milner's dream of a British electoral majority. Those who were not favorable to the Empire, on the other hand, wanted union as a way for South Africa's white communities to stand together against interference from the British.

Eight years after the Peace of Vereeniging, the Union of South Africa was formed on May 31, 1910, as an independent dominion within the British Commonwealth of Nations. The four colonies—Natal, the Orange Free State, the Transvaal, and the Cape—joined together to create the Union of South Africa. Great Britain was left with the "native territories" of Bechuana-land, Basutoland, Swaziland, and Northern and Southern Rhodesia.

The Act of Union was geared to protecting the rights and privileges of the white minority. The question of distribution of power within the white community—not the issue of the rights of Africans—had been the most debated problem. Choosing a national language was part of this problem. The decision to treat English and Afri-kaans on an equal basis convinced the Afrikaners that their language and culture could be preserved within South Africa.

## Consequences

The unification and independence of South Africa came at a high price for the Africans of the region, for the Act of Union made white suprem-acy the rule in the government of the new nation. The political and economic price is still being paid by people of color in South Africa.

African political organizations date back to 1882, but the Act of Union gave the National Native Convention a strong push to begin more open protest. As a result, the African National Congress (ANC) was formed in 1912, led by lawyers and journalists such as J. T. Javabu, who had been trained in Great Britain and the United States. The ANC remained the chief voice of South African black nationalism until it was banned in 1960. It operated underground until 1990, when its leaders and the government of F. W. de Klerk began negotiations that would lead to a nonracial constitution and make it possible for an ANC government to be voted into power.

# Portugal Becomes Republic

*With Portugal in an economic crisis as it lost its colonial holdings, Portuguese radicals overthrew the monarchy and called an assembly to draft a republican constitution.*

**What:** Political reform; Coups
**When:** October 5, 1910
**Where:** Lisbon, Portugal
**Who:**
CARLOS I (1863-1908), king of Portugal from 1889 to 1908
MANUEL II (1889-1910), king of Portugal from 1908 to 1910
TEÓFILO BRAGA (1843-1924), provisional president of the Republic of Portugal from 1910 to 1911
AFONSO COSTA (1871-1937), minister of justice in 1910, who became leader of the radical Democratic Party
MANUEL JOSÉ DE ARRIAGA (1842-1917), first constitutional president of the Republic of Portugal, from 1911 to 1915
SIDÔNIO BERNARDINO CARDOSA DA SILVA PAES (1872-1918), a republican who led a coup in 1917, and became president of Portugal from 1917 to 1918
ANTONIO MACHADO DOS SANTOS, leader of the Carbonaria, a secret society of radical republicans

## Bringing Down the Monarchy

Since the late nineteenth century, the Portuguese government had had many problems. Most of the problems had to do with money. Since Portugal had lost control of Brazil it had been unable to balance its national budget, and civil war had been very costly. There had been attempts to develop Portugal's colonies in Africa so that they would be a source of income, but British achievements in South Africa had blocked those efforts.

Much social reform was needed within Portugal. The government was full of corruption, and the results of all elections were decided in advance at the capital, Lisbon. Small socialist and republican movements raised protests against this corruption, but most of the protest was aimed at the Roman Catholic clergy, which held quite a bit of power throughout the country. The republican movement, founded by Teófilo Braga, was especially marked by fiery opposition to the clergy.

Carlos I, who had become king of Portugal in 1889, could not cope with these problems, and his ministers could not solve them. By 1905, the clamor for reform led many people to believe that the monarchy would soon be overthrown, and in 1908, Carlos and his heir were assassinated. His second son became King Manuel II, but he was quite a young man and was unprepared to lead the country.

The republicans increased their attacks on the government. Together with the Carbonaria, a secret society of radicals led by Antonio Machado Dos Santos, they were mainly responsible for the overthrow of the government. These groups forced Manuel into exile and proclaimed the Republic of Portugal. Braga was named provisional president, and a constituent assembly was called to write a republican constitution.

## The Republic Struggles

From the beginning, the Republic of Portugal was plagued by problems. The Republican Party split into factions—a moderate group led by Machado and a radical group that separated and became the Democratic Party, led by Afonso Costa, the minister of justice in the new government. Political fighting between the Democrats and moderate Republicans took up much time and energy that might otherwise have been given to reform.

With the support of his fellow ministers, Costa began a massive campaign against the clergy even before the Cortes Gerais (parliament) drafted a constitution. The church and state were separated, clerical property was put under the control of the state, the Jesuits were expelled, all monastic orders were dissolved, and the state was given full control of all education. After this first rush of activity to limit the power of the clergy, the Republicans decided to defend the Church—and by doing so caused tensions to increase.

In 1911, a constitution was drafted and promulgated. It was a fairly moderate document that set up a republican parliamentary system. Strikes by labor were permitted—a right that was immediately seized by working people across the country, who rallied to protest the government's lack of action to reform labor and the larger social system. Those who favored a return to monarchy also began to mount protests, and plots to reestablish a king caused the Republicans much concern during the early years of the Republic of Portugal.

Manuel José de Arriaga was named president of the republic after the constitution was promulgated, but he was not able to provide political peace. The Portuguese people did not give wide support to their new form of government. In the first republican election in Lisbon, a city of 350,000 people, fewer than 20,000 voted. No party had a strong base of power. Costa's Democrats alternated in power with the Republicans, but the two parties gave most of their energy to debating whether the Church should be supported or further restricted, so that little attention was given to budget problems and needed social reforms.

## Consequences

As the feud between the moderate Republicans and the Democrats prevented them from addressing any other issue but that of the Church's position, the army began to gain support as the only institution with enough prestige and respectability to keep the country from going out of control. In 1916, the Portuguese government joined the Allies in declaring war on Germany, and Portuguese troops were sent to the Western Front. Yet the Portuguese people had no real grudge against Germany, and the soldiers were not well prepared for war. Tensions within the country increased.

In December, 1917, Major Sidônio Bernardino Cardosa da Silva led the army in overthrowing the government and establishing a dictatorship. This desperate act still did not solve Portugal's problems, and da Silva was assassinated the following year.

*José M. Sánchez*

# Boas Lays Foundations of Cultural Anthropology

> *Boas published his views on the differences between cultures, laying the foundations of modern cultural anthropology.*

**What:** Anthropology
**When:** 1911
**Where:** New York
**Who:**
FRANZ BOAS (1858-1942), the father of American anthropology

## The Fight Against Racism

When Franz Boas published *The Mind of Primitive Man* in 1911, it marked the first step in the battle Boas and his followers fought to push racism out of modern anthropology. As his ideas spread, they had a profound impact on Western civilization.

The title of the book does not do justice to its contents. The title of the German edition, *Kultur und Rasse* (culture and race), published in 1914, is better but still does not give a complete picture. In addition to discussing the mind of primitive people and the connections between culture and race, Boas writes of the connections between race, culture, and language; the importance of heredity; the need for anthropologists to look at the whole culture, keeping its history in mind; and the relationship between psychology and culture. Boas's ideas, which he developed over several decades of research and reflection, laid the foundations of modern cultural anthropology.

The ideas against which Boas and his followers were fighting began when Sir Edward Burnett Tylor—a wealthy Englishman and amateur scholar—published the book *Primitive Culture* in 1871. Tylor was the first to define anthropology as the study of human culture and culture as the knowledge and traditions gained by humans as members of society. He used the culture concept as a synonym for "civilization," which he understood as the common human heritage that becomes increasingly complex and refined as human beings move up on a ladder of progress from "primitive" to more advanced stages. Tylor was an evolutionist; that is, he believed that there were great differences between the people of a primitive culture and those of a "civilized" society.

Tylor, and others like him, did valuable research in cultures that had never been studied before, but the problem with their theory was that it allowed certain cultures to feel superior to others. This idea seemed to justify not only the colonization of land on which only "primitive" peoples lived but also the enslavement of those people. Yet, by fitting so well with the theory of evolution, it seemed a very convincing argument at the time.

Boas's ideas evolved from an experience he had in 1883. Born and educated in Germany, he became interested in the natural sciences and, after earning a doctorate in physics, he set out to do some geographical research on Baffin Island, in the Canadian Arctic. There, he met with the "primitive" Eskimos. What struck him about the Eskimos was the complexity and beauty of their customs. This experience called into question the superiority of Western "civilized" people and led him toward a career in anthropology. In 1899, having moved to the United States to study the Kwakiutl and other Native American groups, he accepted a professorship in anthropology at Columbia University and began to attract a large number of gifted students.

## Psychic Unity

The most controversial ideas in *The Mind of Primitive Man*, at least as far as the social evolutionists were concerned (not to mention the Nazis in Germany, who banned the book), were that racial purity does not exist and that there is no provable connection between race and intelligence or personality. This meant that, since genetics could not explain differences between cultures, these differences must come from the historical development of cultural traits within a particular culture. By so arguing, Boas changed the meaning of the word "culture." Rather than labeling civilization in a general way, "culture" for Boas came to mean the specific heritage of a particular social group, which must be understood only within its own history.

Finally, Boas insisted that humankind shares psychic unity. The evolutionists, on one hand, believed that there was a difference between the way of thinking of primitive people and that of civilized Westerners. Boas, on the other hand, believed that differences are produced only by the specific needs and goals of cultures. There are no mental notions that primitive people cannot possibly grasp; if the need for more sophisticated thought processes should arise, Boas believed, then the necessary mental operations would also appear.

## Consequences

Many of Boas's students, including Ruth Benedict and Margaret Mead, combined Boas's ideas on the relationships between culture and personality with the basic Freudian ideas on personality development and defined cultures on the basis of the type of "personality" they seemed to have. While it is generally accepted that Boas was the founder of American anthropology, it is also clear that some of the theories that were built on his work did not entirely agree with his original ideas. This happened not because his ideas were not complete but rather because many of Boas's students wanted to find answers to questions that he had specifically chosen to leave unanswered.

Boas spent little time building theories. Yet his books, such as *The Mind of Primitive Man*, do contain clear theoretical statements. Boas believed that to make anthropology into a science, it was more important to insist that certain methods be strictly followed than that a particular theory be followed. He also believed that good theory could be built only when enough research had been done to support it. Many of the ideas Boas presented in his book are now so taken for granted that their originality is not fully appreciated.

*E. L. Cerroni-Long*

Library of Congress

*Franz Boas.*

# Italy Annexes Tripoli

> *In 1911-1912, Italy was successful in annexing Tripoli, which had been under Turkish control.*

**What:** Military conflict; Political agression
**When:** 1911-1912
**Where:** Tripoli and Cyrenaica (part of modern Libya)
**Who:**
GIOVANNI GIOLITTI (1842-1928), prime minister of Italy from 1911 to 1914
CARLO CANEVA (1845-1922), commander in chief of the Italian expeditionary force to Tripoli in 1911
ALBERTO POLLIO (1852-1914), Italian chief of staff
ALFRED VON KIDERLEN-WÄCHTER (1852-1912), German foreign secretary beginning in 1910
COUNT ALOIS LEXA VON AEHRENTHAL (1854-1912), foreign minister of Austria-Hungary from 1906 to 1912
COUNT FRANZ CONRAD VON HÖTZENDORF (1852-1925), chief of staff of the Austro-Hungarian army from 1906 to 1911 and from 1912 to 1917

## The Italian Ambition

For thirty years, Italian leaders had been ambitious to establish an empire in North Africa. By 1902, Italy had obtained approval from Germany and Austria-Hungary, its partners in the Triple Alliance, and from France and Great Britain for an eventual move into Tripoli. Russia's approval was gained in 1909.

The Italians had several reasons for wanting to expand into North Africa. The reason they used most often was that their population was growing too large for their area, so that new territory was needed. In 1896, Italy had suffered an embarrassing disaster when it tried to conquer Ethiopia; one of its reasons for seeking to annex Tripoli, then, was its need to restore national pride and confidence. During the first decade of the twentieth century, Italy had been partially successful in making economic investments in Tripoli. When the Young Turks came to power in 1908, however, they put a stop to this economic advance.

By 1911, people throughout Italy were demanding that their fiftieth anniversary of national statehood be celebrated by the annexation of Tripoli. Giovanni Giolitti, the Liberal prime minister of Italy, saw no reasons to delay. In 1900, France had promised Italy a free hand in Tripoli, in exchange for Italy's approval of France's control over Morocco. France had achieved its goals in Morocco in 1911, when it had signed a treaty in which Germany had given up its rights to that country. Now Giolitti was ready to seize Tripoli as quickly as possible. The Italians believed that it was important to maintain a balance of power with the French in the Mediterranean area. They also feared that since Germany had been excluded from Morocco, the Germans would move to claim Tripolitan territories before Italy could do so.

## War and Conquest

On September 28, 1911, Italy put a twenty-four-hour ultimatum before the sultan of the Ottoman Empire, threatening war because Turkey had supposedly mistreated Italian citizens and blocked Italian investments. Though the sultan tried to pacify the Italians, they declared war on the Ottoman Empire on September 29. On the first day of formal hostilities, Italy bombarded Turkish ports on the Tripolitan coast.

Alfred von Kiderlen-Wächter, foreign secretary of the German Empire, tried to mediate between the two states but had no success. On November 5, expecting a quick victory, Giolitti had King Victor Emmanuel III of Italy proclaim the formal annexation of all Libya. Yet this proclamation did not mean victory. The Turks fought well, carrying on the struggle for more than a year.

**218**

Italian commander in chief Carlo Caneva and chief of staff Alberto Pollio made three serious mistakes in conducting the war. First, they imagined that the Turks would give up quickly; second, they did not expect that the Arabs would ally themselves with the Turks; third, because they thought they would be waging a conventional war, they took only twenty thousand Italian troops to Tripoli.

Because the Turks used guerrilla strategies, Caneva and Pollio were forced to expand their forces to 100,000 men. Although the Italians did not gain a major victory in Africa, they did manage, in May, 1912, to capture Rhodes and other islands of the Dodecanese group in the Aegean Sea, previously held by Turkey.

It was not Italian strength but the threat of war with the Balkan states that finally forced the Turks to give up the struggle. Peace negotiations were opened in July, 1912. In the Treaty of Lausanne, signed on October 18, 1912, the Italians gained a certain amount of control over Libya, but that region officially remained under the rule of the Turkish sultan.

The Great Powers of Europe had various responses to Italy's war against the Ottoman Empire. Though all the major European states had at one time or another given their blessing to Italy's ambitions to take over Tripoli, several of them were not altogether pleased with the war. Russia, which had the most to gain from a weakening of Turkey, was the only nation to give Italy full support. France supported Italy's action until January, 1912, when Italians boarded two French ships that they suspected of helping the Turkish war effort; after this incident, the French view of Italy's actions became less favorable. Great Britain expressed its concern that the war might lead to the complete collapse of Turkey and to Russia's taking complete control of the passageway between the Mediterranean and the Black Sea.

The strongest objections to Italy's war of conquest came from its allies, Germany and Austria-

*An Italian artillery desert outpost, in Tripoli, in December, 1911.*

Library of Congress

**219**

Hungary. William II, emperor of Germany, feared that Italy's action might lead to a world war. The foreign minister of Austria-Hungary, Count Alois Lexa von Aehrenthal, announced that his government would not let Italy move its war against the Turks into the Balkans. Yet Aehrenthal stood against the demand by Austro-Hungarian chief of staff Count Franz Conrad von Hötzendorf that Austria declare war on Italy. When Conrad insisted, Emperor Francis Joseph I relieved him temporarily of his command.

As the Tripolitan War was drawing to a close, the Balkan states of Serbia, Montenegro, Bulgaria, and Greece were preparing their own strike against the Ottoman Empire to try to gain more territory. The First Balkan War broke out in October, 1912.

## Consequences

Italy's annexation of Tripoli proved to be a very costly action. Though the Italians felt proud of having gained a colony in Africa, this new colony was expensive to maintain and difficult to control. The Italians who were emigrating from their homeland preferred the United States to Tripoli, which was essentially a desert.

The conquest of Tripoli also altered politics within Italy. The moderate parties that had managed to control the government now gave way before the criticism of new nationalist groups, which demanded that Italy continue to make new conquests and spread its influence abroad. Demands such as these continued to be made loudly throughout Italy during and after World War I.

# British Parliament Passes National Insurance Act

*With its provisions for health and unemployment insurance for manual laborers, the National Insurance Act of 1911 paved the way for the establishment of Britain's welfare state in the mid-1940's.*

**What:** Social reform

**When:** May-December, 1911

**Where:** London

**Who:**

HERBERT HENRY ASQUITH (1852-1928), Liberal prime minister of Great Britain from 1908 to 1916

DAVID LLOYD GEORGE (1863-1945), chancellor of the exchequer from 1908 to 1915

WINSTON CHURCHILL (1874-1965), president of the Board of Trade from 1908 to 1910, and home secretary from 1910 to 1911

## Laws for Social Reform

During the second half of the nineteenth century, the spread of the Industrial Revolution in Great Britain and the extension of voting rights combined to change British people's attitudes about the causes of and cures for poverty. Poverty came to be seen as the result of unemployment and other kinds of social and economic change caused by the ups and downs of various industries.

The Poor Law Administration, which up until this time had governed state aid to the poor in Great Britain, came under increasing attack, for it was seen as inadequate in dealing with poverty in an industrial nation. As the working class gained a political voice in the country, a "New Liberalism" developed at the beginning of the twentieth century, combining older Liberal thinking with socialism. A coalition of Liberal and Labour Members of Parliament was able to pass many laws for social reform between 1906

and 1911: the Old Age Pensions Act, the Labour Exchanges Act, the Trade Disputes Act, the Miners' Eight-Hour-Day Act, and the Trade Boards Act.

In 1906, before leaving office, the Conservative government had given a royal commission permission to study the Poor Law Administration to see how it should be reformed and updated. The commission also was to make recommendations about whether a state system of sickness insurance should be established.

Without waiting for the commission's final report, in 1908, David Lloyd George, chancellor of the exchequer, took a small group of officials with him to tour Germany, where there was already a government program for meeting the health-care needs of the poor. Within a year, Lloyd George instructed Treasury officials to write a bill for social insurance. This bill was to set up a national, compulsory system of insurance, funded by contributions from the employee, the employer, and the state. It would make use of the "friendly societies" that already existed. Sponsored by fraternal organizations and labor unions, these societies sold sickness and accident insurance to more than six million Britons.

## The Bill

In May, 1911, the National Health Insurance Bill was introduced in the House of Commons by the Liberal prime minister, Herbert Henry Asquith. It provided insurance benefits for all employed manual laborers, male and female, between the ages of sixteen and sixty-five, who earned less than 160 pounds per year. The sickness and medical benefits included ten shillings per week sickness pay, payment of doctors' fees, medicines, and special benefits such as thirty

Bain Collections, Library of Congress

*The English elections: an open-air reciting at the Portsmouth docks.*

shillings for the expenses of pregnancy and delivery, and sanatorium treatment for those who suffered from tuberculosis. These benefits would be financed by fourpence deducted from the weekly pay of each employee, threepence per week contributed by employers for each employee, and twopence from the government for each employee.

The bill's second part set up a state system of unemployment insurance. Such a system had not been established by any other modern nation. This part of the bill was largely the work of Winston Churchill, who was president of the Board of Trade from 1908 to 1910, and William Henry Beveridge, a young government worker who belonged to the Fabian Society, a socialist organization.

Because the bill was an experiment, and because insuring against industrial unemployment was risky, part 2 of the bill was at first restricted to

building and engineering—trades in which employment was most subject to change. Like the sickness-insurance system, it was to be funded by compulsory contributions from the worker, the employer, and the government. The worker would be assured of receiving seven shillings per week as an unemployment payment for up to fifteen weeks.

The bill became controversial during long weeks of heated debate from May to December, 1911. The Labour Party was seeking more far reaching changes—an extensive state-run institution to care for health needs, and laws to guarantee the right to work. The small "friendly societies" feared interference by both the government and the large industrial insurance companies, which for political reasons Lloyd George had been forced to accept as "approved" societies.

It was British doctors, however, who raised the loudest objections. They were angry that Lloyd

**222**

George planned to continue the "club practice," in which friendly societies contracted for the doctors' services. A "doctors' revolt" organized by the British Medical Association, an organization with quite a bit of political power, was finally ended when Lloyd George agreed to several of the doctors' demands: He approved free choices of doctors, the end of club practice, and increases in certain fees paid to the doctors.

In response to pressure from Scottish and Irish leaders, the plan was set up to be run by four independent "national insurance committees" for England, Ireland, Scotland, and Wales, but with a central National Health Insurance Joint Committee. In 1911, advocates for health insurance believed that this compromise was a mistake that would make the program ineffective.

With these various compromises and amendments, the National Insurance Act became law.

## Consequences

The goal of the National Insurance Act of 1911 was to improve the living conditions of the working classes and, through a combination of government help and self-help, provide welfare services and income during periods of unemployment and sickness. The act was limited in its effectiveness: The insurance benefits were not generous, workers' families were not covered by the insurance, and hospital stays were not covered.

Yet the system was not designed to meet all the needs that existed, but to test whether such state-run social welfare programs could be made to work. The National Insurance Act created the first state-supported insurance in Great Britain, and it prepared the way for the much more extensive social welfare laws of the 1940's.

*H. Christian Thorup*

# Bingham Discovers Lost Inca City of Machu Picchu

*Hiram Bingham discovered Machu Picchu, a previously unknown Inca city in the Andean jungle, extending the known range of Inca settlement and firing the public's imagination.*

**What:** Archaeology
**When:** July 24, 1911
**Where:** Machu Picchu, Peru
**Who:**
HIRAM BINGHAM (1875-1956), a Yale University professor who discovered Machu Picchu
MELCHOR ARTEAGA, a Quechua Indian farmer who guided Bingham to Machu Picchu
CARRASCO, a Peruvian military policeman who accompanied Bingham's expedition

## A Fallen Empire

The great Inca Empire conquered by Francisco Pizarro and the Spaniards was centered in the Andean highlands of Peru, but the edges of the empire extended as far north as southern Colombia and as far south as central Chile, from the Pacific Ocean to the edges of the Brazilian jungle. Despite being only a small part of the empire, the jungle played an interesting part in the history of the Inca. History told of Vitcos and Vilcabamba, the jungle towns where the Inca who rebelled against the Spaniards fled in 1539 and where they held out until 1572. These fabled places fired Hiram Bingham's imagination early in his career; his greatest triumph was accomplished during his attempts to locate these places.

From the beginning, Bingham was more of an explorer in the nineteenth century mold than a scientist of the twentieth century type. He spent most of his efforts organizing teams and leading expeditions, leaving others to perform the de-

tailed collection, analysis, and interpretation. He received his Ph.D. in history from Harvard University, but he had virtually no archaeological training when he began his series of South American explorations. His expedition consisted initially of Professor Harry Foote (a naturalist) and Doctor William Erving (a surgeon), although they would be joined later by Sergeant Carrasco, a military policeman assigned to them as an escort by the Peruvian president.

In early July, 1911, this group set off, heading down the lush valley of the Urubamba River as it rushed from the high Andes into the jungle below. The deep, winding gorge flanked by dense, tropical vegetation made direct observation and search for ruins difficult. Bingham relied primarily on others to lead him to where there were ruins worthy of his attention. By the evening of July 23, the party had reached an open plain called Mandor Pampa, where, while pitching camp, they encountered a local Indian named Melchor Arteaga. Arteaga told of ruins on the tops of twin mountains nearby, Huayna Picchu and Machu Picchu. Bingham decided to visit these sites the following day.

## The "Lost City"

The next morning broke with a dreary, cold drizzle that was to last all day, and only Bingham retained his interest in pressing on to the ruins Arteaga mentioned. In fact, Foote and Erving decided to stay in camp, and Arteaga had to be bribed with triple wages to guide Bingham to the ruins. Finally, Bingham, Arteaga, and Carrasco set out on a several-hour journey that forced them to scale cliffs, cross primitive bridges over the cascading Urubamba, and carve their way through dense jungle vegetation. Ultimately,

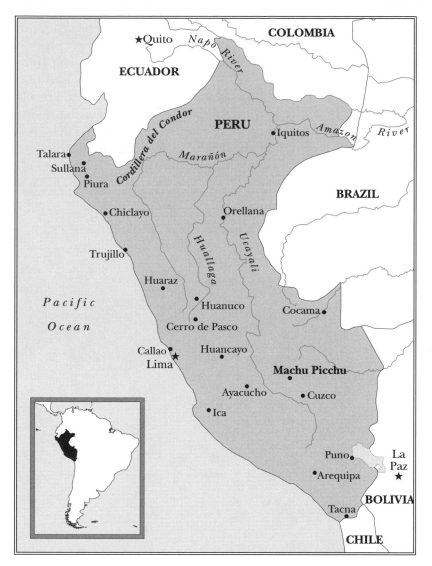

At the top of a steep stone stairway was a cluster of buildings of various styles. Some were temples, many of which had strangely carved rocks within. One was a tower built of huge, carefully decorated stones (now usually interpreted as an observatory), whose circular shape was most unlike typical Incan architecture. Palaces, also made of these large, decorated stones, were found next to the temples and observatory. In another area were modest gabled houses with distinctive trapezoidal doorways, minus their thatched roofs but otherwise largely intact. There also were other gabled buildings, much like the houses but with walls on only three sides, resembling historic buildings where the Inca made *chicha*, a local corn beer. Complex water systems channeled artificial streams through the settlement and produced ornamental streams and fountains. Plazas were marked by open spaces between clusters of buildings. Various types of architecture were scattered throughout the site, some familiar and some exotic, some impressive and some ordinary, but all documenting the complexity and importance of the site.

they found their way to the farm of Richarte and Alvarez, two Quechuan Indian farmers who were using ancient Incan terraced slopes for their plantings. After a rest, the party forged on a short distance to the site of Machu Picchu.

Nearly one hundred ancient agricultural terraces crawled up the slope toward the top of the lower of the two peaks, Machu Picchu. Above the terraces, near the top of the mountain, were dozens of stone buildings, some large and others more modest. Bingham immediately recognized their tremendous importance, though they were overgrown with vegetation. Investigations the following year at the larger peak, Huayna Picchu, revealed a small but significant group of ruins at its top, but Bingham was far more impressed with Machu Picchu.

## Consequences

Bingham was convinced that the had discovered the last capital of the Incas, the site that he had set out to find. He interpreted some of the architecture at the site as part of major defensive works, something to be expected at the last Inca capital. Other archaeologists, however, have not agreed with this interpretation. Although it is clear that the site's location on the top of a mountain would have aided defenders in an attack, there is no clear evidence of defensive works on the scale that Bingham believed.

Refinements in dating techniques have since ruled out the possibility that Machu Picchu was

**225**

the last capital of the Incas. Machu Picchu probably never housed more than a thousand people. It is widely held that a small force of workers and maintenance population lived there year-round, while nobles or other elite came there seasonally for recreation.

Bingham's discovery of Machu Picchu had a profound effect on the public. Bingham's expedition was like something out of a Jules Verne novel, with archaeological treasures and primitive splendor being discovered in the jungle.

Bingham had an engaging style of writing, and his association with the National Geographic Society assured that his story would appear. In an era of literacy, readily available books, and photography, he brought his story before the public with a vividness that sparked the imagination in a way that would not be equaled until the discovery of the tomb of the ancient Egyptian king Tutankhamen in Egypt some ten years later.

*Russell J. Barber*

# British Parliament Limits Power of House of Lords

*Having reached deadlock over a national budget written by members of Great Britain's Liberal Party, the Houses of Parliament passed the Parliament Bill, which put strict limits on the House of Lords' power of veto.*

**What:** Law; Political reform
**When:** August 10, 1911
**Where:** London
**Who:**
HERBERT HENRY ASQUITH (1852-1928), Liberal prime minister of Great Britain from 1908 to 1916
DAVID LLOYD GEORGE (1863-1945), chancellor of the exchequer from 1908 to 1915
ARCHIBALD PHILIP PRIMROSE, EARL OF ROSEBERY (1847-1929), a Liberal leader in the House of Lords
HENRY CHARLES KEITH PETTY-FITZMAURICE, MARQUIS OF LANSDOWNE (1845-1927), a Conservative leader in the House of Lords
ARTHUR JAMES BALFOUR (1848-1930), a Conservative leader in the House of Commons
EDWARD VII (1841-1910), king of Great Britain from 1901 to 1910
GEORGE V (1865-1936), king of Great Britain from 1910 to 1936

## The Budget Battle

The Liberal Party in Great Britain was the majority party in the House of Commons and was led by Herbert Henry Asquith, the prime minister. In 1909, however, though the Liberals had been elected by a landslide vote in 1906, they could not set up their program of new laws without interference or veto from the House of Lords, which was mostly Conservative. A number of important bills passed by the Commons had been thrown out by the Lords. Members of the House of Commons, being answerable to the voters, knew that something needed to change if they were to accomplish any real work.

In planning the national budget of 1909, Asquith and his cabinet faced the unpopular, but necessary, task of raising taxes and finding other sources of income for the government. An extra sixteen million pounds sterling were needed to pay old-age pensions; Great Britain needed to try to stay ahead of Germany in building warships; and the Liberals wanted to begin a new program of government aid to the poor.

David Lloyd George, chancellor of the exchequer, knew that the budget he had put together would upset many Conservatives. Taxes would be raised on liquor and tobacco, there would be new taxes on the sale of gasoline and cars, and income tax and death taxes would increase. The budget also proposed a new "supertax" on incomes over five thousand pounds and other new taxes on land and minerals. Many members of the House of Lords were wealthy landowners who would see these new taxes as a declaration of war against the rich by the Liberal government.

This "People's Budget" did lead to a long struggle between the Houses of Parliament in 1910 and 1911. From April to November, 1909, Parliament was filled with debates on the budget. To go into effect, the budget would have to be passed not only by the House of Commons but also by the House of Lords—and the Lords were determined to exercise their veto. The Conservative leader of the House of Commons, Arthur James Balfour, led fellow Conservatives in opposing the budget without room for compromise. The whole country was engaged in the argument, and some Conservatives organized a Budget Protest League that drew thousands of supporters. Meanwhile, Lloyd George and his young aide Winston Churchill traveled throughout the country to criticize the House of Lords as an old-fashioned "rich men's club."

Many of the Lords believed the new budget to be unconstitutional. To pass it would be politically embarrassing to them, but if they rejected it, the Liberal government would dissolve and the question of the Lords' power of veto would become a political issue fought at the polls. The Lords argued that in rejecting the budget, they were only exercising their ancient rights. The Liberals said that the budget was not included among the Lords' ancient rights, and claimed that the Lords had not rejected a budget for more than 250 years.

### Votes and Negotiations

On November 30, 1909, the Lords voted down the budget 350 to 75. The House of Commons

immediately declared that the rejection of the budget was "a break of the Constitution and a usurpation of the rights of the Commons," and Asquith dissolved his cabinet.

Now matters were in the hands of the voters. In January, 1910, the Liberals were voted back into power. Though their majority in the House of Commons was not quite as large as before, Asquith was determined to keep the House of Lords from imposing its will. The powers of the Lords needed to be cut back if the Liberal program of social reform was to go forward.

In February, 1910, the government had introduced the Parliament Bill, which would severely restrict the Lords' power of veto. Under this bill, money bills would become law when passed by the House of Commons even without the consent of the House of Lords. Other bills, if passed by the Commons in three consecutive sessions and rejected by the Lords, would become law if

Bain Collection, Library of Congress

*Men pause to read election posters on the pavement in front of a polling place.*

two years had passed between the first and third votes by the Commons. Also, the maximum life of a Parliament would be reduced from seven to five years.

As the battle over the Lords' powers began, the unexpected death of King Edward VII on May 6, 1910, suddenly silenced the arguments. The new king, George V, asked Asquith and Balfour to try to settle their differences in private conferences. They met privately many times during a six-month period but failed to resolve the deadlock, and when the debate returned to Parliament, Liberals and Conservatives were still in strong disagreement.

For the second time in less than a year, Asquith dissolved his cabinet and put the issue of the Lords' veto power directly before the voters. The Liberals again won a clear majority.

Now Asquith warned the Lords that if they rejected the Parliament Bill again, he would fill the House of Lords with new Liberal peers. By this threat, he revealed that King George V had promised to use his royal privileges to name as many new peers as necessary to pass the Parliament Bill through the House of Lords.

Ignoring the threat, the Lords insisted on amending the Parliament Bill. The Marquis of Lansdowne (Henry Charles Keith Petty-Fitzmaurice), the Lords' Conservative leader, then received a warning from Asquith: The amendments were unacceptable, and he was ready to ask the king to fulfill his promise to create new Liberal peers. Lansdowne backed down and persuaded the Lords to accept their defeat.

## Consequences

The constitutional crisis ended on August 10, 1911, when the Parliament Bill squeezed through the House of Lords by 131 votes to 114, with most of the peers abstaining. Great Britain's government was thereafter more fully in the hands of the people's elected representatives, rather than those who had inherited titles of nobility. In 1999 further reforms removed most hereditary peers from the House of Lords altogether.

*Henry G. Weisser*

# France and Germany Sign Treaty on Morocco

> *Long-standing conflicts over European interests in Morocco were finally resolved when Germany recognized France's protectorate over Morocco, but many Germans were unhappy with the compromise.*

**What:** International relations
**When:** November 4, 1911
**Where:** Morocco and Germany
**Who:**

WILLIAM II (1859-1941), emperor of Germany from 1888 to 1918

BERNHARD VON BÜLOW (1849-1929), prince and chancellor of Germany from 1900 to 1909

THEOBALD VON BETHMANN-HOLLWEG (1856-1921), chancellor of Germany from 1909 to 1917

FRIEDRICH VON HOLSTEIN (1837-1909), undersecretary of the German foreign ministry from 1905 to 1906

ALFRED VON KIDERLEN-WÄCHTER (1852-1912), German foreign secretary beginning in 1910

## A Long Dispute

In 1904, Great Britain and France formed an agreement called the Entente Cordiale. Great Britain received primary rights to establish its influence in Egypt, while France was given similar rights in Morocco. The French also persuaded Italy and Spain to agree to French supremacy in Morocco. Leaders of Germany, however, were quite alarmed about this agreement. Germany had ambitions to spread its influence in North Africa as well, but France and Great Britain were not sympathetic to these ambitions.

The German response to the British and French agreement on dividing their influence in North Africa was led by Friedrich von Holstein, a politician who had been active in Germany's foreign affairs since the 1870's; in 1905 he became undersecretary in the German Foreign Ministry. Following Holstein's plan, Kaiser William II set off to call on the Sultan of Morocco, and on March 31, 1905, in a bombastic speech in Tangier, he announced that Germany was interested in supporting Moroccan independence.

Chancellor Bernhard von Bülow then suggested publicly that Germany might go to war with France over Morocco. The Germans had decided that this threat was a safe one to make, for they believed that Russia, which was then at war with Japan, would not be able to come to France's aid, though Russia and France were allies under the terms of the Double Entente of 1894. France agreed to Germany's demand for an international conference to decide what was to be done about Morocco. The French people were angry that their leaders had given in to Germany on this matter, but Bülow was later made a prince for his handling of the situation.

The international conference was held at Algeciras, Spain, from January to April, 1906. At this conference, no other nation supported Germany's position. Even Austria-Hungary, Germany's partner in the Triple Alliance, was not interested in promoting German influence over Morocco. Spain was not very involved in the discussions, while Italy sided with France. Russia, no longer in the war with Japan, supported France, as did the United States. At the conference's end, the Act of Algeciras upheld France's right to colonize Morocco, and the end result was that the alliance between France and Great Britain became all the stronger. The conference's effect had been opposite to what the Germans had intended, and as a result Undersecretary Holstein was fired by William II.

In 1907, discussions between Britain and Russia brought an end to the quarrels they had had over control of certain regions of Asia. The Triple Entente of Great Britain, France, and Russia was growing up to stand against Germany. In 1909, Germany signed an agreement with France to divide their influence in Morocco: France would have a free hand in the north of that country, while Germany would be restricted to the south. France also promised to help Germany build a railroad from the German Cameroons to East Africa, across the French and Belgian Congos. Yet that help was never forthcoming.

## Conflict and Compromise

Bülow resigned as chancellor in 1909, and was replaced by Theobald von Bethmann-Hollweg, who appointed Alfred von Kiderlen-Wächter as foreign secretary in the summer of 1910. It was Kiderlen-Wächter who handled the next Moroccan crisis, in 1911. He reasoned that the government needed to achieve some important success in colonization abroad, to distract the German people's attention from the growth of the Social Democratic Party in the German Reichstag, or parliament. Germany should seize a couple of ports in Morocco and insist that France agree to new terms there. William II approved the plan.

Because of uprisings in Morocco, the French moved into the interior of the country and occupied Fez on May 21, 1911. This intrusion contradicted the terms of the Algeciras agreement. Now Germany had its opportunity to assert equal claim to Morocco with France. The German gunboat *Panther* was sent to Agadir Harbor, and within Germany, the government started a campaign to make the Moroccan cause a popular one. Almost the whole country, except the Social Democrats, joined to call for German protection of Morocco's mineral ores.

Yet other European nations were even more opposed to German aims than they had been in 1905. Austria-Hungary did not come to its ally's aid. British prime minister David Lloyd George gave his famous Mansion House speech, promising that his nation would stand in solidarity with France over Morocco. In August, 1911, the British, Russians, and French met for discussions about possible military cooperation.

On July 15, 1911, Kiderlen-Wächter had demanded the whole of the French Congo (French Equatorial Africa with its capital, Brazzaville) in exchange for Germany's renunciation of all claims to Morocco. Two days later, the French prime minister rejected this demand. Meanwhile, Kaiser William II began to back down from Kiderlen-Wächter's threats. A council of advisers told him that war over Morocco would not be a good idea, since Germany was still trying to build up its navy, and since alliances among the other European powers were strong. William instructed Bethmann-Hollweg to bring about a negotiated settlement—a decision that was much attacked by some German business groups.

On November 4, 1911, representatives of France and Germany signed a treaty on Morocco. Germany recognized the French protectorate over Morocco and was allowed equal opportunity in that country. About 270,000 square kilometers in the French Congo, with more than a million inhabitants, were transferred to the German Cameroons.

## Consequences

During the crisis, the people of Germany had, for the most part, become upset at what they saw as Germany's loss of power. Great Britain, which remained loyal to France, was criticized for trying to stop Germany's colonial and economic development. The Germans were not satisfied with the exchange of Morocco for land in central Africa, and many wanted war. Just after the signing of the treaty with France, there were furious debates in the Reichstag, and Bethmann-Hollweg and Kiderlen-Wächter were bitterly attacked.

In a speech on November 9, 1911, Bethmann-Hollweg defended his actions, arguing that Germany had won "free competition" in the Moroccan mines and had made secure its openings to the Congo and Ubangi rivers in central Africa. The treaty, he said, had improved relations between Germany and Great Britain.

Many German political parties and large sections of the press still were not satisfied. Insisting that Germany had been insulted by the outcome of the Moroccan affair, they called on Kaiser William and his government to prepare for war in the future. These attitudes among the German people were certainly a factor in the outbreak of World War I in 1914.

# First Diesel Locomotive Is Tested

*Successful testing of the first diesel locomotive led to the development of the diesel engines used to power modern trains, automobiles, ships, and factories.*

**What:** Transportation
**When:** 1912
**Where:** Winterthur, Switzerland
**Who:**
RUDOLF DIESEL (1858-1913), a German engineer and inventor
SIR DUGOLD CLARK (1854-1932), a British engineer
GOTTLIEB DAIMLER (1834-1900), a German engineer
HENRY FORD (1863-1947), an American automobile magnate
NIKOLAUS OTTO (1832-1891), a German engineer and Daimler's teacher

## A Beginning in Winterthur

By the beginning of the twentieth century, new means of providing society with power were needed. The steam engines that were used to run factories and railways were no longer sufficient, since they were too heavy and inefficient. At that time, Rudolf Diesel, a German mechanical engineer, invented a new engine. His diesel engine was much more efficient than previous power sources. It also appeared that it would be able to run on a wide variety of fuels, ranging from oil to coal dust. Diesel first showed that his engine was practical by building a diesel-driven locomotive that was tested in 1912.

In the 1912 test runs, the first diesel-powered locomotive was operated on the track of the Winterthur-Romanston rail line in Switzerland. The locomotive was built by a German company, Gesellschaft für Thermo-Lokomotiven, which was owned by Diesel and his colleagues. Immediately after the test runs at Winterthur proved its efficiency, the locomotive—which had been designed to pull express trains on Germany's Berlin-Magdeburg rail line—was moved to Berlin and put into service. It worked so well that many additional diesel locomotives were built. In time, diesel engines were also widely used to power many other machines, including those that ran factories, motor vehicles, and ships.

## Diesels, Diesels Everywhere

In the 1890's, the best engines available were steam engines that were able to convert only 5 to 10 percent of input heat energy to useful work. The burgeoning industrial society and a widespread network of railroads needed better, more efficient engines to help businesses make profits and to speed up the rate of transportation available for moving both goods and people, since the maximum speed was only about 48 kilometers per hour. In 1894, Rudolf Diesel, then thirty-five years old, appeared in Augsburg, Germany, with a new engine that he believed would demonstrate great efficiency.

The diesel engine demonstrated at Augsburg ran for only a short time. It was, however, more efficient than other existing engines. In addition, Diesel predicted that his engines would move trains faster than could be done by existing engines and that they would run on a wide variety of fuels. Experimentation proved the truth of his claims; even the first working motive diesel engine (the one used in the Winterthur test) was capable of pulling heavy freight and passenger trains at maximum speeds of up to 160 kilometers per hour.

By 1912, Diesel, a millionaire, saw the wide use of diesel locomotives in Europe and the United States and the conversion of hundreds of ships to diesel power. Rudolf Diesel's role in the story ends here, a result of his mysterious death in 1913—believed to be a suicide by the authori-

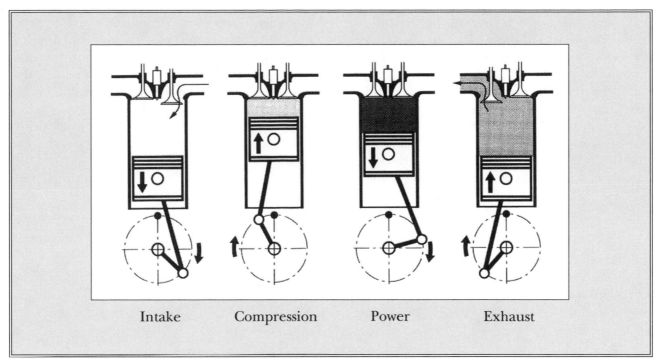

| Intake | Compression | Power | Exhaust |

*The four strokes of a diesel engine.*

ties—while crossing the English Channel on the steamer *Dresden*. Others involved in the continuing saga of diesel engines were the Britisher Sir Dugold Clerk, who improved diesel design, and the American Adolphus Busch (of beer-brewing fame), who bought the North American rights to the diesel engine.

The diesel engine is related to automobile engines invented by Nikolaus Otto and Gottlieb Daimler. The standard Otto-Daimler (or Otto) engine was first widely commercialized by American auto magnate Henry Ford. The diesel and Otto engines are internal-combustion engines. This means that they do work when a fuel is burned and causes a piston to move in a tight-fitting cylinder. In diesel engines, unlike Otto engines, the fuel is not ignited by a spark from a spark plug. Instead, ignition is accomplished by the use of high-temperature compressed air.

In common "two-stroke" diesel engines, pioneered by Sir Dugold Clerk, a starter causes the engine to make its first stroke. This draws in air and compresses the air sufficiently to raise its temperature to 900 to 1,000 degrees Fahrenheit. At this point, fuel (usually oil) is sprayed into the cylinder, ignites, and causes the piston to make its second, power-producing stroke. At the end of

that stroke, more air enters as waste gases leave the cylinder; air compression occurs again; and the power-producing stroke repeats itself. This process then occurs continuously, without restarting.

**Consequences**

Proof of the functionality of the first diesel locomotive set the stage for the use of diesel engines to power many machines. Although Rudolf Diesel did not live to see it, diesel engines were widely used within fifteen years after his death. At first, their main applications were in locomotives and ships. Then, because diesel engines are more efficient and more powerful than Otto engines, they were modified for use in cars, trucks, and buses.

At present, motor vehicle diesel engines are most often used in buses and long-haul trucks. In contrast, diesel engines are not as popular in automobiles as Otto engines, although European automakers make much wider use of diesel engines than American automakers do. Many enthusiasts, however, viewed diesel automobiles as the wave of the future. This optimism is based on the durability of the engine, its great power, and the wide range and econom-

**233**

ical nature of the fuels that can be used to run it. The drawbacks of diesels include the unpleasant odor and high pollutant content of their emissions.

Modern diesel engines are widely used in farm and earth-moving equipment, including balers, threshers, harvesters, bulldozers, rock crushers, and road graders. The recent construction of the Alaskan oil pipeline relied heavily on equipment driven by diesel engines. Diesel engines are also commonly used in sawmills, breweries, coal mines, and electric power plants.

Diesel's brainchild has become a widely used power source, just as he predicted. It is likely that the use of diesel engines will continue and will expand, as the demands of energy conservation require more efficient engines and as moves toward fuel diversification require engines that can be used with various fuels.

*Sanford S. Singer*

# Fischer Discovers Chemical Process for Color Films

> *Rudolf Fischer discovered a chemical process that paved the way for modern color film.*

**What:** Photography
**When:** 1912
**Where:** Berlin, Germany
**Who:**
RUDOLF FISCHER (1881-1957), a German chemist
H. SIEGRIST (1885-1959), a German chemist and Fischer's collaborator
BENNO HOMOLKA (1877-1949), a German chemist

## The Process Begins

Around the turn of the twentieth century, Arthur-Louis Ducos du Hauron, a French chemist and physicist, proposed a tripack (three-layer) process of film development in which three color negatives would be taken by means of superimposed films. This was a subtractive process. (In the "additive method" of making color pictures, the three colors are added in projection—that is, the colors are formed by the mixture of colored light of the three primary hues. In the "subtractive method," the colors are produced by the superposition of prints.) In Ducos du Hauron's process, the blue-light negative would be taken on the top film of the pack; a yellow filter below it would transmit the yellow light, which would reach a green-sensitive film and then fall upon the bottom of the pack, which would be sensitive to red light. Tripacks of this type were unsatisfactory, however, because the light became diffused in passing through the emulsion layers, so the green and red negatives were not sharp.

To obtain the real advantage of a tripack, the three layers must be coated one over the other so that the distance between the blue-sensitive and red-sensitive layers is a small fraction of a thousandth of an inch. Tripacks of this type were suggested by the early pioneers of color photography, who had the idea that the packs would be separated into three layers for development and printing. The manipulation of such systems proved to be very difficult in practice. It was also suggested, however, that it might be possible to develop such tripacks as a unit and then, by chemical treatment, convert the silver images into dye images.

## Fischer's Theory

One of the earliest subtractive tripack methods that seemed to hold great promise was that suggested by Rudolf Fischer in 1912. He proposed a tripack that would be made by coating three emulsions on top of one another; the lowest one would be red-sensitive, the middle one would be green-sensitive, and the top one would be blue-sensitive. Chemical substances called "couplers," which would produce dyes in the development process, would be incorporated into the layers. In this method, the molecules of the developing agent, after becoming oxidized by developing the silver image, would react with the unoxidized form (the coupler) to produce the dye image.

The two types of developing agents described by Fischer are paraminophenol and paraphenylenediamine (or their derivatives). The five types of dye that Fischer discovered are formed when silver images are developed by these two developing agents in the presence of suitable couplers. The five classes of dye he used (indophenols, indoanilines, indamines, indothiophenols, and azomethines) were already known when Fischer did his work, but it was he who discovered that the photographic latent image could be used to promote their formulation from "coupler" and "developing agent." The indoaniline and azomethine types have been found to possess the necessary properties, but the other three suffer from serious defects. Because only p-phenylenediamine and its derivatives can be used to form the indoaniline and azomethine dyes, it has become the most widely used color developing agent.

## Consequences

In the early 1920's, Leopold Mannes and Leopold Godowsky made a great advance beyond the Fischer process. Working on a new process of color photography, they adopted coupler development, but instead of putting couplers into the emulsion as Fischer had, they introduced them during processing. Finally, in 1935, the film was placed on the market under the name "Kodachrome," a name that had been used for an early two-color process.

The first use of the new Kodachrome process in 1935 was for 16-millimeter film. Color motion pictures could be made by the Kodachrome process as easily as black-and-white pictures, because the complex work involved (the color development of the film) was done under precise technical control. The definition (quality of the image) given by the process was soon sufficient to make it practical for 8-millimeter pictures, and in 1936, Kodachrome film was introduced in a 35-millimeter size for use in popular miniature cameras.

Soon thereafter, color processes were developed on a larger scale and new color materials were rapidly introduced. In 1940, the Kodak Re-search Laboratories worked out a modification of the Fischer process in which the couplers were put into the emulsion layers. These couplers are not dissolved in the gelatin layer itself, as the Fischer couplers are, but are carried in small particles of an oily material that dissolves the couplers, protects them from the gelatin, and protects the silver bromide from any interaction with the couplers. When development takes place, the oxidation product of the developing agent penetrates into the organic particles and reacts with the couplers so that the dyes are formed in small particles that are dispersed throughout the layers.

In one form of this material, Ektachrome (originally intended for use in aerial photography), the film is reversed to produce a color positive. It is first developed with a black-and-white developer, then reexposed and developed with a color developer that recombines with the couplers in each layer to produce the appropriate dyes, all three of which are produced simultaneously in one development.

In summary, although Fischer did not succeed in putting his theory into practice, his work still forms the basis of most modern color photographic systems. Not only did he demonstrate the general principle of dye-coupling development, but the art is still mainly confined to one of the two types of developing agent, and two of the five types of dye, described by him.

*Genevieve Slomski*

# Leavitt Measures Interstellar Distances

*Henrietta Swan Leavitt discovered that the pulsating brightness of a Cepheid variable star is directly proportional to the star's brightness.*

**What:** Astronomy; Photography
**When:** 1912
**Where:** Cambridge, Massachusetts
**Who:**
HENRIETTA SWAN LEAVITT (1868-1921), an American astronomer at Harvard College Observatory
HARLOW SHAPLEY (1885-1972), an American astronomer at Mount Wilson Observatory
EDWIN POWELL HUBBLE (1889-1953), an American astronomer at Mount Wilson Observatory
EJNAR HERTZSPRUNG (1873-1967), a Danish astronomer

## Variable Stars

The modern view of the universe is an expanding sphere of approximately one trillion galaxies. Each galaxy consists of one hundred billion to one trillion stars. Each star, including the sun, is an immense thermonuclear furnace composed mostly of the elements hydrogen and helium. The sun is one of approximately four hundred billion stars in the Milky Way galaxy.

By 1900, astronomers had firmly established that Earth, other planets, asteroids, and comets revolved about the sun and that the sun was only one of billions of stars in the Milky Way galaxy. Nevertheless, there were many unresolved problems. Some astronomers believed that the sun was located at the center of the Milky Way. Others correctly theorized that the sun was an average star not located at the galactic center. Yet evidence was difficult to find. The existence of other galaxies had not been clearly demonstrated, and there were no completely reliable methods for measuring the enormous distances between stars.

In 1902, Henrietta Swan Leavitt became a permanent observatory staff member at Harvard College Observatory. She studied variable stars, stars that change their luminosity (brightness) in a fairly predictable pattern over time. During her tenure at Harvard, Leavitt observed and photographed nearly 2,500 variable stars.

Variable stars can be of three principal types: eclipsing binaries, novas, and Cepheid variables. Cepheid variables and their similar RR Lyrae stars are stars that have exhausted their hydrogen fuel and have switched to helium. This causes them to be unstable, periodically increasing, then decreasing their energy and light output. Each Cepheid variable star has a predictable, repeating cycle of brightening and dimming. They are named after Delta Cephei, the first variable star discovered in the constellation Cepheus in 1768. Polaris, the North Star, is also a Cepheid.

## Measuring the Distance

At the time, the principal means of determining stellar distances was a trigonometric method known as parallax. Parallax is the apparent change in position of an object produced by a real change in position of the viewer. Perhaps the most familiar example of parallax is that objects near a road appear to move quickly as one drives past, while objects farther from the road seem to move more slowly. In a similar fashion, as Earth moves around the sun, nearby stars appear to move as seen from Earth. Measuring how much the stars seem to move allows one to find the distance to nearby stars. The very close stars Alpha Centauri and Barnard's star are 4.27 and 5.97 light-years distant, respectively. Sirius is 8.64 light-years distant. During the twentieth century, more than ten thousand stellar distances have been obtained using this method. Yet if one considers the universe as having a radius of perhaps twenty billion light-years, parallax fails at relatively short astronomical distances (for example, fifty thousand light-years).

Equipped with photographs of the Large and Small Magellanic Clouds (very small galaxies visible in the Southern Hemisphere), Leavitt measured the luminosities of variable stars over time. The Small Magellanic Cloud contained seventeen Cepheid variables having very predictable periods ranging from 1.25 days to 127 days. She carefully measured the changes in brightness and collected photographs of other Cepheids.

While Leavitt was studying Cepheids, Ejnar Hertzsprung of the Leiden University in the Netherlands and Henry Norris Russell of the Mount Wilson Observatory in Pasadena, California, independently discovered a relationship between a star's luminosity and its spectral class (that is, its color and temperature). Together, their results produced the Hertzsprung-Russell diagram, the astronomical equivalent of chemistry's periodic table. Thinking along the same lines as Hertzsprung and Russell, Leavitt carefully measured the luminosities and cyclic periods for many Cepheid variables from the Magellanic Clouds. She discovered that a Cepheid's apparent luminosity is directly proportional to the length of its period, or the time it takes to complete one cycle of brightening and dimming. A faint Cepheid variable has a very short period, usually ranging from one to four days. A more luminous Cepheid has a longer cyclic period, usually twenty to thirty days or more.

Harlow Shapley, an astronomer at the Mount Wilson Observatory, combined Leavitt's plot with parallax data for Cepheid distances and constructed a Cepheid period-absolute luminosity curve. With this curve, one can find a Cepheid's absolute luminosity by measuring its period. Knowing both the apparent and absolute lumi-

nosities of a star makes it possible to calculate its distance and, therefore, the distances of all the stars in that particular star cluster.

**Consequences**

Leavitt's discovery of the Cepheid variable period-luminosity relationship, reported in 1912 in the *Harvard College Observatory Circular,* was an important achievement in twentieth century astronomy. Her period-luminosity relationship established Cepheid variables as standard reference points for measuring distances between stars and galaxies.

Immediate applications of her work appeared in the studies of Shapley and Hubble. From RR Lyrae stars, Cepheid-like pulsating stars located within the globular clusters that surround our galaxy, Shapley produced an approximate distance map for the entire Milky Way galaxy. He demonstrated that the Milky Way is a flattened spiral disk with a thickened center. He also measured the approximate diameter of the Milky Way and estimated that the sun is about fifty thousand light-years (later corrected to about thirty thousand light-years) from galactic center. The sun is located in one of the Milky Way's spiral arms.

Hubble used the work of Leavitt and Shapley to measure the distances to RR Lyrae stars located within the Andromeda galactic disk, obtaining an approximate intergalactic distance of 750,000 light-years (later recalibrated to one million light-years), showing Andromeda to be one of the closest galaxies to Earth. Hubble applied this technique to other galaxies as well. This work contributed to his later studies of galactic redshift velocities, which led to his monumental astronomical discovery that the universe is expanding.

*David Wason Hollar, Jr.*

# United States Establishes Public Health Service

*What had begun as the Marine Hospital Service in 1798 grew to offer an increasing number of services to all Americans; in 1912, it was renamed the U.S. Public Health Service.*

**What:** Social reform; Health
**When:** 1912
**Where:** Washington, D.C.
**Who:**
ALEXANDER HAMILTON (1757-1804), U.S. secretary of the treasury from 1789 to 1795
EDWARD LIVINGSTON (1764-1836), congressman from New York from 1795 to 1801
JOHN M. WOODWORTH (fl. 1870), a surgeon

## The Early Years

The United States Public Health Service has its origins in England in the sixteenth century, when the English people, out of gratitude for their navy's successes against the Spanish, established a seamen's hospital. Throughout much of the American colonial period, the British government collected small taxes from its sailors to care for the sick and disabled.

After the United States was founded, Alexander Hamilton urged that a similar program be established; he reasoned that a good merchant marine was important for promoting American trade. In 1798, New York congressman Edward Livingston pushed through Congress a bill that created the Marine Hospital Service. The law provided that the Treasury Department should collect twenty cents per month from each merchant seaman to support this service. The following year, the law was amended to include the navy and marines.

The first temporary U.S. Marine Hospital was established in Boston in 1799, and the first government-owned hospital was built in Norfolk County, Virginia. In 1802, the law was broadened to cover the crews on boats and rafts sailing down the Mississippi River to New Orleans. As the nation expanded toward the west, Congress, in 1837, provided for new hospitals in the Mississippi River Valley and the Great Lakes region.

During these years, the Marine Hospital Service was a responsibility of the Treasury Department, with local collectors of customs being assigned to oversee the collection of fees. Much of the medical care was provided on a contract basis, and the contracts were often handed out as political favors. The collector of customs was traditionally a political appointee as well. As a result, the quality of services patients received in the first seventy years of the Marine Hospital Service varied widely.

As the nineteenth century advanced, the service began to build more hospitals, but it became clear that their locations were chosen for political reasons. In New Orleans, work was started on a marine hospital in 1837. Delayed for many years, the project was finally completed in 1851 at a total cost of $123,000—an enormous amount of money in those days. Though this building contract obviously had been mismanaged, an even larger hospital project was begun in 1855. This one was built in a swamp, and one of the walls sank two feet before the building was completed. This structure, which eventually cost half a million dollars, was never even used by the Marine Hospital Service.

## Reform and Growth

Because of complaints about the marine hospitals, Congress appointed a commission to investigate them in 1849. Seven years later, the commission's recommendations for improving procedures in the hospitals were put into effect.

**239**

*Children line up to be vaccinated in 1910, two years before the creation of the U.S. Public Health Service.*

There was some improvement as a result, but political interference still kept most of the hospitals from providing the best health care at a low cost.

In 1870, Congress completely reorganized the Marine Hospital Service, making it a bureau of the Treasury Department, headed by a supervising surgeon. The first officer to hold this position was John M. Woodworth, a very able doctor and administrator. The law of 1870 also increased the hospital tax to forty cents per month. The monthly charge remained the same until 1884, when it was replaced with a tonnage tax. Finally, in 1906, the tonnage tax was replaced with direct appropriations by Congress.

Under Woodworth's leadership, the service improved dramatically. In 1873 he introduced a personnel policy by which health-service workers would be hired for their skill, not for political reasons. In this and other ways, he gradually raised the quality of the staff. The next advance came in 1889, when a commissioned-officer corps was created, giving the service a force of highly qualified health experts who could move around the country to help take care of specific needs and problems.

Woodworth was a supporter of the emerging public-health movement. He worked for a national quarantine system to help keep contagious diseases from spreading, and he helped draft the first federal quarantine law in 1878. After this law was passed, the Marine Hospital Service cooperated with state and city governments to improve their quarantine agencies. When Congress strengthened the National Quarantine Law in 1893, the Marine Hospital Service began taking over state and local quarantine stations, completing the work in 1921, when it took over the quarantine facilities of the Port of New York.

**240**

The responsibilities of the Marine Hospital Service were increased further by the Immigration Law of 1891, which required the service's medical officers to examine all immigrants. In 1887, a one-room bacteriological laboratory was set up in the Marine Hospital on Staten Island, New York. This laboratory was the forerunner of the Hygienic Laboratory for Bacteriological Research, which was established four years later in Washington, D.C.

Because of the growth and development of the Marine Hospital Service, Congress changed its name in 1902, to the United States Public Health and Marine Hospital Service. In 1912, the name was shortened to the United States Public Health Service.

## Consequences

In the twentieth century, the U.S. Public Health Service has continued to provide new services. Its research laboratories have done outstanding work on epidemiology—the spread of diseases—and the service has played an important part in helping state and local health boards to work well. Under the Social Security Act of 1935, the Public Health Service was given responsibility for distributing eight million dollars each year for grants to public health agencies around the country. In the years since, the budget of the Public Health Service has increased dramatically.

The Public Health Service is now a division of the Department of Health, Education, and Welfare. It is headed by the U.S. surgeon general, and its divisions include the National Library of Medicine, the Bureau of State Services, the Bureau of Medical Services, and the National Institutes of Health.

*John Duffy*

**241**

# Bohr Develops Theory of Atomic Structure

*Bohr applied Max Planck's quantum theory to Ernest Rutherford's nuclear model of the atom, providing a theoretical explanation for a large number of atomic phenomena.*

**What:** Physics
**When:** 1912-1913
**Where:** Copenhagen, Denmark
**Who:**
NIELS BOHR (1885-1962), a Danish physicist
ERNEST RUTHERFORD (1871-1937), the founder of nuclear physics
JOSEPH JOHN THOMSON (1856-1940), the discoverer of the electron

## Theories of the Atom

At the beginning of the twentieth century, physicists were learning about the structure of the atom. In 1897, Sir Joseph John Thomson had discovered the part of the atom that became known as the electron, and he developed a model (or picture) of the atom as consisting of a central sphere, or "nucleus," that carried a positive electrical charge; around this sphere orbited the electrons, which carried a negative charge. All the electrons traveled in the same planar "shell"; moreover, the electrons accounted for a large portion of the atom's mass. This model was accepted by most scientists.

In 1911, both Niels Bohr and Thomson were at Cambridge University, where Bohr was completing his doctoral dissertation on the electron theory of metals. Bohr then went to Manchester, where he joined Ernest Rutherford, who had recently proposed a different model of the atom—one in which the positively charged nucleus was much smaller than the atom as a whole. At the same time, even though it was very small, the mass of this nucleus was very great.

There appeared to be a problem with this model, however: It was not stable according to the laws of physics that were known at that time, which were governed by Sir Isaac Newton's laws

of mechanics. Bohr became interested in this problem after reading a paper by Charles Galton Darwin and noticing some errors in it. Since classical mechanics could not explain how the atom could remain stable, Bohr decided to turn to Max Planck's quantum theory. He assumed that an electron must have a certain, exact amount of energy in order to maintain a stable orbit around a nucleus. This amount of energy, which was defined by a ratio discovered by Planck, is called "Planck's constant." Bohr called such stable orbits "stationary states."

This theory allowed Bohr to explain the relationship of many different chemical elements to one another and to position them on the Periodic Table of Elements. He suggested that the radioactive properties of an element (which determine how unstable it is) depend on the atom's nucleus, and the chemical properties of an element (which determine how it combines with other elements to form molecules, for example) depend on the number of electrons in the atom. Bohr also considered how atoms would act in different energy states and explained some of the phenomena that had been observed concerning the spectral lines emitted by atoms.

## Physicists Accept Bohr's Model

Bohr discovered that electrons could travel in various stable orbits, or stationary states, around an atomic nucleus, not only in a single stable orbit, as Thomson had thought. It was possible for an electron to move from a higher-energy orbit farther from the nucleus to a lower-energy orbit closer to the nucleus, or vice versa. The electron could move to a higher-energy orbit if it received energy from outside the atom, and it could move to a lower-energy orbit if it gave off energy. The energy given off or received by the electron could be seen as a "spectral line," a light or dark line in the spectrum of light given off by the atom. Such a line would be either an "absorption line" (if the electron received the energy) or an "emission line" (if the electron gave off the energy).

Bohr presented his ideas about the atom in a trilogy of papers published over the course of the year 1913. Reactions were mixed, because most physicists at the time still doubted that Planck's quantum theory could have any effects on observed physical phenomena. This changed, however, when Bohr's theory began to explain the details of the spectra emitted by atoms, which had not been satisfactorily explained before. More physicists began to accept the new atomic model.

## Consequences

One part of Bohr's theory, called the "correspondence principle," became especially important in the overall development of quantum theory, which in turn shaped all of modern physics. According to the correspondence principle, the results of quantum mechanics do not conflict with those of classical mechanics in the realm of physical phenomena, where classical laws are valid. Bohr's original theory was therefore extended to other areas of physics.

Bohr's theory was able to make remarkably accurate predictions for atoms with a single electron; it was less reliable, however, when it was applied to atoms with more than one electron. In 1920, Bohr focused on this problem and presented an improved and consistent theory.

Bohr's groundbreaking trilogy of 1913, although flawed, paved the way for quantum mechanics, which would ultimately dominate twentieth century physics. Bohr himself not only

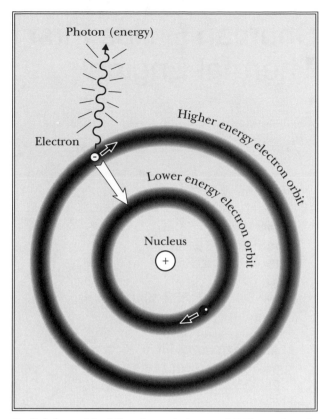

*In Bohr's model of the atom, electrons orbit the nucleus in discrete energy states. When an electron moves from a higher-energy orbit to a lower-energy orbit, the extra energy is released as a photon of light.*

continued to contribute to physics but also educated a new generation of physicists who went on to develop quantum mechanics. Although he never became as famous as Albert Einstein, his work is among the most important in the history of physics.

*Roger Sensenbaugh*

# Shuman Builds First Practical Solar Thermal Engine

> *Frank Shuman proved that a large-scale solar plant could operate as cheaply as a coal-fired plant.*

**What:** Energy
**When:** 1912-1913
**Where:** Meadi, Egypt
**Who:**
FRANK SHUMAN (1862-1918), an American inventor
JOHN ERICSSON (1803-1889), an American engineer
AUGUSTIN MOUCHOUT (1825-1911), a French physics professor

## Power from the Sun

According to tradition, the Greek scholar Archimedes used reflective mirrors to concentrate the rays of the sun and set afire the ships of an attacking Roman fleet in 212 B.C.E. The story illustrates the long tradition of using mirrors to concentrate solar energy from a large area onto a small one, producing very high temperatures.

With the backing of Napoleon III, the Frenchman Augustin Mouchout built, between 1864 and 1872, several steam engines that were powered by the sun. Mirrors concentrated the sun's rays to a point, producing a temperature that would boil water. The steam drove an engine that operated a water pump. The largest engine had a cone-shaped collector, or "axicon," lined with silver-plated metal. The French government operated the engine for six months but decided it was too expensive to be practical.

John Ericsson, the American famous for designing and building the warship *Monitor*, built seven steam-driven solar engines between 1871 and 1878. In Ericsson's design, rays were focused onto a line rather than a point. Long mirrors, curved into a parabolic shape, tracked the sun.

The rays were focused onto a water-filled tube mounted above the reflectors to produce steam. The engineer's largest engine, which used an $11 \times 16$-foot trough-shaped mirror, delivered nearly 2 horsepower. Because his solar engines were ten times more expensive than conventional steam engines, Ericsson converted them to run on coal to avoid financial loss.

Frank Shuman, a well-known inventor in Philadelphia, Pennsylvania, entered the field of solar energy in 1906. The self-taught engineer believed that curved, movable mirrors were too expensive. His first large solar engine was a hot-box, or flat-plate, collector. It lay flat on the ground and had blackened pipes filled with a liquid that had a low boiling point. The solar-heated vapor ran a 3.5-horsepower engine.

Shuman's wealthy investors formed the Sun Power Company to develop and construct the largest solar plant ever built. The site chosen was in Egypt, but the plant was built near Shuman's home for testing before it was sent to Egypt.

When the inventor added ordinary flat mirrors to reflect more sunlight into each collector, he doubled the heat production of the collectors (see figure 1). The 572 trough-type collectors were assembled in twenty-six rows. Water was piped through the troughs and converted to steam. A condenser converted the steam to water, which reentered the collectors. The engine pumped 3,000 gallons of water per minute and produced 14 horsepower per day; performance was expected to improve 25 percent in the sunny climate of Egypt.

British investors requested that professor C. V. Boys review the solar plant before it was shipped to Egypt. Boys pointed out that the bottom of each collector was not receiving any direct solar energy; in fact, heat was being lost through the bottom. He suggested that each row of flat mirrors be replaced by a single parabolic reflector (see figure 2), and Shuman agreed. Shuman thought Boys's idea was original, but he later realized it was based on Ericsson's design.

The company finally constructed the improved plant in Meadi, Egypt, a farming district on the Nile River. Five solar collectors, spaced 25 feet apart, were built in a north-south line. Each was about 200 feet long and 10 feet wide. Trough-shaped reflectors were made of mirrors held in place by brass springs that expanded and contracted with changing temperatures. The parabolic mirrors shifted automatically so that

the rays were always focused on the boiler. Inside the 15-inch boiler that ran down the middle of the collector, water was heated and converted to steam. The engine produced more than 55 horsepower, which was enough to pump 6,000 gallons of water per minute.

The purchase price of Shuman's solar plant was twice as high as that of a coal-fired plant, but its operating costs were far lower. In Egypt, where coal was expensive, the entire purchase price would be recouped in four years. Afterward, the plant would operate for practically nothing. The first practical solar engine was now in operation, providing enough energy to drive a large-scale irrigation system in the floodplain of the Nile River.

By 1914, Shuman's work was enthusiastically supported, and solar plants were planned for India and Africa. Shuman hoped to build 20,000 reflectors in the Sahara Desert and generate energy equal to all the coal mined in one year, but the outbreak of World War I ended his dreams of large-scale solar developments. The Meadi project was abandoned in 1915, and Shuman died before the war ended. Powerful nations lost interest in solar power and began to replace coal with oil. Rich oil reserves were discovered in many desert zones that were ideal locations for solar power.

### Consequences

Although World War I ended Frank Shuman's career, his breakthrough proved to the world that solar power held great promise for the future. His ideas were revived in 1957, when the Soviet Union planned a huge solar project for Siberia. A large boiler was fixed on a platform 140 feet high. Parabolic mirrors, mounted on 1,300 railroad cars, revolved on circular tracks to focus light on the boiler. The full-scale model

*Figure 1.* Trough-shaped collectors with flat mirrors produced enough solar thermal energy to pump 3,000 gallons of water per minute. *Figure 2.* Trough-shaped collectors with parabolic mirrors produced enough solar thermal energy to pump 6,000 gallons of water per minute.

**245**

was never built, but the design inspired the solar power tower.

In the Mojave desert near Barstow, California, an experimental power tower, Solar One, began operation in 1982. The system collected solar energy to deliver steam to turbines that produced electric power. The 30-story tower was surrounded by more than 1,800 mirrors that adjusted continually to track the sun. Solar One generated about 10 megawatts per day, enough power for 5,000 people.

Solar One and its successor, Solar Two, were expensive, but future power towers will generate electricity as cheaply as fossil fuels can. If the costs of the air and water pollution caused by coal burning were considered, solar power plants would already be recognized as cost effective. Meanwhile, Frank Shuman's success in establishing and operating a thoroughly practical large-scale solar engine continues to inspire research and development.

# Abel Develops First Artificial Kidney

*John Jacob Abel developed a way to take waste end-products and poisons out of the blood when the kidneys are not working properly.*

**What:** Medicine
**When:** 1912-1914
**Where:** Baltimore, Maryland
**Who:**

JOHN JACOB ABEL (1857-1938), a pharmacologist and biochemist known as the "father of American pharmacology"

WILLEM JOHAN KOLFF (1911-　　), a Dutch American clinician who pioneered the artificial kidney and the artificial heart

## Cleansing the Blood

In the human body, the kidneys are the dual organs that remove waste matter from the bloodstream and send it out of the system as urine. If the kidneys fail to work properly, this cleansing process must be done artifically—such as by a machine.

John Jacob Abel was the first professor of pharmacology at The Johns Hopkins University School of Medicine. Around 1912, he began to study the by-products of metabolism that are carried in the blood. This work was difficult, he realized, because it was nearly impossible to detect even the tiny amounts of the many substances in blood. Moreover, no one had yet developed a method or machine for taking these substances out of the blood.

In devising a blood filtering system, Abel understood that he needed a saline solution and a membrane that would let some substances pass through but not others. Working with Leonard Rowntree and Benjamin B. Turner, he spent nearly two years figuring out how to build a machine that would perform dialysis—that is, remove metabolic by-products from blood. Finally their efforts succeeded.

The first experiments were performed on rabbits and dogs. In operating the machine, the blood leaving the patient was sent flowing through a celloidin tube that had been wound loosely around a drum. An anticlotting substance (hirudin, taken out of leeches) was added to blood as the blood flowed through the tube. The drum, which was immersed in a saline and dextrose solution, rotated slowly. As blood flowed through the immersed tubing, the pressure of osmosis removed urea and other substances, but not the plasma or cells, from the blood. The celloidin membranes allowed oxygen to pass from the saline and dextrose solution into the blood, so that purified, oxygenated blood then flowed back into the arteries.

Abel studied the substances that his machine had removed from the blood, and he found that they included not only urea but also free amino acids. He quickly realized that his machine could be useful for taking care of people whose kidneys were not working properly. Reporting on his research, he wrote, "In the hope of providing a substitute in such emergencies, which might tide over a dangerous crisis . . . a method has been devised by which the blood of a living animal may be submitted to dialysis outside the body, and again returned to the natural circulation." Abel's machine removed large quantities of urea and other poisonous substances fairly quickly, so that the process, which he called "vividiffusion," could serve as an artificial kidney during cases of kidney failure.

For his physiological research, Abel found it necessary to remove, study, and then replace large amounts of blood from living animals, all without dissolving the red blood cells, which carry oxygen to the body's various parts. He realized that this process, which he called "plasmaphaeresis," would make possible blood banks, where blood could be stored for emergency use.

**247**

Library of Congress

*John Jacob Abel.*

In 1914, Abel published these two discoveries in a series of three articles in the *Journal of Pharmacology and Applied Therapeutics*, and he demonstrated his techniques in London, England, and Groningen, The Netherlands. Though he had suggested that his techniques could be used for medical purposes, he himself was interested mostly in continuing his biochemical research. So he turned to other projects in pharmacology, such as the crystallization of insulin, and never returned to studying vividiffusion.

## Refining the Technique

Georg Haas, a German biochemist working in Giessen, West Germany, was also interested in dialysis; in 1915, he began to experiment with "blood washing." After reading Abel's 1914 writings, Haas tried substituting collodium for the celloidin that Abel had used as a filtering membrane and using commercially prepared heparin instead of the homemade hirudin Abel had used to prevent blood clotting. He then used this machine on a patient and found that it showed promise, but he knew that many technical problems had to be worked out before the procedure could be used on many patients.

In 1937, Willem Johan Kolff was a young physician at Groningen. He felt sad to see patients die from kidney failure, and he wanted to find a way to cure others. Having heard his colleagues talk about the possibility of using dialysis on human patients, he decided to build a dialysis machine.

Kolff knew that cellophane was an excellent membrane for dialyzing, and that heparin was a good anticoagulant, but he also realized that his machine would need to be able to treat larger volumes of blood than Abel's and Haas's had. During World War II (1939-1945), with the help of the director of a nearby enamel factory, Kolff built an artificial kidney that was first tried on a patient on March 17, 1943. Between March, 1943, and July 21, 1944, Kolff used his secretly constructed dialysis machines on fifteen patients, of whom only one survived. He published the results of his research in *Acta Medica Scandinavica*. Even though most of his patients had not survived, he had collected information and developed the technique until he was sure dialysis would eventually work.

Kolff brought machines to Amsterdam and The Hague and encouraged other physicians to try them; meanwhile, he continued to study blood dialysis and to improve his machines. In 1947, he brought improved machines to London and the United States. By the time he reached Boston, however, he had given away all of his machines. He did, however, explain the technique to John P. Merrill, a physician at the Harvard Medical School, who soon became the leading American developer of kidney dialysis and kidney-transplant surgery.

Kolff himself moved to the United States, where he became an expert not only in artificial kidneys but also in artificial hearts. He helped develop the Jarvik-7 artificial heart (named after its chief inventor, Robert Jarvik), which was implanted in a patient in 1982.

## Consequences

Abel's work showed that the blood carried some substances that had not been previously known and led to the development of the first dialysis machine for humans. It also encouraged interest in the possibility of organ transplants.

After World War II, surgeons had tried to transplant kidneys from one animal to another, but after a few days the recipient began to reject the kidney and die. In spite of these failures, researchers in Europe and America transplanted kidneys in several patients, and they used artificial kidneys to take care of the patients who were waiting for transplants. In 1954, Merrill—to whom Kolff had demonstrated an artificial kidney—successfully transplanted kidneys in identical twins.

After immunosuppressant drugs (used to prevent the body from rejecting newly transplanted tissue) were discovered in 1962, transplantation surgery became much more practical. After kidney transplants became common, the artificial kidney became simply a way of keeping a person alive until a kidney donor could be found.

*Thomas P. Gariepy*

# Braggs Invent X-Ray Crystallography

> *The Braggs founded the science of X-ray crystallography, which allowed them to deduce the crystal structures of many substances.*

**What:** Physics; Chemistry; Earth science; Photography
**When:** 1912-1915
**Where:** Cambridge, England
**Who:**

SIR LAWRENCE BRAGG (1890-1971), the son of Sir William Henry Bragg and cowinner of the 1915 Nobel Prize in Physics

SIR WILLIAM HENRY BRAGG (1862-1942), an English mathematician and physicist and cowinner of the 1915 Nobel Prize in Physics

MAX VON LAUE (1879-1960), a German physicist who won the 1914 Nobel Prize in Physics

WILHELM CONRAD RÖNTGEN (1845-1923), a German physicist who won the 1901 Nobel Prize in Physics

RENÉ-JUST HAÜY (1743-1822), a French mathematician and mineralogist

AUGUSTE BRAVAIS (1811-1863), a French physicist

## The Elusive Crystal

A crystal is a body that is formed once a chemical substance has solidified. It is uniformly shaped, with angles and flat surfaces that form a network based on the internal structure of the crystal's atoms. Determining what these internal crystal structures look like is the goal of the science of X-ray crystallography. To do this, it studies the precise arrangements into which the atoms are assembled.

Central to this study is the principle of X-ray diffraction. This technique involves the deliberate scattering of X rays as they are shot through a crystal, an act that interferes with their normal path of movement. The way in which the atoms are spaced and arranged in the crystal determines how these X rays are reflected off them while passing through the material. The light waves thus reflected form a telltale interference pattern. By studying this pattern, scientists can discover variations in the crystal structure.

The development of X-ray crystallography in the early twentieth century helped to answer two major scientific questions: What are X rays? and What are crystals? It gave birth to a new technology for the identification and classification of crystalline substances.

From studies of large, natural crystals, chemists and geologists had established the elements of symmetry through which one could classify, describe, and distinguish various crystal shapes. René-Just Haüy, about a century before, had demonstrated that diverse shapes of crystals could be produced by the repetitive stacking of tiny solid cubes.

Auguste Bravais later showed, through mathematics, that all crystal forms could be built from a repetitive stacking of three-dimensional arrangements of points (lattice points) into "space lattices," but no one had ever been able to prove that matter really was arranged in space lattices. Scientists did not know if the tiny building blocks modeled by space lattices actually were solid matter throughout, like Haüy's cubes, or if they were mostly empty space, with solid matter located only at the lattice points described by Bravais.

With the disclosure of the atomic model of Danish physicist Niels Bohr in 1913, determining the nature of the building blocks of crystals took on a special importance. If crystal structure could be shown to consist of atoms at lattice points, then the Bohr model would be supported, and science then could abandon the theory that matter was totally solid.

## X Rays Explain Crystal Structure

In 1912, Max von Laue first used X rays to study crystalline matter. Laue had the idea that irradiating a crystal with X rays might cause diffraction. He tested this idea and found that X rays were scattered by the crystals in various directions, revealing on a photographic plate a pattern of spots that depended on the orientation and the symmetry of the crystal.

The experiment confirmed in one stroke that crystals were not solid and that their matter consisted of atoms occupying lattice sites with substantial space in between. Further, the atomic arrangements of crystals could serve to diffract light rays. Laue received the 1914 Nobel Prize in Physics for his discovery of the diffraction of X rays in crystals.

Still, the diffraction of X rays was not yet a proved scientific fact. Sir William Henry Bragg contributed the final proof by passing one of the diffracted beams through a gas and achieving ionization of the gas, the same effect that true X rays would have caused. He also used the spectrometer he built for this purpose to detect and measure specific wavelengths of X rays and to note which orientations of crystals produced the strongest reflections. He noted that X rays, like visible light, occupy a definite part of the electromagnetic spectrum. Yet most of Bragg's work focused on actually using X rays to deduce crystal structures.

Sir Lawrence Bragg was also deeply interested in this new phenomenon. In 1912, he had the idea that the pattern of spots was an indication that the X rays were being reflected from the planes of atoms in the crystal. If that were true, Laue pictures could be used to obtain information about the structures of crystals. Bragg developed an equation that described the angles at which X rays would be most effectively diffracted by a crystal. This was the start of the X-ray analysis of crystals.

Henry Bragg had at first used his spectrometer to try to determine whether X rays had a particulate nature. It soon became evident, however, that the device was a far more powerful way of analyzing crystals than the Laue photograph method had been. Not long afterward, father and son joined forces and founded the new science of X-ray crystallography. By experimenting with this technique, Lawrence Bragg came to believe that if the lattice models of Bravais applied to actual crystals, a crystal structure could be viewed as being composed of atoms arranged in a pattern consisting of a few sets of flat, regularly spaced, parallel planes.

Diffraction became the means by which the Braggs deduced the detailed structures of many crystals. Based on these findings, they built three-dimensional scale models out of wire and spheres that made it possible for the nature of crystal structures to be visualized clearly even by nonscientists. Their results were published in the book *X-Rays and Crystal Structure* (1915).

## Consequences

The Braggs founded an entirely new discipline, X-ray crystallography, which continues to grow in scope and application. Of particular im-

*Sir William Henry Bragg.*

**251**

portance was the early discovery that atoms, rather than molecules, determine the nature of crystals. X-ray spectrometers of the type developed by the Braggs were used by other scientists to gain insights into the nature of the atom, particularly the innermost electron shells. The tool made possible the timely validation of some of Bohr's major concepts about the atom.

X-ray diffraction became a cornerstone of the science of mineralogy. The Braggs, chemists such as Linus Pauling, and a number of mineralogists used the tool to do pioneering work in deducing the structures of all major mineral groups. X-ray diffraction became the definitive method of identifying crystalline materials.

Metallurgy progressed from a technology to a science as metallurgists became able, for the first time, to deduce the structural order of various alloys at the atomic level. Diffracted X rays were applied in the field of biology, particularly at the Cavendish Laboratory under the direction of Lawrence Bragg. The tool proved to be essential for deducing the structures of hemoglobin, proteins, viruses, and eventually the double-helix structure of deoxyribonucleic acid (DNA).

*Edward B. Nuhfer*

# Wegener Articulates Theory of Continental Drift

*Alfred Lothar Wegener proposed that all lands were once part of the supercontinent of Pangaea, which then fragmented and whose pieces drifted apart to form present-day continents.*

**What:** Earth science
**When:** January, 1912
**Where:** Frankfurt, Germany
**Who:**
ALFRED LOTHAR WEGENER (1880-1930), a German meteorologist and earth scientist
FRANK BURSLEY TAYLOR (1860-1938), an American student of geology and astronomy
ALEXANDER LOGIE DU TOIT (1878-1948), a South African geologist
ARTHUR HOLMES (1890-1965), a British geologist

## A Giant Jigsaw Puzzle

The concept of continental drift was developed, at least in part, to explain the striking parallelism between the coasts of continents bordering the Atlantic Ocean, which seem as though they could fit together as pieces of a giant jigsaw puzzle. In particular, the fit between the eastern coast of South America and the Western coast of Africa is very striking.

The idea that continents were once joined together as part of a single landmass has been around for several centuries. As early as 1620, the English philosopher and author Sir Francis Bacon had discussed the possibility that the Western Hemisphere had once been joined to Africa and Europe. In 1668, a scientist by the name of Placet expressed similar ideas. Antonio Snider-Pellegrini in his book *La Création et ses mystères dévoilés* (1859; creation and its mysteries revealed) recognized the similarities between American and European fossil plants of the Carboniferous period (about 300 million years ago) and pro-

posed that all continents were once part of a single landmass. By the end of the nineteenth century, Austrian geologist Eduard Suess had noticed the close correspondence between geological formations in the lands of the Southern Hemisphere and had fitted them together into a single landmass he termed "Gondwanaland." In 1908, Frank Bursley Taylor of the United States, and in 1910, Alfred Lothar Wegener of Germany, independently suggested mechanisms that could account for large, lateral displacements of the earth's crust and, therefore, how continents could be driven apart. Wegener's work became the center of the debate that has lasted until the present.

## Folding Mountain Ranges

The concept of continental drift was best expressed by Wegener in his book *Die Entstehung der Kontinente und Ozeane* (1912; *The Origin of Continents and Oceans*, 1924). He based the theory not only on the shape of the continents but also on geologic evidence found around the world. Wegener specifically cited similarities in fossil fauna and flora (extinct animals and plants) found in Brazil and Africa. A series of maps were developed to show three stages of the drift process, and the original supercontinent was named "Pangaea" (a word meaning "all lands").

Wegener believed that the continents, composed of light-density granitic rocks, were independently propelled and plowed through the denser basalts of the ocean floor driven by forces related to the rotation of the earth. He provided evidence based on detailed correlations of geological features and fossils indicating a common historical record on both sides of the Atlantic. He also proposed that the supercontinent of Pangaea existed before the beginning of the Mesozoic era (about 200 million years ago).

**253**

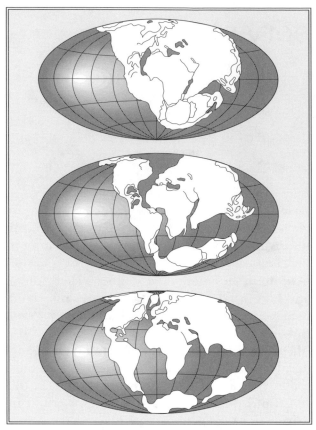

*Wegener theorized that the continents began as one great landmass more than 200 million years ago (top) which drifted apart beginning about 190 million years ago (middle), eventually resembling the world as we know it today (bottom).*

The split of Pangaea was visualized as beginning during the Jurassic period (about 190 million years ago), with the southern continents moving westward and toward the equator. South America and Africa began to drift apart during the Cretaceous period (70 million years ago). The opening of the north Atlantic was accomplished during the Pleistocene epoch (approximately 2.5 million years ago). Greenland and Norway started to separate as recently as 1.5 million years ago.

The Indian peninsula drifted northward, colliding with the Asian continent and giving rise to the folded mountains of the Himalayas. Similarly, the European Alps and the Atlas Mountains of North Africa were explained as a westward extension of the Himalayan chain. Wegener also suggested that as the drifting continents met the resistance of the ocean floor, their leading edges were compressed and folded into mountains. In this way, he also explained the Western Cordillera of the Americas and the mountains of New Zealand

and New Guinea. The tapering ends of Greenland and South America and the island arcs of the Antilles and East Asia were visualized as stragglers trailing behind the moving continents. Periods of glaciation found in the southern part of South America, Africa, Australia, peninsular India, and Madagascar provided further evidence of drift.

Detailed studies by the South African geologist Alexander Logie Du Toit provided strong support to Wegener's concepts. Du Toit postulated two continental masses rather than the single entity of Pangaea. He visualized the northern supercontinent of Laurasia and its southern counterpart Gondwanaland separated by a seaway called "Tethys." Du Toit was also the first to propose that the continental masses of the Southern Hemisphere had moved relative to the position of the South Pole. His ideas were published in *Our Wandering Continents* (1937), a book he dedicated to Wegener.

## Consequences

Although Wegener and Du Toit had provided compelling evidence in favor of the drift theory, one monumental problem remained: What forces could be strong enough to rupture, fragment, and cause the continents to drift? Arthur Holmes of the University of Edinburgh was the originator of the now popular concept of thermal convection in the earth's mantle as the main cause of drift. Holmes's model, published in 1931, is very similar to that presently used in the widely accepted theory of plate tectonics (the modern version of Wegener's theory). Holmes was also the first to introduce the idea that the continents are being carried along by a moving mantle in a sort of conveyor-belt motion.

Although appealing, Wegener's theory of continental drift remained controversial and was not widely accepted until the American geologist Harry Hammond Hess and the American geophysicist Robert Sinclair Dietz introduced the theory of seafloor spreading in the early 1960's.

Continental drift and its modification into the concept of seafloor spreading and plate tectonics remains as one of the most significant theories in earth science. Although not devoid of problems, it is the most complete theory of global tectonics in existence.

*Robert G. Font*

# China's Manchu Dynasty Is Overthrown

*The Chinese Revolution of 1911 succeeded in ousting the Manchu Dynasty, but revolutionary leaders were not able to unify China.*

**What:** Political reform; Coups
**When:** February 12, 1912
**Where:** China
**Who:**
SUN YAT-SEN (1866-1925), a Chinese revolutionary leader who conspired against the Manchus
CIXI (TZ'U-HSI; 1835-1908), dowager empress of China from 1861 to 1908
HUANG XING (HUANG HSING; 1873-1916), a Chinese revolutionary conspirator who cooperated with Sun Yat-sen
CHUN (CH'UN; 1883-1951), prince regent of China from 1908 to 1912
HENRY PUYI (P'U-YI; 1906-1967), the last Manchu emperor of China, from 1908 to 1912
YUAN SHIKAI (YÜAN SHIH-K'AI; 1859-1916), a military officer under the Manchus, and president of China from 1912 to 1916

## Call for Reform

The most important cause of the Chinese Revolution of 1911 was the impact of Western nations on China. Ever since the disastrous Opium War of 1839-1842, the Chinese, who had long believed their civilization to be far superior to all others, had watched helplessly as other nations chipped away at their sovereignty. These other nations included not only the Great Powers of the West but, after 1895, Japan as well. Intent on gaining special trade privileges, these nations were caught up in a fierce rivalry. Their competition at least had the benefit of keeping China from becoming the colony of any single European state. The Manchu Dynasty, established in the seventeenth century, seemed unable to respond to the foreign threats.

After China was defeated in the Sino-Japanese War of 1895, the Chinese people rose up against foreign influences in the Boxer Rebellion. Yet, in 1900, a multinational army assembled by the Great Powers was able to defeat this revolt. After this humiliation, thoughtful Chinese increasingly began calling for an end to the ineffective Manchu leadership. Some young revolutionaries said that the monarchy must be altogether abolished in order to save China. One of the earliest and most prominent of these revolutionaries was Sun Yat-sen.

Born in the southern Chinese province of Guangdong (Kwangtung) in 1866, Sun had emigrated to Hawaii as a teenager. There he had received a Western education and had been converted to Christianity. Although Sun wanted China to be strong enough to stand against Western imperialism, he also wanted to see the nation copy what he saw as the good aspects of modern Western civilization.

In October, 1895, Sun and a group of fellow revolutionaries, using Hong Kong as a base, tried to begin a rebellion in Canton, the capital of Guangdong Province. When this rebellion was defeated, Sun was forced to leave China. He then spent long years as a political exile in Japan, Southeast Asia, and the West.

In the summer of 1905, Sun Yat-sen, Huang Xing (another anti-Manchu exile), and others founded the Revolutionary Alliance, or Tougmenhui (T'ung-Meng-hui). This revolutionary coalition, led by Sun, brought together all the foes of the Manchu regime. Overseas Chinese (Chinese people who had emigrated to Southeast Asia, and their descendants) began to give financial support to the coalition. Until 1911, however, all Sun's schemes for stirring up revolt in China came to nothing.

After 1900, Manchu Dowager Empress Cixi, who had once strongly opposed reform, suddenly had a change of heart. There were many changes in the government during the first decade of the twentieth century. Locally elected provincial assemblies were established, and a constitutional parliamentary government was promised for the future. Public officials began to be selected on the basis of their training in specific skills rather than on the basis of their knowledge of the Confucian classics. A new educational system was established to train leaders, and certain students were sent abroad for further training. Also, the army was modernized and made more efficient.

Yet all these changes did not make the Chinese people more content with the rule of the Manchus. Instead, dissent increased. The new standards for the selection of government officials actually made it harder for poor or middle-class young men to find a place in government bureaucracy, since more schooling was now required, and education was expensive. Instead, many of these ambitious young men entered the army,

where they continued to oppose the Manchu regime. In the new schools and colleges, students developed a questioning, rebellious attitude toward the Manchus. Similarly, the provincial assemblies became places where local leaders could push for a greater say in running their own affairs.

## Uprising and Revolution

After the dowager empress's death in 1908, there was no strong ruler able to keep a lid on the boiling pot of discontent. The year 1911 was marked by famine, floods, and economic disaster. In May of that year, Prince Regent Chun, who governed in the name of the "boy emperor" Puyi, became quite unpopular because he delayed calling a parliament, created a cabinet made up mostly of Manchus, and tried to get a foreign loan to pay for his plan to nationalize the railways. Protests arose in various parts of China. Normally, the army would have been able to crush dissent, but rebellion was beginning to spread within its ranks as well.

Hulton Archive

*A member of the Manchu imperialist army prepares to behead a revolutionary kneeling at the side of his own grave.*

In October, a revolutionary organization was founded in the garrison of Wuchang, capital of the province of Hubei. Instead of submitting to discipline, the soldiers revolted on October 10, kidnapped their commander, and forced him, under threat of death, to lead their rebellion against the Manchu Dynasty. This day became known as the "Double Ten" among Chinese patriots.

Across the nation, the response to this mutiny was sudden and strong. Assembly after assembly in the provinces declared independence from the Manchu Dynasty. Soon a revolutionary regime was established in the central Chinese city of Nanjing. All that remained for the Manchus was the capital city, Beijing, and several provinces in northern China.

Sun Yat-sen returned from exile in December, and on December 29, 1911, he was elected provisional president of the Republic of China by the Nanjing government.

Yet the revolutionary victory was not complete. Yuan Shikai, an able military leader, had come out of retirement to defend the Manchus. His forces were far too strong for Nanjing to defeat, but far too weak to defeat the new Nanjing government. After some bargaining, the ambitious Yuan struck a deal with the revolutionaries. Yuan would persuade the Manchus to abdicate; in return, he would become president of China.

On February 12, 1912, the Manchus formally announced the end of their dynasty. Three days later, Sun Yat-sen resigned the presidency, recommending that Yuan be named as his successor. On March 10, 1912, Yuan Shikai was formally sworn in as provisional president of the Republic of China. In response to his urging, the provisional parliament voted on April 1 to transfer the capital to Beijing. China was now a united republic.

## Consequences

The revolution did not secure liberal democracy in China. On March 20, 1913, Song Jiaoren (Sung Chiao-jen), a liberal who had drafted the new constitution, was assassinated. In July and August, Yuan took power away from the provinces and banned all organized opposition to his regime. Sun Yat-sen was once again forced into exile in Japan.

After Yuan died in June, 1916, China soon disintegrated into a number of feuding states ruled by warlords. Sun died in 1925, and internal peace did not come to China until 1949, when the Communists, led by Mao Zedong, secured power over the mainland, while the Nationalists, led by Chiang Kai-shek, were confined to the island of Taiwan. Both the Communist regime and the Nationalist government claimed to be the exclusive heir to the glorious tradition of Sun Yat-sen and the Chinese Revolution of 1911.

*S. M. Chiu*

# Rutherford Presents Theory of the Atom

*Ernest Rutherford discovered the nucleus of the atom from experiments with radioactive elements and in the process deduced the true nature of the atom.*

**What:** Physics
**When:** March 7, 1912
**Where:** Manchester, England
**Who:**
ERNEST RUTHERFORD (1871-1937), an English physicist who won the 1908 Nobel Prize in Chemistry
SIR JOSEPH JOHN THOMSON (1856-1940), an English physicist who won the 1906 Nobel Prize in Physics
HANS GEIGER (1882-1945), a German physicist
ERNEST MARSDEN, Rutherford's assistant
NIELS BOHR (1885-1962), a Danish physicist who won the 1922 Nobel Prize in Physics

## From Philosophy to Science

Until the beginning of the twentieth century, nearly all atomic theory had come straight from philosophers. The early Greeks assumed that one could divide matter no farther than into tiny particles they called *atomos*, meaning indivisible. In English, *atomos* became "atom." Atomic theory was made more or less respectable by the British physicist and chemist Robert Boyle in the seventeenth century, but it was still philosophy without hard scientific evidence.

On April 29, 1897, atomic science was transformed from philosophy to hard science. That evening, Sir Joseph John Thomson announced that he had discovered tiny subatomic particles, which he called "corpuscles," that were much smaller than the atom. (They were later renamed "electrons.") Thomson's discovery confirmed not only that atoms existed but also that they were probably made up of even smaller particles. According to Thomson's theory, the atom was made up of a positively charged fluidized inte-

rior of great volume, compared with the tiny negatively charged electron enclosed in the fluid. Thomson described the aggregate as similar to "plum pudding."

In 1907, Ernest Rutherford accepted a position at Manchester, England. He had been experimenting with the particles that appeared to emanate from the radioactive atom. These efforts had been narrowed to a series of experiments that he hoped would finally identify these particles and their nature. Assisting him were Hans Geiger and an undergraduate student named Ernest Marsden.

Enough data from Thomson's work had filtered in so that Rutherford and his assistants knew that the electron was a piece of the atom and that it was both considerably lighter and smaller than the whole atom. They also knew its charge was negative, so that whatever was left of the atom had to be much heavier and have a net positive charge. Yet the particles that were emitted by the radioactive material Rutherford was examining (called "alpha" particles) were much heavier than electrons but still smaller than a whole atom. The question that perplexed Rutherford was whether, like the electron, these were also a part of the atom.

## Throwing Out the "Plum Pudding"

The experimental apparatus that Rutherford set up was quite simple, compared with the multi-million-dollar devices used by physicists a century later. It consisted of a glass tube that held an alpha particle emitter at one end. At the other end was a target of gold foil and beyond that a fluorescent screen that acted as the detector.

The theory behind Rutherford's experiments was that the alpha particle from the radioactive element would race down the tube from its source and strike the atoms in the gold foil. If the atoms were made up of Thomson's "plum pudding," then as the massive alpha particle struck the electrons, they would be deflected only slightly or not at all. By measuring where the tiny blips of light struck the gold foil, Rutherford could calculate the angle of deflection and indirectly determine the mass of whatever the alpha particle had struck on its way down the tube. He reasoned that the deflections of the more massive alpha particles striking tiny electrons would be minimal, but that if, by the most bizarre of circumstances, one of these particles should encounter a series of electrons on its way through an atom, the deflection might register as much as 45 degrees.

The experiments began in 1910 with Geiger assisting Marsden, counting the almost invisible flashes of light on the fluorescent screen through a magnifying lens in a completely blackened laboratory. They immediately found an astonishing effect. One out of about eight thousand alpha particles was deflected at an angle, varying from greater than 45 to 180 degrees.

It was obvious to Rutherford that plum pudding could never account for such wild deflections. He considered that perhaps the nucleus held a charge vastly greater than any hypothesized and that the alpha particle was being whipped around the interior of the atom like a comet tossed back into the deep solar system by the sun. The only other plausible explanation, which he eventually accepted, was that the atom contained a tiny, pinpoint nucleus that occupied only a minuscule portion of the total volume of the atom but, at the same time, itself contained nearly all the atom's mass. The electrons, he supposed, orbited like tiny, flyweight particles at huge distances from the densely packed core. On March 7, 1912, Rutherford presented his theory at the Manchester Literary and Philosophical Society.

Rutherford had the correct idea of the nucleus. Electrons do not "orbit" in the classical sense, however; they "exist" in a quantum state, as Niels Bohr would later prove. In the process, Bohr would change the face of physics. Rutherford's discovery would be the last major finding of classical physics. By 1913, Rutherford's vision would be replaced by Bohr's quantum view.

## Consequences

From the time of classical Greece, people had viewed matter as made up of tiny, indivisible particles. The notion was nothing more than an educated guess. It held through thousands of years not because of its inherent accuracy but because of the lack of technology to prove otherwise. This idea became so firmly implanted that it became a kind of theology of reason without implicit cause. When Rutherford proved this notion wrong, he was met with immediate disbelief. At least Thomson's atom had substance; according to Rutherford, however, atoms were made up mostly of space.

Bohr would soon redefine the atom in new and innovative terms. Bohr described everything equal in size or smaller than the atom in terms of "quantum mechanics," which deals with the interaction of matter and radiation, atomic structure, and so forth. Rutherford explored as deeply inside the atom as one could go within the framework of knowledge current at that time. A new science had to be developed to go even deeper. Quantum mechanics would join with Einstein's work on relativity to reorder physics and redefine the nature of all matter and energy.

*Dennis Chamberland*

# Home Rule Stirs Debate in Ireland and Great Britain

*As Irish nationalists pressed for the reestablishment of the Irish parliament, the issue of Ireland's ties to Britain became so controversial that the United Kingdom nearly broke out in civil war.*

**What:** Political reform

**When:** April 11, 1912-September 15, 1914

**Where:** London and Ireland

**Who:**

HERBERT HENRY ASQUITH (1852-1928), Liberal prime minister of Great Britain from 1908 to 1916

WINSTON CHURCHILL (1874-1965), first lord of the admiralty from 1911 to 1915, and Liberal leader in the House of Commons

BONAR LAW (1858-1923), Conservative leader in the House of Commons

JOHN EDWARD REDMOND (1856-1918), leader of the Irish Nationalist Party

EDWARD HENRY CARSON (1854-1935), leader of the Ulster Unionists

EOIN MACNEILL (1867-1945), founder of the Nationalist Volunteers and leader of the Gaelic League

PADHRAIC PEARSE (1879-1916), leader of the Irish Republicans (Sinn Féin)

## The Drive for Home Rule

The Act of Union of 1800, which established the United Kingdom of Great Britain and Ireland, had never been popular in Ireland. Throughout the nineteenth century, many Irish people had called for the restoration of the Irish parliament, or Home Rule. The issue became important in British politics in the 1880's, when the Irish members of Parliament, led by Charles Stewart Parnell, used their eighty-five votes to keep the Liberals in power, in exchange for Home Rule.

As a result, William Ewart Gladstone's Liberal government introduced the First Home Rule Bill in 1886, but it was defeated by thirty votes, and part of the Liberal Party, the Liberal Unionists (who opposed Home Rule), broke away. Gladstone's Second Home Rule Bill passed the House of Commons by thirty-four votes in 1893, but it was soundly defeated in the House of Lords by a vote of 419 to 41.

In the following years, the Conservative and Unionist governments of Great Britain tried to distract the Irish from the issue of Home Rule by making economic reforms to benefit the Irish. John Edward Redmond, an eloquent speaker, became leader of the united Irish Nationalist Party in the House of Commons in 1900. Yet since neither the Liberals nor the Conservatives needed Irish votes at that time, no progress was made until 1910. In the campaign of that year, the Liberals supported Home Rule. They lost so many seats in Parliament that they needed the votes of the Irish Nationalists to form a cabinet and ensure that their proposed program could go forward.

Now that they needed the help of the Irish Nationalists, the Liberals were obligated to put together a new Home Rule bill. The House of Lords would no longer be an obstacle, since the Parliament Act of 1911 had reduced its power.

## Debate over the Bill

The Third Home Rule Bill was introduced into the House of Commons on April 11, 1912, by the Liberal prime minister, Herbert Henry Asquith. The bill was managed by the First Lord of the Admiralty, Winston Churchill, who was then Liberal leader in the House of Commons and a strong supporter of Home Rule. Like the two previous Home Rule bills, this one would set up an Irish parliament in Dublin, but control over foreign affairs customs duties, and defense would

*John Edward Redmond, leader of the Irish Nationalist Party (right member of walking pair), and Irish Nationalist John Dillon leave Buckingham Palace after a conference on home rule.*

remain with the British parliament at Westminster. The number of Irish members of Parliament at Westminster was to be reduced by half, to forty-two.

Redmond and the Irish Nationalists praised the bill and argued that it would strengthen the relationship between Great Britain and Ireland. The Liberal leaders praised it as a way of providing justice and greater independence for Ireland. Another Liberal argument was that Home Rule would relieve the British parliament of the burden of having to deal with local affairs in Ireland.

Yet the Unionists—both the Conservative Unionists, led by Bonar Law, and the Ulster Unionists, led by Edward Henry Carson—opposed the Home Rule Bill bitterly. Their chief concern was the fate of Ulster, present-day Northern Ireland. The mostly Protestant Scottish-Irish residents of Ulster saw themselves as loyal Britons who were

being sacrificed to the interests of the Roman Catholic residents of southern Ireland, which was more rural. They were convinced that the Catholic Irish would persecute them religiously and economically.

Neither the North nor the South wished to divide Ireland into two separate states. There was a small but influential Protestant minority in the South, and the Ulster Protestants did not want to abandon them to the domination of Catholics. The Irish Nationalists—predominately Catholics—wanted Ireland to remain one nation.

After vigorous debate, the Home Rule Bill passed the House of Commons on January 16, 1913. As had been expected, the House of Lords rejected it by a huge majority, which meant that the bill would have to be passed again by the Commons at its next two sessions to overcome the Lords' veto. Because the same Parliament

**261**

would still be in session at that time, however, it was almost certain to pass.

Carson began to call on fellow Ulsterites to rise up against Great Britain. At first, the Liberal government did not take this threat too seriously. Then, however, a "covenant" opposing Home Rule gained a total of 470,000 signatures, and the Ulster Unionists began to drill a militia, the Ulster Volunteers. Conservative leaders gave open support to this threatened rebellion.

The southern Irish were quite upset by Ulster's defiance and British weakness in the face of these threats. While Redmond continued to support Asquith and Parliament, other Irish people became convinced of the need for stronger measures than Home Rule. Groups such as the Irish Republicans, or Sinn Féin, the Gaelic League, and certain labor unions urged the South to arm itself just as Ulster had. In November, 1913, the Nationalist Volunteers, led by Eoin MacNeill and by Padhraic Pearse, leader of the Sinn Féin, began to drill in the South.

Because of these threats, when the Home Rule Bill passed the Commons for the third and last time in March, 1914, Asquith added an Amending Bill. This compromise excluded Ulster from Home Rule for six years. Redmond reluctantly accepted it, but the radical Irish Nationalist groups objected.

Meanwhile, German rifles were smuggled into Ulster for the Ulster Volunteers. The Nationalist Volunteers in the South were also growing in influence and becoming restless. Redmond unwillingly took command of them in May to keep his ties with MacNeill, Pearse, and Sir Roger David Casement, a rebel who was later hanged as a traitor.

The Home Rule Bill passed the Commons on May 25, and the Amending Act passed the Commons on June 23. The House of Lords then amended the Amending Act enough to destroy the compromise Asquith had worked out. On July 12, Carson proclaimed the establishment of the "Ulster Provisional Government."

Negotiations were not working, and on July 26, British troops fired on an angry crowd in Bachelors' Walk, Dublin. Three people were killed and many others wounded. The Irish Nationalists became even angrier, and civil war seemed about to begin.

The next day, however, Austria declared war on Serbia, and by August 4, Great Britain was plunged into World War I. In this crisis, Asquith, Redmond, Carson, and Law agreed to postpone the Amending bill, and the threat of civil war passed. On September 15, the Home Rule Bill became law, but Parliament agreed not to put it into effect until after the war.

## Consequences

With the outbreak of World War I, Home Rule was no longer a real possibility. The radical Irish Nationalists soon split from Redmond to gain more followers. Eventually they were responsible for the Easter Rebellion of 1916 and for civil war after 1918.

The battle over the Home Rule Bill had many serious consequences for England and Ireland. The Conservatives' preaching of revolution had seriously threatened Great Britain's parliamentary system of government. The failure to enforce Home Rule helped to break the bonds between Great Britain and Ireland, preparing the way for the Irish Rebellion and eventual independence. Furthermore, Ulster opposition to Home Rule helped to cause the permanent split of Northern Ireland from the rest of the island.

*Eugene S. Larson*

# Iceberg Sinks "Unsinkable" Ocean Liner
## *Titanic*

*Despite being advertised as "unsinkable," the fabulous ocean liner RMS* Titanic *took nearly fifteen hundred people to watery graves in the North Atlantic after hitting an iceberg. Instead of achieving fame as the ultimate in ocean travel, the name* Titanic *became synonymous with maritime disaster.*

**What:** Transportation; Disasters
**When:** April 14-15, 1912
**Where:** North Atlantic Ocean, off Newfoundland
**Who:**
J. Bruce Ismay (1862-1937), chairman of the White Star Line
Edward John Smith (1850-1912), captain of the *Titanic*

## Fanfare

The *Titanic* was designed to carry the world's elite on transatlantic voyages that would provide them with an experience unlike any other. That is exactly what it delivered on its maiden and only voyage. Part hype, part sound shipbuilding methods, it was built on a colossal scale and equipped with all the latest technological advances. Its builders believed that bigger was better and that the ship's very size would attract passengers to the White Star Line, which was competing with the Cunard Line. Cunard's recent acquisition of the giant luxury liners *Mauretania* and *Lusitania* had opened a new era in shipbuilding and transatlantic travel. The Cunard ships were not only bigger and faster than earlier liners, they also had the latest in creature comforts. White Star Line chairman J. Bruce Ismay believed that unless his line acquired bigger and better ships, it might go out of business.

White Star commissioned the building of sister ships of monumental proportions and christened them the *Olympic* and *Titanic* to symbolize their grandeur. Just over 882 feet in length, the *Titanic* displaced 46,328 gross tons and was powered by engines that combined to produce 46,000 horsepower. It had a cruising speed of twenty-one knots and top speed of twenty-five knots. However, while the ship had berths for 3,547 passengers and crew, its meager supply of twenty lifeboats was capable of carrying less than a third of that number.

Scheduled for mid-April, 1912, the *Titanic*'s maiden voyage generated enormous public attention. Many wealthy people booked passage on the Titanic simply because they wanted to be a part of its maiden voyage, not because they needed to cross the Atlantic. Families such as the Guggenheims, Astors, and Strausses reserved palatial and elegantly furnished suites. A gymnasium, lounge, swimming pool, and vast dining room helped separate first-class passengers from others. Commercial manufacturers from soap companies to porcelain makers wanted to be associated with the event and published advertisements touting products that were to be used aboard the ship. The idea that "if it's good enough for the *Titanic*, it's good enough for me" pervaded society.

Accommodations for second-class passengers were so fine that it was said that second-class accommodations on the *Titanic* matched the first-class accommodations of any other ship of its time. However, the accommodations for several hundred third-class passengers in steerage were located deep in the bowels of the ship and were segregated by sex, on the apparent assumption that lower-class people had to be treated differently.

## The First and Last Voyage

After an unsettling near-collision with another ship before it even left the southern English port of Southampton, the *Titanic* set sail on Wednesday, April 10, 1912. Within hours, it took on additional passengers at Cherbourg in Normandy and then dropped anchor two miles off Queenstown, where another 120 passengers boarded, and 1,385 sacks of mail were loaded. As a Royal Mail Ship, the *Titanic* was to be a major mail carrier between Europe and the United States. Finally, at 1:30 P.M. of April 11, the ship set out for the Atlantic Ocean, destination: New York City.

Everything went smoothly as the ship glided past milemarkers. Between Friday and Sunday, April 14, it covered 1,065 miles. White Star chairman Ismay, who was aboard as a passenger, wanted the ship to break the transatlantic speed record recently set by Cunard's *Mauretania*. The veteran captain Edward James Smith—who delayed his retirement to command the *Titanic's* maiden voyage—thought it wiser to save the speed test for the return voyage, but Ismay overruled him, and the ship plowed ahead at a high speed.

Radio warnings of icebergs had been issued in the North Atlantic for more than a week, but they were apparently not taken seriously on the *Titanic* until Captain Smith set an iceberg watch on Sunday night. Ismay was so obsessed with setting a speed record that he pocketed an iceberg warning for more than five hours before showing it to Captain Smith.

At about 11:30 P.M., lookouts in the crow's nest reported seeing something that looked like an iceberg in a misty haze. Only after ten fateful minutes did they realize it was an a mountain of ice so immense that its mere tip projected well over fifty feet above the water. By then it was too late for the helmsman to steer away from it. Moving at 22.5 knots, the *Titanic* struck the iceberg hard enough to open a long gash in its hull. Water then began flooding into its supposedly watertight compartments. Shortly after midnight, Captain Smith ordered the lifeboats readied for lowering

At the moment of collision, most passengers felt only a slight jarring and were not alarmed—until the crew ordered them to put on lifejackets

*The sinking of the* Titanic *is front-page news in the April 16, 1912, edition of* The New York Times.

| *TITANIC* SURVIVORS | | | |
|---|---|---|---|
| *People* | *Aboard* | *Survived* | *%* |
| First-class passengers | 324 | 201 | 62.0 |
| Second-class passengers | 277 | 118 | 42.6 |
| Third-class passengers | 708 | 181 | 25.6 |
| Crew | 885 | 212 | 24.0 |
| Postmen/musicians | 13 | 0 | 0 |
| Total | 2,207 | 712 | 32.3 |

*Source:* Encyclopedia Titanica: Statistics of *Titanic* sinking.
   (http://www.encyclopedia-titanica.org)

and to go to the lifeboats on the upper decks. In pajamas and fur coats, first-class passengers were first to step out into subfreezing air temperatures. Most second-class passengers also made it outside, but steerage passengers had trouble climbing out of the lower decks because of the poor directions they received and the confusing labyrinthine design of passageways. Few of them would survive.

As radio distress signals alerted nearby ships to the *Titanic*'s peril, flares lit up the night sky, but no ships were near enough to offer immediate assistance. Meanwhile, confusion reigned on the *Titanic* as the ship slowly sank, bow first. Some of the first lifeboats lowered to the sea carried less than half their capacities. Calls for women and children to go first were ignored, and even White Star chairman Ismay himself got aboard a lifeboat. The last distress signal went out at 2:17 A.M., the ship's lights failed one minute later.

In the ship's final moments, its stern rose high above the water, then slowly slipped back down, until the ship's immense weight caused it nearly to snap in two. Finally, the forward section sank and pulled the stern down after it shortly after 2:20 A.M. Several hours later, another ship, the *Carpathia*, arrived and began collecting survivors. The sea was calm, but the water was so frigid that almost no one who failed to get into a lifeboat survived.

**Impact**

Nearly 1,500 people went down with the *Titanic*. Had the ship carried more lifeboats, the loss of life would almost surely have been less. Hearings on the disaster began less than a week after the sinking. In these and later inquiries, Captain Smith's name was cleared. Ismay was also cleared of liability, but he was never the same afterward. He retired from the White Star Line a year later. New maritime safety regulations came out of the inquiries. Ships were eventually required to carry enough lifeboats to accommodate all passengers, lifeboat drills became mandatory, watertight decks became standard in ship construction, and lookouts had to have regular eye examinations.

Meanwhile, the *Titanic* itself has become the subject of legend. The tragic story of its sinking has fascinated people worldwide and inspired books, films, and museum exhibitions. In 1985, the discovery of the ship's resting place and subsequent release of poignant underwater pictures of the wreck spurred new interest in the *Titanic*. Public fascination reached a peak in 1997 when a film taking its title from the ship achieved worldwide success.

*Noelle K. Heenan*

# U.S. Marines Go to Nicaragua to Quell Unrest

> *By sending marines into Nicaragua, the United States prevented the Nicaraguan people from choosing their own government.*

**What:** Political aggression
**When:** August 3, 1912
**Where:** Nicaragua
**Who:**
José Santos Zelaya (1845-1919), dictator of Nicaragua from 1893 to 1909
Juan J. Estrada (1871-1947), leader of the 1909 revolt that overthrew Zelaya
Adolfo Díaz (1874-1964), president of Nicaragua from 1913 to 1917 and from 1926 to 1928
Augusto César Sandino (1895-1934), a legendary guerrilla leader
Anastasio Somoza García (1896-1956), dictator of Nicaragua from 1936 to 1956

### Rebellion in Nicaragua

In the nineteenth century, political leaders struggled to gain control of Nicaragua, and there were many abuses of human rights: election fraud, political corruption, destruction and confiscation of property, exile, execution, and murder. These problems decreased toward the end of the century, as José Santos Zelaya began to gain power.

Zelaya was a dictator, but he was also a Liberal who expanded education, improved the armed forces, and encouraged trade. Still, the United States was not pleased with his rule. At the beginning of the twentieth century, American leaders began planning to build a canal across Central America, and Nicaragua was their first choice of location. Zelaya eagerly wanted a canal, but he insisted that Nicaragua's sovereignty must be guarded. As a result, the Americans de-

cided to construct a canal through Panama instead.

Along with many other Nicaraguans, Zelaya resented this loss and hoped that another world power would decide to build a canal through Nicaragua. In angry public speeches, he pushed for a canal agreement with Germany and Japan. This possibility was alarming to American leaders, for they were determined to keep European powers out of the Americas.

On October 10, 1909, a revolt against Zelaya broke out in the Atlantic port of Bluefields. This uprising was led by General Juan J. Estrada, who was governor over the Atlantic coastal region of Nicaragua. Soon Zelaya's forces caught two American mercenaries laying mines in the San Juan River. These two men were quickly tried, condemned, and executed by a firing squad.

Though these two soldiers had become officers in the revolutionary movement and in doing so had given up their rights as American citizens, the United States took the opportunity to break diplomatic relations with Zelaya's government. On December 1, 1909, Zelaya resigned and went into exile.

### The Americans Interfere

The United States had already sent gunboats to the Bluefields area, and after the revolution had begun American Marines had landed in Nicaragua, supposedly to protect American lives and property. Declaring Bluefields a neutral zone, the United States kept the Nicaraguan government forces from shelling the revolutionaries. Since the revolutionaries had control of the port, they were able to receive customs money as well as weapons shipments from abroad.

The Nicaraguan congress chose José Madriz, a respected Liberal judge, to replace Zelaya, but the United States refused to recognize him. Though the Liberals were a majority in Nicaragua, the United States insisted that the Conservatives were the true representatives of the people.

In 1910, the United States supported Estrada as provisional president of Nicaragua. He put together a government that included both Liberals and Conservatives, but when the marines left, the coalition fell apart. In May, 1911, Estrada resigned and gave his place to the vice president, Adolfo Díaz (another Conservative).

When the Liberals, led by Benjamín Zeledón, revolted, Díaz asked for U.S. help. On August 3, 1912, twenty-seven hundred U.S. Marines landed and took over the main railroad and principal cities. The Conservatives quickly hunted down the Liberal leaders and displayed Zeledón's body as a prize.

For almost all the next twenty years, U.S. Marines would stay in Nicaragua. Usually there were not more than one hundred at a time, but that was enough to keep the peace and protect the political leaders whom the United States wished to support.

The United States also became deeply involved in Nicaragua's finances. After Zelaya fell in 1909, the country's finances had become chaotic, with many debts to foreign creditors. To prevent European countries from moving into Nicaragua to recover the money that was owed to

*Nicaraguans line up on a dock, waiting for U.S. Marines to arrive.*

Library of Congress

**267**

them, the United States arranged for new loans from New York bankers. To pay back the new loans, however, Nicaragua had to mortgage its future customs duties and its income from rail and steamship transportation. American bankers set up the National Bank of Nicaragua, whose board of directors met in New York City rather than in Nicaragua.

In the Bryan-Chamorro Treaty, which was signed in 1914 and ratified in 1916, the United States bought the right to construct a canal through Nicaragua for only three million dollars. Having finished the Panama Canal in 1914, the United States had no intention of building another in Nicaragua; the only purpose of the treaty was to prevent any other world power from building a Central American canal.

## Consequences

Realizing that the Nicaraguans were not happy with U.S. interference in their affairs, the United States withdrew the marines in 1925 and trained the Nicaraguan National Guard as a replacement. A new government of Conservatives and Liberals was formed, but the Conservatives were not cooperative, and the Liberals soon rebelled again. U.S. Marines returned to Nicaragua in 1926; by the next February, there were fifty-four hundred marines and eleven destroyers and cruisers in Nicaraguan cities and ports.

Finally, the United States tried to come to an agreement with the Liberals—but one Liberal general, Augusto César Sandino, had had enough. He demanded that the United States leave Nicaragua immediately and let it solve its own problems. The National Guard tried to hunt Sandino down, but he and his supporters used guerrilla warfare to escape capture.

For six years, the United States had two thousand soldiers fighting Sandino and helping the National Guard. Bombs were dropped on Nicaraguan villages, killing many civilians, but Sandino and his movement endured. Finally the United States pulled the marines out in January, 1933, leaving behind Anastasio Somoza García as head of the National Guard.

Somoza found a way to assassinate Sandino on February 23, 1934, and with the support of the National Guard Somoza became president of Nicaragua in 1936. Until his own assassination in 1956, he ruled Nicaragua as a dictator, using beatings, torture, imprisonment, and murder against his opponents. After his death, his two sons took power and carried on the tradition of dictatorship. Educated in the United States, the younger Somozas kept close ties with the U.S. government.

In 1961, the Sandinista Front of National Liberation (FSLN), named in Sandino's honor, was founded and began an eighteen-year guerrilla struggle against the Somozas. On July 19, 1979, the last Somoza was brought down and the Sandinistas held power in Nicaragua until voted out in 1990.

*Maurice P. Brungardt*

# Hess Determines Extraterrestrial Origins of Cosmic Rays

*Victor Franz Hess pioneered dangerous high-altitude balloon experiments that indicated the existence of cosmic rays originating outside Earth and its solar system.*

**What:** Physics; Astronomy
**When:** August 7 and 12, 1912
**Where:** Aussig, Austria
**Who:**

VICTOR FRANZ HESS (1883-1964), an Austrian experimental physicist who was a cowinner of the 1936 Nobel Prize in Physics

ROBERT ANDREWS MILLIKAN (1868-1953), an American physicist who won the 1923 Nobel Prize in Physics

CARL DAVID ANDERSON (1905-1991), an American physicist who discovered the positron and was a cowinner with Hess of the 1936 Nobel Prize in Physics

## Nuisance or New Science?

The beginning of the twentieth century presented scientists with a puzzling fact. A tightly sealed electroscope (a device that detects the presence of minute electrical charges) slowly developed an electric charge even when all sources of electrical leakage were carefully eliminated. Scientists wondered if the strange charging was merely a nuisance or if it pointed to some new science awaiting discovery. The charging diminished slightly with thick shielding over the electroscope. If the charging were real, a penetrating radiation might produce it by colliding with the trapped air inside the sealed electroscope and releasing electrons. The charging was as strong over the oceans as over the land; therefore, radioactivity from rocks could not be the source of the penetrating rays. Since the land and oceans were not the source, perhaps the rays came from the skies. Measurements were made on the Eiffel

Tower and in primitive balloon flights with little change in the charging.

In 1910, Victor Franz Hess, an assistant at the Institute of Radium Research of the Vienna Academy of Science, began to study the problem. Hess had calculated that heights greater than one and one-half that of the Eiffel Tower would be needed to show any real difference in charging dependent upon altitude and that much greater altitudes would be necessary to distinguish a sky source different from ground radiation. He approached the Austrian Aeroclub to request high-altitude balloon flights. The Aeroclub agreed. Hess gathered daring companions, and in 1911, he began his series of dangerous balloon experiments.

## Catching Rays

Hess began his next-to-last flight with two companions on August 7, 1912, from the vicinity of Aussig, Austria. Hess monitored his three electroscopes, one companion navigated, and the other checked altitude and temperature. After rising for two and one-half hours, the balloon drifted between 4,000 and 5,000 meters for one hour more, covering 200 kilometers to Pieskew, Germany, near Berlin. At the highest altitude, Hess's electroscopes acquired a charge four times faster than at ground level. He believed the radiations responsible for the charging effect were coming from outside Earth.

Hess's next ascent was to determine whether the extraterrestrial rays originated in the sun. He scheduled the balloon flight for August 12 to coincide with a solar eclipse. He took the balloon to 3,000 meters and found no reduction in the intensity of the rays as the sun disappeared from view. This eliminated the sun as a major source of the cosmic rays. The August 12 flight marked the

The Nobel Foundation

*Victor Franz Hess.*

end of the historic set of ten flights, half of which were carried out at night. Hess summarized his conclusions in a paper published that November: "The results of my observations are best explained by the assumption that a radiation of very great penetrating power enters our atmosphere from above."

Confirmation by other scientists came slowly. Werner Kohlhörster found that the unknown rays were twelve times as intense at 10 kilometers altitude as on the earth's surface. Hess continued his research and found that a small daily variation in cosmic ray intensity was accounting for a very slight amount of the overall cosmic ray intensity of the sun. His accumulated data indicated the origin of cosmic rays to be beyond the galaxy. Nevertheless, general acceptance was still

slow. Many scientists could not envision rays coming from outer space. It was not until 1925 that Robert Andrews Millikan acknowledged his full acceptance of Hess's pioneering efforts by labeling the unknown radiation as "cosmic rays."

## Consequences

With the recognition that cosmic rays were real, physicists became aware that they had an extraordinary tool for investigating not only outer space but also the innermost part of matter. Evidence showed that cosmic rays possess extraordinary energy. Carl David Anderson was one of the first to put these extreme energies to use.

Physicists often measure radiation in volts. This unit measures the energy of an electron or a proton (which has the same value of charge as the electron but is positive rather than negative). For example, visible light requires a particle energy of several volts, while a particle energy of 50,000 volts can generate penetrating X rays.

Anderson knew that cosmic rays possess extraordinary energies, although he did not know the full extent of these energies. He enlisted the rays in tracking down the positron, the positively charged twin of the tiny electron. He set a vapor-filled vessel in a large magnetic field, exposed the contents to Hess's ever-present cosmic rays, and photographed the results. His discovery of the positron came in a cloud chamber picture of the collision debris left when the antiparticle was hit by a 500-million-volt cosmic ray. No other instrument at the time could generate such energies.

A wide range of new experimental techniques pointed to supernovas as a source of cosmic rays. Supernovas throw off energetic particles with very high energies; these particles can acquire more energy in their long journey within galaxies. Gamma rays, which are very energetic X rays, are formed when cosmic rays encounter magnetic fields in space. The cosmic rays generated by supernovas have almost unimaginable energies and pervade the galaxy. Whatever the exact source, the cosmic rays, particles and gammas, are space relics that can tell scientists much about the unknowns of outer space.

*Peter J. Walsh*

# Balkan Wars Divide European Loyalties

> *Two wars fought in southeastern Europe increased tensions among the various European states and set the stage for the eruption of World War I.*

**What:** Military conflict
**When:** October 18, 1912
**Where:** The Balkans
**Who:**
COUNT LEOPOLD VON BERCHTOLD (1863-1942), foreign minister of Austria-Hungary from 1912 to 1915
SERGEI DMITRIEVICH SAZONOV (1861-1927), Russian minister of foreign affairs from 1910 to 1916
SIR EDWARD GREY (1862-1933), foreign secretary of Great Britain from 1905 to 1916
FERDINAND I OF SAXE-COBURG (1861-1948), king of Bulgaria from 1908 to 1918

## The First War

Since 1908, when Austria had annexed the Turkish provinces of Bosnia and Herzegovina, Russia had wanted to keep Austria from advancing any farther in the Balkans, especially against Serbia. So from 1909 to 1913, Russia worked for the establishment of a Balkan League, based on separate agreements between Bulgaria on the one hand and Serbia, Greece, and Montenegro on the other, along with an accord between Montenegro and Serbia.

The league finally came into existence in 1912, but the leaders of its various nations were not interested in simply maintaining things as they were in the Balkans in order to please Russia. Austria's annexation of Bosnia and Herzegovina in 1908 and Italy's annexation of Tripoli in 1912 encouraged the members of the Balkan League to try to drive the Turks out of Europe altogether and divide the conquered lands among themselves.

None of the Great Powers wanted to see such a war, which might easily spread beyond the Balkans. Austria and Russia were particularly concerned. On October 8, 1912, Foreign Minister Count Leopold von Berchtold of Austria-Hungary and Russian minister of foreign affairs Sergei Dmitrievich Sazonov, speaking for all the European powers, made a declaration to the Balkan states warning them not to make war on the Turks. The declaration warned that if the Balkan nations did make war and won (a possibility that was considered very unlikely), they would not be allowed to annex any territory.

Yet the warning came too late. On the same day it was issued, Montenegro boldly declared war on the Ottoman Empire. War began in earnest on October 18, when Bulgaria, Serbia, and Greece entered the conflict.

Six weeks later, on December 3, the badly beaten Turks called for an armistice. At the same time, a serious international crisis had arisen over Serbia's occupation of a stretch of the northern Adriatic coast of Albania. Austria and Italy were completely opposed to Serbia's taking control of any part of the Adriatic coastline, for that would allow Serbia and Russia to challenge the naval supremacy of the Austrians and Italians. Rather reluctantly, Germany agreed to support Austria if Austria were attacked while defending its interests.

Sazonov, afraid that Austria might pose a threat to Serbian interests (as had happened in 1908), spoke up for Serbian claims. Realizing, however, that Russia was not yet ready to become involved in a major war, Sazonov worked to resolve the conflict through negotiation. He helped to organize the London Peace Conference, which opened on December 16, 1912, under the chairmanship of Edward Grey, foreign secretary of Great Britain.

Early in 1913, Austria and Italy were able to gain some acceptance from other European powers for the creation of an enlarged Albania, which would safely keep Serbia from having access to the Adriatic. Meanwhile, hostilities had broken out again between members of the Balkan League and the Turks, on January 30, 1913. The Turks, however, were again defeated, and the Treaty of London of May 30, 1913, brought the First Balkan War to a close.

## The Second War

None of the Balkan states was satisfied, however, with the terms it had been forced to accept in London. Because Serbia had been denied an outlet to the Adriatic, it demanded a substitute: Bulgaria should give it a larger slice of Macedonian territory than the Treaty of London had assigned. Bulgaria indignantly refused. Bulgaria also did not want to give in to Greece's claims to the Thessalonica area of Macedonia or Romania's claims to the Dobruja, an area near the mouth of the Danube River.

Surrounded by hostile nations, Ferdinand I of Saxe-Coburg, who had been the independent king of Bulgaria since 1908, decided to remove the Serbs and the Greeks from Macedonia. Bulgaria attacked on June 29, 1913, and in response

*Montenegrins attack a town in Serbia in October, 1912.*

Bain Collection, Library of Congress

Serbia and Greece declared war on Bulgaria. They were soon joined by Montenegro and Romania, and then by their former enemy, the Ottoman Empire.

Bulgaria suffered a serious defeat. On August 10 in Bucharest, it agreed to terms of peace with the Balkan states and gave back the territories it had wanted to take over. In a separate peace treaty with Turkey, signed at Constantinople on September 29, Bulgaria gave Turkey the greater part of Thrace (which it had gained in the First Balkan War), including the important city of Adrianople.

## Consequences

The Balkan Wars were over, but the peace of southeastern Europe was in ruins, and the stability of all Europe remained in danger. Relations between Austria and Serbia became even more bitter. The combined might of Austria-Hungary and Germany had, once again, forced Russia to back down in its support of Serbia. This was quite humiliating for Russia. Bulgaria and Turkey were dissatisfied with the results of the Balkan Wars; when World War I broke out, these two nations ended up joining forces with Austria-Hungary and Germany.

*Edward P. Keleher*

**273**

# Wilson Is Elected U.S. President

*With the Republican Party split between two candidates, Democrat Woodrow Wilson won the U.S. presidency and brought a new era of social and political reform.*

**What:** National politics
**When:** November, 1912
**Where:** United States
**Who:**
WOODROW WILSON (1856-1924), president of the United States from 1913 to 1921
WILLIAM HOWARD TAFT (1857-1930), president of the United States from 1909 to 1913
THEODORE ROOSEVELT (1858-1919), president of the United States from 1901 to 1909, and leader of the new Progressive Party
JAMES BEAUCHAMP "CHAMP" CLARK (1850-1921), speaker of the House of Representatives from Missouri from 1911 to 1919
OSCAR UNDERWOOD (1862-1929), an influential congressman from Alabama
LOUIS DEMBITZ BRANDEIS (1856-1941), a lawyer who later became a member of the Supreme Court

## The Republican Disaster

When Theodore Roosevelt left the United States presidency in 1909, he was succeeded by William Howard Taft. Taft, who had been supported by Roosevelt, was expected to continue Roosevelt's progressive reforms. Roosevelt went off to Africa on a hunting expedition, believing that the White House would continue to operate much as it had during his presidency.

Through Roosevelt's ability to negotiate and to compromise between liberal and conservative policies, he had been able to hold together an unusual balance of supporters in Congress, including Eastern conservatives and Midwestern and urban progressives. Taft lacked both the physical energy and the political know-how to keep this coalition of support operating. He was a conservative at heart and, unable to inspire a commitment to unity, let the Republican Party fall apart into squabbling factions.

When Roosevelt returned home in 1910, the party was in a state of disarray, and the Democrats had scored important gains in the local and congressional elections of that year. By 1912, the division among Republicans had become even greater. Roosevelt marched out of the Republican convention in Chicago on June 22, 1912, to form the new Progressive Party. With this split in Republican loyalties, the election of a Democrat to the presidency was almost certain.

## Wilson Campaigns and Wins

When the Democrats gathered in Baltimore on June 25, 1912, the party was at historic crossroads. Woodrow Wilson had been the frontrunner for the presidential nomination until the Republican split made the Democratic nomination much more valuable. The new leader for the nomination was Champ Clark, the Speaker of the House. He was a typical old-line politician who had broadened his base of support from his home state of Missouri to attract many in the West who had at one time supported William Jennings Bryan. Another leading candidate, Oscar Underwood of Alabama, had strong backing in the South. After a bitter convention fight, on the forty-sixth ballot Wilson finally won the nomination.

Wilson had campaigned for and won his first political office, the governorship of New Jersey, only in 1910. Before his entrance into politics, Wilson had had a career as a respected scholar. He had been graduated from Princeton University in 1879 and had practiced law for a short time; then he finished a doctorate in political science and history at The Johns Hopkins Univer-

274

sity in 1886. In 1902, he had been elected president of Princeton University. By that time, he had already written three books on the U.S. system of government.

Some of his educational reforms eventually led to a dispute at Princeton, and just at that time (1910) he had the opportunity to enter politics as a candidate for the governorship of New Jersey. As governor, Wilson began certain basic reforms that changed New Jersey's government from a corrupt regime run by political bosses into a reform-oriented state. His successes had brought him to national attention as a leader in the Democratic Party.

The election became a confrontation between two progressive philosophies that reflected the new realities of an industrial society. Roosevelt's New Nationalism argued for a strong federal government. Roosevelt wanted to strengthen and regulate large corporations and to launch a program of government-supported social welfare.

In responding, Wilson obtained the help of a prominent Massachusetts lawyer, Louis D. Brandeis. Wilson's New Freedom platform emphasized giving power to state and local governments, breaking up large corporations, and encouraging small businesses.

In the election, Wilson won with 6,293,454 votes to Roosevelt's 4,119,538, Taft's 3,484,980, and socialist Eugene V. Debs's 900,672. Even though Wilson's share of the popular vote was only 42 percent, he won an overwhelming victory in the electoral college, with 435 votes to Roosevelt's 88 and Taft's 8.

## Consequences

As a newcomer in politics, Woodrow Wilson was not burdened with political debts like the favors Roosevelt owed to Eastern Republican business interests. Wilson also was fortunate to deal with a sympathetic Congress that was ready to cooperate in bringing about progressive reforms.

Wilson strengthened the presidency even further than Roosevelt had. He became the leader of the people as well as the Congress, and he dominated the government in both his terms. In his first term he influenced Congress to pass important legislative reforms in tariffs and banking and the breaking up of business monopolies. Later he added to his own program almost all the proposals Roosevelt had championed in the great debate of 1912. When the United States was drawn into World War I, Wilson became the first of the powerful war presidents of the twentieth century.

The effects of the 1912 campaign and of Wilson's election to the presidency have continued in American liberal politics. The New Deal of Franklin D. Roosevelt contained many echoes of Wilson's thinking, and the expansion of power Wilson brought about for the executive branch allowed the later programs of social welfare to be carried out. In international affairs, Wilson's Fourteen Points, which promised a peace without victory in World War I, expressed some of the ideals that are still being debated in American foreign policy.

*Richard H. Collin*

Library of Congress

*Woodrow Wilson.*

# Burton's Thermal Cracking Process Improves Gasoline Production

*The development of the Burton process for cracking petroleum increased the gasoline yield from petroleum and stimulated the development of the petroleum industry.*

**What:** Chemistry
**When:** 1913
**Where:** United States
**Who:**
WILLIAM M. BURTON (1865-1949), an American chemist
ROBERT E. HUMPHREYS, an American chemist

## Gasoline, Motor Vehicles, and Thermal Cracking

Gasoline is a liquid mixture of hydrocarbons (chemicals made up of only hydrogen and carbon) that is used primarily as a fuel for internal combustion engines. It is produced by petroleum refineries that obtain it by processing petroleum (crude oil), a naturally occurring mixture of thousands of hydrocarbons, the molecules of which can contain from one to sixty carbon atoms.

Gasoline production begins with the "fractional distillation" of crude oil in a fractionation tower, where it is heated to about 400 degrees Celsius at the tower's base. This heating vaporizes most of the hydrocarbons that are present, and the vapor rises in the tower, cooling as it does so. At various levels of the tower, various portions (fractions) of the vapor containing simple hydrocarbon mixtures become liquid again, are collected, and are piped out as "petroleum fractions." Gasoline, the petroleum fraction that boils between 30 and 190 degrees Celsius, is mostly a mixture of hydrocarbons that contain five to twelve carbon atoms.

Only about 25 percent of petroleum will become gasoline via fractional distillation. This amount of "straight run" gasoline is not suffi-

cient to meet the world's needs. Therefore, numerous methods have been developed to produce the needed amounts of gasoline. The first such method, "thermal cracking," was developed in 1913 by William M. Burton of Standard Oil of Indiana. Burton's cracking process used heat to convert complex hydrocarbons (whose molecules contain many carbon atoms) into simpler gasoline hydrocarbons (whose molecules contain fewer carbon atoms), thereby increasing the yield of gasoline from petroleum. Later advances in petroleum technology, including both an improved Burton method and other methods, increased the gasoline yield still further.

## More Gasoline!

Starting in about 1900, gasoline became important as a fuel for the internal combustion engines of the new vehicles called automobiles. By 1910, half a million automobiles traveled American roads. Soon, the great demand for gasoline—which was destined to grow and grow—required both the discovery of new crude oil fields around the world and improved methods for refining the petroleum mined from these new sources. Efforts were made to increase the yield of gasoline—at that time, about 15 percent—from petroleum. The Burton method was the first such method.

At the time that the cracking process was developed, Burton was the general superintendent of the Whiting refinery, owned by Standard Oil of Indiana. The Burton process was developed in collaboration with Robert E. Humphreys and F. M. Rogers. This three-person research group began work knowing that heating petroleum fractions that contained hydrocarbons more complex than those present in gasoline—a process called "coking"—produced kerosene, coke

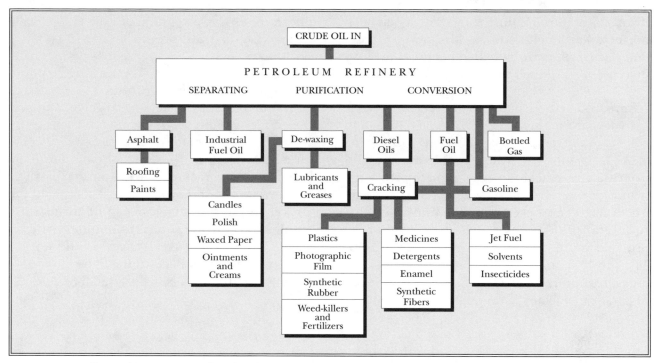

*Burton's process contributed to the development of petroleum refining, shown in this diagram.*

(a form of carbon), and a small amount of gasoline. The process needed to be improved substantially, however, before it could be used commercially.

Initially, Burton and his coworkers used the "heavy fuel" fraction of petroleum (the 66 percent of petroleum that boils at a temperature higher than the boiling temperature of kerosene). Soon, they found that it was better to use only the part of the material that contained its smaller hydrocarbons (those containing fewer carbon atoms), all of which were still much larger than those present in gasoline. The cracking procedure attempted first involved passing the starting material through a hot tube. This hot-tube treatment vaporized the material and broke down 20 to 30 percent of the larger hydrocarbons into the hydrocarbons found in gasoline. Various tarry products were also produced, however, that reduced the quality of the gasoline that was obtained in this way.

Next, the investigators attempted to work at a higher temperature by bubbling the starting material through molten lead. More gasoline was made in this way, but it was so contaminated with gummy material that it could not be used. Continued investigation showed, however, that moderate temperatures (between those used in the hot-tube experiments and that of molten lead) produced the best yield of useful gasoline.

The Burton group then had the idea of using high pressure to "keep starting materials still." Although the theoretical basis for the use of high pressure was later shown to be incorrect, the new method worked quite well. In 1913, the Burton method was patented and put into use. The first cracked gasoline, called Motor Spirit, was not very popular, because it was yellowish and had a somewhat unpleasant odor. The addition of some minor refining procedures, however, soon made cracked gasoline indistinguishable from straight run gasoline. Standard Oil of Indiana made huge profits from cracked gasoline over the next ten years. Ultimately, thermal cracking subjected the petroleum fractions that were utilized to temperatures between 550 and 750 degrees Celsius, under pressures between 250 and 750 pounds per square inch.

## Consequences

In addition to using thermal cracking to make gasoline for sale, Standard Oil of Indiana also profited by licensing the process for use by other gasoline producers. Soon, the method was used

**277**

throughout the oil industry. By 1920, it had been perfected as much as it could be, and the gasoline yield from petroleum had been significantly increased. The disadvantages of thermal cracking include a relatively low yield of gasoline (compared to those of other methods), the waste of hydrocarbons in fractions converted to tar and coke, and the relatively high cost of the process.

A partial solution to these problems was found in "catalytic cracking"—the next logical step from the Burton method—in which petroleum fractions to be cracked are mixed with a catalyst (a substance that causes a chemical reaction to proceed more quickly, without reacting itself). The most common catalysts used in such cracking were minerals called "zeolites." The wide use of catalytic cracking soon enabled gasoline pro-ducers to work at lower temperatures (450 to 550 degrees Celsius) and pressures (10 to 50 pounds per square inch). This use decreased manufacturing costs because catalytic cracking required relatively little energy, produced only small quantities of undesirable side products, and produced high-quality gasoline.

Various other methods of producing gasoline have been developed—among them catalytic reforming, hydrocracking, alkylation, and catalytic isomerization—and now about 60 percent of the petroleum starting material can be turned into gasoline. These methods, and others still to come, are expected to ensure that the world's needs for gasoline will continue to be satisfied—as long as petroleum remains available.

*John A. Heitmann*

# Edison Develops Talking Pictures

*Thomas Alva Edison developed a system of talking pictures by using a phonograph linked to a film projector.*

**What:** Communications; Photography; Entertainment
**When:** 1913
**Where:** West Orange, New Jersey
**Who:**
THOMAS ALVA EDISON (1847-1931), an American inventor
WILLIAM KENNEDY LAURIE DICKSON (1860-1937), an American mechanic and inventor
MILLER REESE HUTCHISON (1876-1944), the American inventor of the Klaxon, or car horn
LEE DE FOREST (1873-1961), an American inventor
GEORGE EASTMAN (1854-1932), an American inventor and industrialist

## Combining Sight and Sound

Thomas Alva Edison began experimenting with a device that would show moving pictures because he wanted a visual image to accompany the music reproduced by his phonograph. He said that moving pictures would do for the eye what the phonograph had done for the ear. He thought that both sound and visual images could be stored on a phonograph cylinder and then reproduced at any time by the user. He selected William Kennedy Laurie Dickson from his staff to work on the project at Edison's new laboratory in West Orange, New Jersey.

Dickson was an accomplished experimenter with a strong interest in photography. He started work in 1887. He fixed small photographs onto a phonograph cylinder and viewed them through a microscope as they rotated. He found that sequential images of a moving object have the impression of movement when passed rapidly through the viewer's line of sight. Edison and

Dickson had examined several devices that exploited the phenomenon of persistence of vision—when the eye retains the image a short time after the image has disappeared. It was well known that a sequence of images could be manipulated to give the visual illusion of movement. Edison and Dickson were confident that a series of small photographs could achieve this effect. This was the conceptual beginning of motion pictures.

By 1888, Dickson had built a device that could reproduce both sight and sound. This "moving view" apparatus contained a phonograph cylinder and a drum covered with a series of small photographs. Both were centered on a common axle. When the axle was turned, the viewer watched the moving pictures through a microscope and listened to the accompanying sound track through earphones connected to a mechanism on the cylinder. This device was the first attempt to show talking pictures.

Further experiments proved that this approach was not feasible because the sound and the vision elements could not be synchronized. The phonograph cylinder had to revolve with a continuous motion to ensure good reproduction of the sound track, while the photograph drum needed an intermittent motion: The image had to be moved into the line of sight, fixed there for a fraction of a second so that the eye could record it, and then rapidly moved out of the line of sight. Despite months of experiments, the problem remained insurmountable, forcing Edison to drop the idea of combining sight and sound in one device.

Once released from the demands of talking pictures, Dickson experimented on the best method of manipulating sequential photographs to give the impression of movement. His research produced the important innovation of the perforated filmstrip that replaced the photo-

**279**

Hulton Archive

*Thomas Alva Edison produced the first talking motion pictures, using this building.*

graphs on a revolving drum. The edges of the filmstrip were punched with holes to fit the teeth of the ratchet that moved the strip rapidly past the lens. The film camera, patented by Edison in 1891, was a light-proof box containing a lens, shutter, electric motor, and length of perforated filmstrip.

## Advancing Technology

Although moving-picture technology was developed after the phonograph, synchronized sound reproduction proved to be the greatest obstacle. The best format for a long-playing phonograph was the disc rather than the cylinder. Invented by American Emil Berliner, the disc-playing talking machine had become a serious competitor to Edison's phonograph. The playing time of a disc record could easily be extended to around seven minutes, and oversized discs of-

fered much longer playing times. The average playing time of an Edison phonograph cylinder at this time was about two minutes. Edison found it difficult to abandon the format he had invented and concentrated instead on increasing the size of the cylinder. He developed cylinders of about 11 centimeters in diameter and 19 centimeters in length that could play for about six minutes.

Once the playing time had been increased, there still remained the task of amplifying the volume of the playback to ensure that the sound track could be heard from every seat in the theater. In 1908, Daniel Higham of Massachusetts was hired to work on sound amplification. Higham designed an oversized reproducer assembly that contained a special device that did a better job of picking up the sound vibrations made by the reproducing needle. The result was much louder playback.

A close associate of Edison, Miller Reese Hutchison, developed methods of sound recording with numerous horns in order to achieve a more balanced sound track. Hutchison also tried a variety of controlling devices to coordinate the film projector at the rear of the auditorium with the phonograph behind the screen. After rejecting a system of electrical controls, Hutchison finally decided on a simple mechanical system. A clutch mechanism was used to adjust the speed of the revolving cylinder of the phonograph to keep it in time with the film. The film projectionist worked the clutch with a cord that ran on pulleys. This device was formally introduced in 1913, with Edison's claim that he had perfected talking pictures.

## Consequences

Edison's device, called the "kinetophone," had a profound effect on motion-picture audiences. When it was first publicly demonstrated in New York City, the newspapers reported that gasps of astonishment could be heard when sound came from the screen.

The fame of the kinetophone was short-lived, however, for after the novelty had worn off, movie audiences became impatient with its shortcomings. The playing time of the kinetophone could not accommodate the longer, epic silent films that were popular in 1913. Audiences wanted their entertainment to last hours, not minutes. Problems with synchronization continued to plague the kinetophone. After another year of unsuccessful experiments and theatrical failures, Edison gave up on the kinetophone and made no further attempts to perfect talking pictures.

Yet Edison had demonstrated the appeal of the talkies to film audiences. Another ten years were to pass before electronic recording and amplification made talking pictures a commercial reality.

*Andre Millard*

**281**

# Geothermal Power Is Produced for First Time

*Hot springs in northern Italy were used to provide steam to power an electric generator, thus inaugurating the first geothermal power installation.*

**What:** Energy
**When:** 1913
**Where:** Larderello, Italy
**Who:**
PIERO GINORI CONTI (1865-1939), Italian prince and industrialist
SIR CHARLES PARSONS (1854-1931), an English engineer
B. C. MCCABE, an American businessman

## Developing a Practical System

The first successful use of geothermal energy was at Larderello in northern Italy. The Larderello geothermal field, located near the city of Pisa about 240 kilometers northwest of Rome, contains many hot springs and fumaroles (steam vents). In 1777, these springs were found to be rich in boron, and in 1818, Francesco de Larderel began extracting the useful mineral borax from them. Shortly after 1900, Prince Piero Ginori Conti, director of the Larderello borax works, conceived the idea of using the steam for power production. An experimental electrical power plant was constructed at Larderello in 1904 to provide electric power to the borax plant. After this initial experiment proved successful, a 250-kilowatt generating station was installed in 1913 and commercial power production began.

As the Larderello field grew, additional geothermal sites throughout the region were prospected and tapped for power. Power production grew steadily until the 1940's, when production reached 130 megawatts; however, the Larderello power plants were destroyed late in World War II (1939-1945). After the war, the generating plants were rebuilt and were producing more than 400 megawatts by 1980.

The Larderello power plants encountered many technical problems that would later concern other geothermal facilities. For example, hydrogen sulfide in the steam was highly corrosive to copper, so the Larderello power plant used aluminum for electrical connections much more than did conventional power plants of the time. Also, the low pressure of the steam in early wells at Larderello presented problems. The first generators simply used steam to drive a generator and vented the spent steam into the atmosphere. A system of this sort, called a "noncondensing system," is useful for small generators but not efficient to produce large amounts of power.

Most steam engines derive power not only from the pressure of the steam but also from the vacuum created when the steam is condensed back to water. Geothermal systems that generate power from condensations, as well as direct steam pressure, are called "condensing systems." Most large geothermal generators are of this type. Condensation of geothermal steam presents special problems not present in ordinary steam engines: There are other gases present that do not condense. Instead of a vacuum, condensation of steam contaminated with other gases would result in only a limited drop in pressure and, consequently, very low efficiency.

Initially, the operators of Larderello tried to use the steam to heat boilers that would, in turn, generate pure steam. Eventually, a device was developed that removed most of the contaminating gases from the steam. Although later wells at Larderello and other geothermal fields produced steam at greater pressure, these engineering innovations improved the efficiency of any geothermal power plant.

## Expanding the Idea

In 1913, the English engineer Sir Charles Parsons proposed drilling an extremely deep (12-kilometer) hole to tap the earth's deep heat. Power from such a deep hole would not come from natural steam as at Larderello but would be generated by pumping fluid into the hole and generating steam (as hot as 500 degrees Celsius) at the bottom. In modern terms, Parsons proposed tapping "hot dry-rock" geothermal energy.

The first use of geothermal energy in the United States was for direct heating. In 1890, the municipal water company of Boise, Idaho, began supplying hot water from a geothermal well. Water was piped from the well to homes and businesses along appropriately named Warm Springs Avenue. At its peak, the system served more than four hundred customers, but as cheap natural gas became available, the number declined.

Although Larderello was the first successful geothermal electric power plant, the modern era of geothermal electric power began with the opening of the geysers geothermal field in California. Early attempts began in the 1920's, but it was not until 1955 that B. C. McCabe, a Los Angeles businessman, leased 14.6 square kilometers in the geysers area and founded the Magma Power Company. The first 12.5-megawatt generator was installed at the geysers in 1960, and production increased steadily from then on. The geysers surpassed Larderello as the largest producing geothermal field in the 1970's, and more than 1,000 megawatts were being generated by 1980. By the end of 1980, geothermal plants had been installed in thirteen countries, with a total capacity of almost 2,600 megawatts, and projects with a total capacity of more than 15,000 megawatts were being planned in more than twenty countries.

## Consequences

Geothermal power has many attractive features. Because the steam is naturally heated and under pressure, generating equipment can be simple, inexpensive, and quickly installed. Equipment and installation costs are offset by savings in fuel. It is economically practical to install small generators, a fact that makes geothermal plants attractive in remote or underdeveloped areas. Most important to a world faced with a variety of technical and environmental problems connected with fossil fuels, geothermal power does not deplete fossil fuel reserves, produces little pollution, and contributes little to the greenhouse effect.

Despite its attractive features, geothermal power has some limitations. Geologic settings suitable for easy geothermal power production are rare; there must be a hot rock or magma body close to the surface. Although it is technically possible to pump water from an external source into a geothermal well to generate steam, most geothermal sites require a plentiful supply of natural underground water that can be tapped as a source of steam. In contrast, fossil-fuel generating plants can be at any convenient location.

*Steven I. Dutch*

# Gutenberg Discovers Earth's Mantle-Outer Core Boundary

*Beno Gutenberg expanded the use of seismographs to the global scale upon his discovery of the boundary between the earth's outer core and the lower mantle.*

**What:** Earth science
**When:** 1913
**Where:** Göttingen, Germany
**Who:**
BENO GUTENBERG (1889-1960), a German-born geologist
JOHN MILNE (1850-1913), a seismologist who invented the first seismograph in 1883
ANDRIJA MOHOROVIČIĆ (1857-1936), a Yugoslavian meteorologist

## A Model of the Earth

Based on evidence gained through indirect and secondary research and experiments, scientists have been able to determine that the earth consists of layers. The uppermost layer is the crust, on or in which most life exists. The structure of the remainder of the earth's interior and the size and composition of the deeper layers have been learned through various methods. The original source of information that founded modern understanding of the earth's structure is seismology, the study of earthquakes and vibrations of the earth.

In 1883, one of the pioneers in the field of seismology, John Milne, perfected the clockwork-powered Milne seismograph, which produced a record of earth vibrations on light-sensitive film. Milne discovered that the vibrations from a distant earthquake arrived in a series of separate vibrations traveling at different speeds. The greater the separation between the waves of different types, the farther away the earthquake. The method was identical to that used to calculate the distance of a storm from the time between the arrival of the lightning and the sound of the thunder.

The seismograph allowed for the identification of three different kinds of waves. The fastest waves are those of compression, the pressure waves or P waves, which move through the air, water or earth by compression and expansion. Another type are shear waves, or S waves, the secondary waves; these waves move only through solids. Finally, surface waves move along boundaries such as the boundary between the rock and the air or the water. The surface waves are a varied group that behave in a way similar to waves in the sea. Milne's early seismograph allowed for the base recording of earthquake waves.

Later research in the field of seismology used the records (seismograms) from Milne and other seismologists to learn the composition of the earth. The results of the research revealed that the innermost section of the earth is composed possibly of pure iron. Surrounding the inner core is the molten outer core, with a radius of approximately 3,500 kilometers and composed of an iron alloy. The mantle of the earth, 3,000 kilometers thick, is the area between the outer core and crust. Probably the most active area in the layers of the earth, the mantle contains large amounts of the mineral olivine.

## The Shadow Zone

Beno Gutenberg's method of research was first used by Andrija Mohorovičić, a Yugoslavian meteorologist, who was one of the pioneers in the science of seismology. The key event that spurred Mohorovičić's inspiration was a minor earthquake in 1909 in Zagreb, Yugoslavia. Upon examining the seismographic records of the earthquake, Mohorovičić found two P and two

S waves recorded for each tremor. The separate groups of P and S waves appeared to be traveling at different velocities. Because the type of material as well as its density changes the rate, and sometimes the direction, of the wave's speed, the seismogram indicated the presence of a layer of material under the earth's outer crust that was dense enough to alter the velocity of the second group of P and S waves.

Because the second wave group reached the recording stations before the first wave group, Mohorovičić deduced that the boundary—now called the "Mohorovičić Discontinuity"—was denser than the crust. Along with the discovering of the boundary between the earth's outer crust and upper mantle, Mohorovičić showed how seismology can be used to explore the interior structure of the earth.

Gutenberg, a graduate student in Germany, was encouraged to begin his own studies in seismology two years after the discovery of the Mohorovičić Discontinuity. Gutenberg's area of concentration was the mysterious "shadow zone" of seismology. For some unknown reason, P waves disappeared when passing through a 4,400-kilometer-wide area on the side of the earth opposite the focus, or epicenter, of the earthquake.

Gutenberg began a mathematical investigation to explain the temporary disappearance of P waves. The most likely answer was suggested by the work of Richard D. Oldham, a geologist, and Emil Wiechert, a geophysicist. Their independently formed theories proposed that the center of the earth contained a large, dense, and perhaps partially molten core.

Assuming they were correct, Gutenberg made mathematical models of the effects a dense core would have on P waves. He positioned his hypothetical core at various depths in the earth, then calculated the course and behavior of P waves in each. After comparing his models to actual seismograph readings, he discovered one that confirmed his work. The model that eventually matched the real graphs was based on a core

2,900 kilometers below the surface of the earth. The shadow zone was the boundary between the lower mantle and outer core. Later research that continued with Gutenberg's original outline altered the accuracy of his placement only slightly and revealed that the outer edge of the core must be molten. Gutenberg's discovery of the boundary between the lower mantle and the outer core is called the "Gutenberg Discontinuity."

## Consequences

Gutenberg's confirmation of the presence of the earth's core and his discovery of the boundary between the outer core and lower mantle in 1913 solved the mystery of the shadow zone and raised new questions regarding the composition of these areas. While the discovery of the boundary between the earth's lower mantle and outer core was significant in itself, the research techniques applied in the study opened new venues for tackling old problems, as well as providing a built-in continuation of the original research.

Later research indicated that the Gutenberg Discontinuity also marked the lowest edges of the continental plates. Through the use of seismographs, the speed of earthquake waves can be applied to learning the composition of a material through which a random wave will pass. Because the waves travel more slowly in the lower part of the mantle, Gutenberg postulated that the lower mantle was softer than the upper mantle. This theory shed new light on the study of plate tectonics.

Plate tectonics is the theory that the outer shell of the earth is formed by several large plates. As these plates move toward or away from one another, they create earthquakes, volcanoes, mountains, and other geological features of the earth's surface. Gutenberg's theory helped to form this theory of plate tectonics: If the lower mantle is the layer in which the plates extend the deepest and if it is partially molten, the movement of plates is more easily explained.

*Earl G. Hoover*

# Salomon Develops Mammography to Detect Breast Cancer

Using a technique from X-ray photography, Albert Salomon developed the first X-ray procedure for detecting and diagnosing breast cancer.

**What:** Medicine; Photography
**When:** 1913
**Where:** Germany
**Who:**
ALBERT SALOMON, the first researcher to use X-ray technology instead of surgery to identify breast cancer
JACOB GERSHON-COHEN (1899-1971), a breast cancer researcher

### Studying Breast Cancer

Medical researchers have been studying breast cancer for more than a century. At the end of the nineteenth century, however, no one knew how to detect breast cancer until it was quite advanced. Often, by the time it was detected, it was too late for surgery; many patients who did have surgery died. So after X-ray technology first appeared in 1896, cancer researchers were eager to experiment with it.

The first scientist to use X-ray techniques in breast cancer experiments was Albert Salomon, a German surgeon. Trying to develop a biopsy technique that could tell which tumors were cancerous and thereby avoid unnecessary surgery, he X-rayed more than three thousand breasts that had been removed from patients during breast cancer surgery. In 1913, he published the results of his experiments, showing that X rays could detect breast cancer. Different types of X-ray images, he said, showed different types of cancer.

Though Salomon is recognized as the inventor of breast radiology, he never actually used his technique to diagnose breast cancer. In fact, breast cancer radiology, which came to be known as "mammography," was not taken up quickly by other medical researchers. Those who did try to reproduce his research often found that their results were not conclusive.

During the 1920's, however, more research was conducted in Leipzig, Germany, and in South America. Eventually, the Leipzig researchers, led by Erwin Payr, began to use mammography to diagnose cancer. In the 1930's, a Leipzig researcher named W. Vogel published a paper that accurately described differences between cancerous and noncancerous tumors as they appeared on X-ray photographs. Researchers in the United States paid little attention to mammography until 1926. That year, a physician in Rochester, New York, was using a fluoroscope to examine heart muscle in a patient and discovered that the fluoroscope could be used to make images of breast tissue as well. The physician, Stafford L. Warren, then developed a stereoscopic technique that he used in examinations before surgery. Warren published his findings in 1930; his article also described changes in breast tissue that occurred because of pregnancy, lactation (milk production), menstruation, and breast disease. Yet Stafford's technique was complicated and required equipment that most physicians of the time did not have. Eventually, he lost interest in mammography and went on to other research.

### Using the Technique

In the late 1930's, Jacob Gershon-Cohen became the first clinician to advocate regular mammography for all women to detect breast cancer before it became a major problem. Mammography was not very expensive, he pointed out, and it was already quite accurate. A milestone in breast cancer research came in 1956, when Gershon-Cohen and others began a five-year study of more than 1,300 women to test the accuracy of mammography for detecting breast cancer. Each woman studied was screened once every six months. Of the 1,055 women who finished the study, 92 were diagnosed with benign tumors and 23 with malignant tumors. Remarkably, out of all these, only one diagnosis turned out to be wrong.

During the same period, Robert Egan of Houston began tracking breast cancer X rays. Over a span of three years, one thousand X-ray photographs were used to make diagnoses. When these diagnoses were compared to the results of surgical biopsies, it was confirmed that mammography had produced 238 correct diagnoses of cancer, out of 240 cases. Egan therefore joined the crusade for regular breast cancer screening.

Once mammography was finally accepted by doctors in the late 1950's and early 1960's, re-

searchers realized that they needed a way to teach mammography quickly and effectively to those who would use it. A study was done, and it showed that any radiologist could conduct the procedure with only five days of training.

In the early 1970's, the American Cancer Society and the National Cancer Institute joined forces on a nationwide breast cancer screening program called the "Breast Cancer Detection Demonstration Project." Its goal in 1971 was to screen more than 250,000 women over the age of thirty-five.

Since the 1960's, however, some people had argued that mammography was dangerous because it used radiation on patients. In 1976, Ralph Nader, a consumer advocate, stated that women who were to undergo mammography should be given consent forms that would list the dangers of radiation. In the years that followed, mammography was refined to reduced the amount of

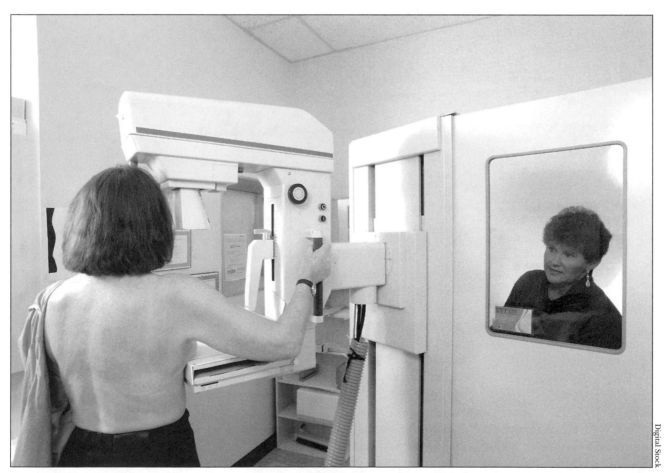

*A woman has a mammogram. Albert Salomon's use of X-ray technology to detect breast cancer was the foundation of modern mammography.*

radiation needed to detect cancer. It became a standard tool for diagnosis, and doctors recommended that women have a mammogram every two or three years after the age of forty.

## Consequences

Radiology is not a science that concerns only breast cancer screening. While it does provide the technical facilities necessary to practice mammography, the photographic images obtained must be interpreted by general practitioners, as well as by specialists. Once Gershon-Cohen had demonstrated the viability of the technique, a means of training was devised that made it fairly easy for clinicians to learn how to practice mammography successfully. Once all these factors—accuracy, safety, simplicity—were in place, mammography became an important factor in the fight against breast cancer.

The progress made in mammography during the twentieth century was a major improvement in the effort to keep more women from dying of breast cancer. The disease has always been one of the primary contributors to the number of female cancer deaths that occur annually in the United States and around the world. This high figure stems from the fact that women had no way of detecting the disease until tumors were in an advanced state.

Once Salomon's procedure was utilized, physicians had a means by which they could look inside breast tissue without engaging in exploratory surgery, thus giving women a screening technique that was simple and inexpensive. By 1971, a quarter million women over age thirty-five had been screened. Twenty years later, that number was in the millions.

*Michael S. Ameigh*

# Schick Introduces Test for Diphtheria

> *Béla Schick developed the Schick test, which is performed on the skin to find out how susceptible a person is to diphtheria.*

**What:** Medicine; Health

**When:** 1913

**Where:** Vienna, Austria

**Who:**

BÉLA SCHICK (1877-1967), a Hungarian microbiologist and pediatrician

EDWIN KLEBS (1834-1913) and FRIEDRICH LÖFFLER (1852-1915), German microbiologists who identified the bacteria that cause diphtheria

ÉMILE ROUX (1853-1933), a French microbiologist

ALEXANDRE YERSIN (1863-1943), a Swiss microbiologist

EMIL VON BEHRING (1854-1917), a German microbiologist who discovered a diphtheria antitoxin

## A Killer Disease

Diphtheria is a serious disease of the upper respiratory tract—the mouth, nose, and pharynx. The person with this disease may have a fever, a sore throat, and pain all over the body. If the disease is not treated with antibiotics, the infection spreads and causes tissue damage in the heart or kidneys and the victim will eventually die.

Diphtheria is caused by *Corynebacterium diphtheriae*, a rod-shaped species of bacteria. The bacterium can be spread from one person to another by touching or by droplets (for example, from sneezes). Once it enters a person's body, the bacterium releases protein toxins that destroy the membranes and inner structures of cells.

In the 1800's, this disease was not yet understood, but the first steps toward that task were taken. Louis Pasteur, Robert Koch, and other microbiologists (scientists who study organisms too small to be seen by the naked eye) established the germ theory of disease, showing that infectious diseases are carried by microorganisms, usually bacteria or viruses. Diphtheria, typhoid fever, scarlet fever, tuberculosis, and several other diseases were major killers in the nineteenth century, especially among patients in hospitals. Microbiologists were determined to discover the microorganisms that caused these diseases.

In 1883, the German microbiologists Edwin Klebs and Friedrich Löffler raised guinea pigs infected with diphtheria. Under microscopes, they observed rod-shaped bacteria growing in blood samples from the infected animals. When this bacterium was injected into healthy guinea pigs, they became ill with diphtheria, too. In this way, the scientists proved that this rod-shaped bacterium was the cause of diphtheria.

Löffler believed that these bacteria hurt their victims by releasing a toxin (a chemical that damages cells). In 1888, microbiologists Émile Roux and Alexandre Yersin, working together at the Pasteur Institute in Paris, separated the diphtheria bacteria from the serum in which they were being grown. Roux and Yersin then injected the bacteria-free serum into healthy animals. The animals soon came down with diphtheria, even though they had not been exposed to the bacteria. The scientists realized that a toxin was being released by the *Corynebacterium diphtheriae* into the growth serum, and that it was this toxin that made people and animals ill.

## Fighting Diphtheria

Now microbiologists could go to work finding a vaccine and designing methods of diagnosing and treating diphtheria victims. In 1890, German microbiologist Emil von Behring discovered that the blood of animals infected with diphtheria produced an antitoxin, a chemical that binds to a toxin and makes it harmless.

Behring realized that the antitoxin might be helpful in producing a vaccine to protect people against diphtheria. He injected animals with weakened diphtheria toxin—just enough so that their immune systems would create antitoxin but not enough to hurt the animals. Unfortunately, this diphtheria toxin was too dangerous for using on humans.

Yet Behring's work led to the later use of diphtheria antitoxin produced in horses as a treatment for human victims of the disease. In 1923, a formalin-treated toxin was used to vaccinate people against diphtheria and was found to be safe.

In 1908, Béla Schick, a pediatrician and microbiologist from Boglár, Hungary, became an assistant to Theodor Escherich at the University of Vienna, Austria. These two scientists began studying diseases caused by bacteria, including diphtheria and scarlet fever.

In 1913, Schick used Behring's work with antitoxins to develop a test that would show how susceptible a person was to catching diphtheria. The result was the Schick test, which proved to be simple and reliable. About 0.1 milliliter of a weakened toxin solution is injected just under the skin inside a patient's arm. The toxin is treated so that it will lead to a bit of swelling in susceptible persons without hurting them. If the patient is susceptible to diphtheria, a reddened, swollen rash (caused by damaged skin cells) will appear around the injection site within a few days. A person who is not susceptible will have no reaction, because the toxin is not causing damage.

Those who test positive with the Schick test should be immunized. People who are already suffering from diphtheria can be treated with a combination of antibiotics and horse serum antitoxin. Antibiotics destroy the *Corynebacterium diphtheriae* bacteria, while the horse serum antitoxin destroys the diphtheria toxin until the victim's body is strong enough to make enough of its own antitoxin.

## Consequences

Schick's test for diphtheria became a valuable tool for identifying the disease and which people most needed immunization. For his findings, Schick was named Extraordinary Professor of Children's Diseases at the University of Vienna in 1918. His test saved thousands of lives, especially among children, who tend to be susceptible to diphtheria. In the middle-to-late 1920's, when the first successful toxoid vaccine was available, the number of cases of diphtheria around the world dropped dramatically.

During the first two decades of the twentieth century, before the test and vaccine were available, there were between 150,000 and 200,000 diphtheria cases every year in the United States alone. By the 1970's, the number of diphtheria cases had dropped to ten a year.

The work of Schick and others also helped show how microorganisms are present everywhere in the environment and can cause disease once they are inside the human body. This led to a better understanding of the importance of sterilization. Before the 1900's, surgical instruments were kept clean, but they were never sterile (clear of all microorganisms); as a result, many patients died after surgery. Microbiological research in Schick's day led to the sterilization of surgical equipment, antiseptic treatment to keep all hospital rooms and equipment clean, and the sanitation of water.

*David Wason Hollar, Jr.*

# Sixteenth Amendment Legalizes Income Tax

> *With the passing of the Sixteenth Amendment to the Constitution, Congress gained the power to collect taxes on incomes, without apportionment among the states and without regard to the census.*

**What:** Law; Economics
**When:** February 25, 1913
**Where:** Washington, D.C.
**Who:**
JOSEPH WELDON BAILEY (1863-1929), senator from Texas from 1901 to 1913
ALBERT CUMMINS (1850-1926), senator from Iowa from 1908 to 1926
NELSON ALDRICH (1841-1915), senator from Rhode Island from 1881 to 1911
WILLIAM HOWARD TAFT (1857-1930) president of the United States from 1909 to 1913

## The Tax Question

The history of income tax in the United States dates back to the War of 1812, when Treasury Secretary Alexander J. Dallas recommended on January 21, 1815, that Congress adopt a tax on incomes to raise funds for waging the war. Although the war had been ended with the signing of the Treaty of Ghent on December 24, 1814, this fact was probably not yet known in Washington in January, 1815. Once it was known that the war was over, however, Congress did not act on Secretary Dallas's request.

The income-tax question did not arise again until the Civil War, when the United States' first income tax was levied by the Union government. The Internal Revenue Act of 1862 provided for a tax on incomes and also said that this tax would be progressive—that is, the rates would be higher on higher incomes. For the first time in American history, people were to be taxed according to their ability to pay. The Act of 1862 called for a 3 percent tax on incomes up to ten thousand dollars and a 5 percent tax on incomes over that amount. The income tax remained in effect—

though the rates changed—until 1872, when Congress let the law expire because of pressure from business groups.

In the late 1880's, however, the Populists and other groups began to call for a new income tax. In fact, the national platform of the Populist Party in 1892 contained a demand for a graduated, or progressive, income tax. Two years later, Congress passed the Wilson-Gorman Tariff Act, which, among other things, set up an income tax of 2 percent on all incomes over four thousand dollars. In 1895, however, the Supreme Court declared the income tax unconstitutional in *Pollock v. Farmers' Loan and Trust*, since the Constitution prohibited direct taxation.

Those who favored the income tax did not agree with the Supreme Court's interpretation of the Constitution, and they continued to fight for the tax. Wealth was being concentrated more and more in the hands of a few, and many Americans considered this a direct threat to democracy. The income tax seemed a way of reversing that trend, of redistributing some wealth to those who had less.

One of the leaders of the struggle for an income tax was Senator Joseph W. Bailey, a Democrat from Texas. In April, 1909, Bailey proposed an income-tax amendment to the Payne-Aldrich Tariff Bill, which Congress was considering at that time. Bailey suggested a 3 percent tax on all incomes over five thousand dollars. His idea was soon replaced, however, with a more radical proposal by Albert Cummins, senator from Iowa: a graduated income tax ranging from 2 percent on incomes over five thousand dollars to 6 percent on incomes over $100,000.

## The Amendment

Conservative Democrats and Republicans, led by Senator Nelson Aldrich of Rhode Island, did

*This* Puck *magazine cartoon, published in 1895, the year that the Supreme Court ruled against the income tax, expresses disapproval of this tax.*

Although those who favored the income tax feared that the amendment was the conservatives' attempt at defeating or at least delaying any income tax, they finally agreed to the idea of a constitutional amendment. In July, 1909, the amendment was put before the states.

The road to ratification was not easy. Throughout 1910 and 1911, Republican and Democratic conservatives attacked the amendment, saying that it would lead the United States down the path to socialism. Yet in the state governments the forces of Progressivism were strong, and in the end, only five states did not ratify the amendment. On February 25, 1913, Secretary of State Philander C. Knox certified that the Sixteenth Amendment was officially part of the Constitution.

## Consequences

Soon after the ratification of the Sixteenth Amendment, Congress passed the first income-tax law as part of the Underwood-Simmons Tariff Act. Although by modern standards the rate of taxation was low, the bill set an important precedent by making the income tax progressive. The Sixteenth Amendment had made no mention of graduated rates of taxation, but from that time forward they were accepted as a fair way of structuring the income tax. In this way, more of the tax burden was shifted to those who were best able to pay.

The income tax did reverse the trend toward concentration of wealth. After 1913, the poorer classes carried a smaller share of taxation in the United States.

*Fredrick J. Dobney*

not think that taxes should be used to try to redistribute wealth. When they realized that the income-tax amendment to the tariff bill might have enough votes to pass, Senator Aldrich and President William Howard Taft suggested that a constitutional amendment should be passed to make sure that the income tax was legal.

# Ford Develops Assembly Line

*In his manufacturing plants in Detroit, Michigan, Henry Ford and a team of engineers refined the concept of the assembly line to allow for a much faster production of automobiles.*

**What:** Economics
**When:** March 1, 1913-January 5, 1914
**Where:** Detroit, Michigan
**Who:**
HENRY FORD (1863-1947), the head of the Ford Motor Company
CHARLES SORENSEN (1882-1968), a technician who contributed to the moving assembly line
WILLIAM C. KLANN, a technician who adapted the idea to the assembly of motors
CARL EMDE, a German technician who built machinery for Ford
JAMES COUZENS (1872-1936), Ford's business manager

## New Methods

Mass production and the assembly line are often thought of in connection with Henry Ford, but these methods were not actually his original ideas. The three basic ideas of mass production—standardization, simplification, and interchangeability—date back to the eighteenth century. Before 1913, there were a number of companies that produced such things as telephone sets, bicycles, typewriters, and cash registers in large quantities. Ford himself had an assembly-line system in his original plant; it produced a car every thirteen hours.

After the opening of the Ford Motor Company in 1903, Ford decided that he would produce only one type of automobile, the Model T. Rival companies, which were manufacturing a number of different models, believed this decision would doom Ford's business to failure. Ford's idea was to keep his car's design standard and invest most of his money, time, and effort on the equipment and machinery to produce it. He wanted to "build a motor car for the great multitude"—for average middle-class Americans—and this goal could be achieved only if he produced large numbers of cars and was able to keep their prices low.

Early in 1907, the plant slowed down as it switched over to the production of Model T's only. This plant, in the Piquette area, had within it a number of assembly-line techniques: Work was brought to the workers, and the men, machines, and materials were placed in a logical arrangement that followed the order in which the cars were assembled. About twelve thousand cars were produced in this plant in 1909.

In the following year, Ford moved into a plant in the Highland Park area and greatly expanded his factory. There, between 1912 and 1913, the continuous assembly line was developed. The new techniques were aimed at increasing "power, accuracy, economy, system, continuity, speed, repetition." Ford's assistants in the design of this project included Carl Emde, who created the necessary machinery; William C. Klann, who worked with the system to make it useful for the assembly of motors; and Charles Sorensen, who added the idea of the continuous conveyor belt and who completed the assembly line.

Although there is no detailed record of the steps in the designing and building of the continuous assembly line, it is known that it was in operation by March 1, 1913. The main assembly line had forty-five operations, and it quickly increased the speed of production. A motor that had previously taken one person 600 minutes to assemble could now be put together in 226 minutes. A chassis that had taken one person 12 hours and 28 minutes to complete could now be finished in 1 hour and 33 minutes. The plant that had produced 78,440 automobiles in 1911-1912 was by 1916-1917 producing 730,041 automobiles—or 2,000 each day.

## The Human Factor

The Ford Motor Company grew very rapidly and was on its way to becoming the largest manufacturer of cars in the United States. Yet its methods began to be criticized. The use of machines and the division of each job into many small, repetitive tasks meant that most factory work required very little real skill. (The exceptions were the jobs of a few top engineers who designed the product and process.)

Sociologists and social reformers began to protest the dreary monotony of the endless repetition of work. As if to confirm their complaints, workers began to leave Ford at an alarming rate.

Beginning in 1913, the Ford Motor Company introduced "the most advanced labor policies yet known in large-scale American industry." Safety measures were improved and extended, the workday was reduced to eight hours, and the factory was converted to run on three shifts instead of two. Aptitude tests, sick-leave allowances, an English-language school for immigrants, a technical school, better medical care for the injured, and an improved factory environment were all soon introduced. The company even made special efforts to hire persons with disabilities and former convicts.

The most remarkable new policy of all, however, was the five-dollar minimum wage. This policy was announced by Henry Ford and his company's business manager, James Couzens, on January 5, 1914. Various reasons have been suggested to explain this decision: to get the pick of Detroit mechanics; to keep workers for a longer time and reduce turnover; to respond to the threat of unionization by the Industrial Workers of the World; and to motivate employees to be agreeable to a new speedup of work at the plant.

Library of Congress

*The Ford Motor Company's Model-T assembly line in its Highland Park factory.*

The reason Ford himself gave was "profit sharing and efficiency engineering." The company was enjoying a very large net income and wanted to share some of it with the workers. Furthermore, a higher wage would mean more loyal, better disciplined, and more efficient and productive workers.

To qualify for the five-dollar minimum wage, each employee had to meet certain standards: to be a person of good personal habits with a decent home and, if under twenty-two years of age, to prove himself to be "sober, saving, steady, industrious." To find out which employees met these standards, and to help those who did not, the company created a "Sociological Department."

## Consequences

There is no question that assembly lines have greatly increased the efficiency of manufacturing and packaging. Yet criticism of continuous assembly lines has continued throughout the twentieth century. Assembly-line work is seen as numbing to the mind, and increasingly it has been recognized that continuously repetitive movements can be physically harmful.

In the second half of the twentieth century, there has been experimentation with other styles of manufacturing. Some research indicates that workers' satisfaction with their work increases greatly when their job is set up so that they can see a product develop from beginning to end. Another option has been to design machines that can take over tiresomely repetitive tasks. Yet regardless of whether modern factories use a version of Ford's continuous assembly line, it is clear that his values of speed and efficiency continue to influence the way manufacturing is carried out.

*Russell M. Magnaghi*

# Congress Passes Federal Reserve Act

*The Federal Reserve Act established a system designed to promote a stable dollar and orderly growth in the economy of the United States.*

**What:** Economics
**When:** December, 1913
**Where:** Washington, D.C.
**Who:**
NELSON ALDRICH (1841-1915), Republican senator from Rhode Island from 1881 to 1911
WILLIAM JENNINGS BRYAN (1860-1925), a Democratic leader from Nebraska
CARTER GLASS (1858-1946), a conservative Democratic congressman from Virginia from 1902 to 1919
WILLIAM GIBBS MCADOO (1863-1941), secretary of the treasury from 1913 to 1918
ROBERT LATHAM OWEN (1856-1947), Democratic senator from Oklahoma
HENRY PARKER WILLIS (1874-1937), a bank expert
WOODROW WILSON (1856-1924), president of the United States from 1913 to 1921

## Need for Banking Reform

By the beginning of the twentieth century the American economy was the most powerful in the world, yet the country's banking system was old-fashioned and inadequate. The United States had two types of commercial banks: those chartered by the states and those having national charters from the federal government. Under the National Bank Acts of 1863 and 1864, national banks were required to buy government securities and to be regulated by the federal government, but they could issue national bank notes, a fairly stable paper currency.

The nation's banking system, made up of state and national banks, had many problems. Banking laws varied from one state to the next, and some states were lax in regulating banking activity. State bank charters could be obtained so easily that there were many banks that were mismanaged and lost the depositors' money. National banks tended to be more stable than state banks, but at critical moments national banks often found it difficult to get needed cash.

The major problems all banks faced was that the currency supply was "inelastic"—that is, neither banks nor the government had the power to expand or contract the money supply to meet the seasonal needs of industry and farming, or to prevent financial panics.

In the late nineteenth century, various reform measures were proposed to meet these needs. Among the most popular were the Greenback movement of the 1870's and the Populists' proposal for a Subtreasury Plan and for the free and unlimited coinage of silver in the 1890's. Many who were more conservative believed it would be helpful to establish a central bank—rather like the Bank of England—to hold government deposits and central banking reserves and to have full responsibility for the issuing of paper money.

The financial panic of 1907 clearly showed the weakness of the banking system. Throughout the country, depositors hurried to their banks to withdraw their savings; even banks that had been well managed were unable to meet the flood of demands. This crisis led to more calls for banking reform.

## The Act Is Passed

In 1908, Congress created the National Monetary Commission, made up of members of Congress, to make a plan for revision of the banking system. Headed by Senator Nelson Aldrich, the commission drew up the so-called Aldrich Plan. It called for a voluntary system headed by a central bank, the National Reserve Association. With its branch banks, the National Reserve would issue currency, hold the deposits of the federal government, and furnish reserve credit to member banks.

Some large banks supported the Aldrich Plan, but by the time it was submitted to Congress in 1912 it faced strong opposition. Many Progressives believed that the plan would allow large financial institutions even more control over the nation's money. They were relieved when Woodrow Wilson was elected president in 1912.

Although the Democrats controlled the new Congress, they were divided on the issue of banking. On one side were the Southern and Western radicals who followed William Jennings Bryan of Nebraska. On the other were conservatives headed by Congressman Carter Glass of Virginia, chairman of the House Banking Committee. Bryan, who was joined by William McAdoo, soon to be secretary of the treasury, and Senator Robert L. Owen of Oklahoma, insisted that any new banking system be under the government's full control, and that the government control and guarantee the currency supply. Glass and his followers opposed any plan for a central bank; they wanted to see a loose, disconnected system of regional reserve banks.

In the weeks before Wilson was inaugurated, he received help from Glass and banking expert H. Parker Willis to draft a new banking law. They proposed a privately controlled system of regional reserve banks under a general board that would coordinate and supervise their activities. This was actually a decentralized version of the Aldrich Plan. Glass gave in to certain wishes of Bryan's faction and added to the bill a federal guarantee of the notes issued by the new system.

The plan was put before Congress as the Federal Reserve Bill (also called the Glass-Owen Bill). More changes were made in committee, and the measure was finally passed in December, 1913.

The Federal Reserve Act was intended to establish no more than twelve Federal Reserve banks, to allow for an elastic currency, and to make the supervision of banking in the United States more effective. The twelve "bankers' banks" do not accept deposits from individuals or loan money to them. They are controlled by a board of governors (originally with five members, later seven) appointed by the president of the United States for ten-year terms. The board works with the secretary of the treasury and the comptroller of the currency to supervise the system.

All national banks are required to belong to the system, but state banks and trust companies can also join if they meet certain requirements. Member banks elect six of the nine directors of the district Federal Reserve banks. Although the system is a corporation owned by the member commercial banks, the Federal Reserve is in fact a public agency. It is directly responsible to Congress, some of its officials are appointed by the president, and it has traditionally put the public interest before private profit.

**Consequences**

The Federal Reserve Act created a system that did solve some banking problems in the United States. It allowed for a more elastic currency and system of making loans. Yet it has not been able to eliminate financial crises. There was wild inflation in the United States during World War I, while in 1921, there was a brief but severe financial depression. Throughout the 1920's, many banks failed. The Federal Reserve banks allowed too much money to be invested in risky ventures during that period, and in this way they contributed to the stock market crash of 1929. In the 1920's, the Federal Reserve Bank of New York dominated the system, and as Bryan's followers had feared, New York-based banks and corporations increased their hold on the country's supply of credit.

The nation's worst economic crisis and the "holiday" that closed all the banks of the country took place in the 1930's. Only then were measures taken to insure depositors' money against loss and to regulate the banking system more strictly. In spite of these changes, the Federal Reserve System still exists in the same basic form as when it was created in 1913.

*Merle O. Davis*

# Russell Announces Theory of Stellar Evolution

Henry Norris Russell developed a theory of how stars change over time based on the relationship between a star's brightness and its color.

**What:** Astronomy
**When:** December, 1913
**Where:** Princeton, New Jersey
**Who:**
HENRY NORRIS RUSSELL (1877-1957), an American astronomer
EJNAR HERTZSPRUNG (1873-1967), a Danish astronomer and photographer
SIR JOSEPH NORMAN LOCKYER (1836-1920), a British astronomer

## A Garden of Stars

The British astronomer Sir William Herschel described the starry sky as a garden in which one can see stars, like plants, in various stages of growth. The assumption that stars have different ages and change as they age was an important step in forming theories of how stars evolve. Increasingly sophisticated techniques for identifying and describing stars in the late nineteenth and early twentieth centuries brought a wealth of data from which a theory of stellar evolution could be built.

During the late 1800's, Sir Joseph Norman Lockyer helped formulate a theory about the life cycles of stars. Lockyer used the simple classification systems of the astronomers Angelo Secchi and Hermann Karl Vogel, which placed stars in one of four categories. At the time, physicists believed that stars released heat and light as a result of the force of gravity. Scientists thought that if enough interstellar matter came together in one place as a result of the force of gravity, a star would form. The star would then begin to contract under the force of gravity and to heat up and to shine. Eventually, when the star could collapse no further, it would cool off and die.

When Henry Norris Russell began his study of the heavens, many more stars had been classified. Each star was classified according to its spectrum. (A spectrum is a picture of the light of a star that has been spread out in all its different colors, like a rainbow. Astronomers can use the black lines in a star's spectrum to discover the composition and velocity of the star.)

A star's classification was believed to be linked to its surface temperature and color, and the different types were thought to be the result of differing temperatures; however, no consensus had been reached on the cause of differences in brightness. Russell showed that differences in brightness were directly related to density. In 1913, Russell produced a plot of spectral type versus absolute brightness ("true" brightness, as opposed to how bright a star appears on Earth). Ejnar Hertzsprung made a similar diagram in 1911; this type of plot is known today as a Hertzsprung-Russell, or H-R, diagram. Russell used this plot to view the relationship between brightness (and density) and spectral type (and color and temperature).

## Following the Diagram

Russell first presented his diagram to the Royal Astronomical Society on June 13, 1913, and to the American Astronomical Society on December 30, 1913. He also offered his interpretation of the diagram in terms of stellar evolution. Most stars fell either on a diagonal band stretching across the diagram (the "main sequence") or on a horizontal strip across the top of the diagram (the "giant sequence"). On the main sequence, stars vary in brightness and color, ranging from bright blue to dim red. On the giant sequence, stars have a fairly constant brightness but vary in color. These two sequences were explained by Russell in terms of the ages of stars in each sequence. He asserted that a star's evolution is driven by gravity alone and that a star begins its life cool, red, dim, and diffuse, and grows increasingly dense, bright, and hot (with an associated color change) as it contracts. Once it has contracted so far that gravity can condense it no further, it begins to cool off and becomes less bright and more red.

Russell hypothesized that the large red stars at one end of the giant sequence are the youngest of the stars, which are very diffuse and just beginning to collapse. As a star collapses, it becomes more dense and begins to change color as it moves across the giant sequence; it eventually brightens and leaves the giant sequence for the main sequence. At its hottest point, which Russell believed to be the midpoint of its life, the star is at the top of the main sequence among the brightest and bluest stars. As it then begins to cool, while continuing to become denser, it slides down the main sequence from being a hot blue star to being a yellow star like the sun and finally to being a dim red star, very dense and near the end of its life. Thus, there were two sorts of red stars: young and old. Hertzsprung had demonstrated earlier that the spectra of the two types of red stars were different and thus enabled astronomers to tell which type they were observing.

## Consequences

Russell presented a concise and straightforward scheme of stellar evolution that neatly fit the known data and accommodated the accepted explanation for why stars shine and how they form, exist, and die. He was able to use his diagram to illustrate the life cycle of a star. Although his evolutionary scheme later required revision, it was still an important step in understanding the "garden" of varying stars we see.

The later discovery that nuclear fusion, rather than gravitational collapse, powers stars for most of their lifetimes brought about drastic revisions in Russell's scheme. His use of the diagram, however, was a key step in the developing science of astrophysics. Also, in his explanation of how the H-R diagram reveals the evolution of stars, Russell gave at least a hint of what was to be discovered later about nuclear power fueling the stars. He suggested that perhaps there is a type of energy release related to radioactivity that could counteract the gravitational pull inward and give a star a longer lifetime than it would have had otherwise.

Today, however, it is believed that while a star starts to form because a cloud of material collapses under the influence of gravity, eventually conditions become hot enough in the center of the forming star that nuclear fusion begins to occur. The star then lives out most of its life cycle in one spot on the H-R diagram, its gravitational pull inward balanced by the energy being released in nuclear fusion. Gravity becomes important again at the end of the star's lifetime, when its fate is determined by the amount of mass it contains.

Much has been learned about such exotic objects as "white dwarfs," "neutron stars," and "black holes," which are the end products of evolution for various masses of stars. Without the H-R diagram, and the foundation of knowledge it offers for the understanding of the interrelationships among a star's density, brightness, temperature, and spectral type, astronomers could not have arrived at their current understanding.

*Mary Hrovat*

**299**

# Assassination of Archduke Ferdinand Begins World War I

> *After the heir apparent to the throne of Austria-Hungary was killed by a Serbian nationalist, aggression and mistrust spread quickly beyond Austria and Serbia, so that all Europe became engulfed in war.*

**What:** Assassination; War
**When:** June 28, 1914
**Where:** Europe
**Who:**
ARCHDUKE FRANCIS FERDINAND (1863-1914), heir apparent of Austria-Hungary
COUNT LEOPOLD VON BERCHTOLD (1863-1942), foreign minister of Austria-Hungary from 1912 to 1915
SERGEI DMITRIEVICH SAZONOV (1861-1927), Russian minister of foreign affairs from 1910 to 1916
SIR EDWARD GREY (1862-1933), foreign secretary of Great Britain from 1905 to 1916
THEOBALD VON BETHMANN-HOLLWEG (1856-1921), chancellor of the German empire from 1909 to 1917
GAVRILO PRINCIP, Bosnian terrorist

## Assassination and Ultimatum

On June 28, 1914, Gavrilo Princip, a Bosnian member of a Serbian terrorist organization, shot and killed Archduke Francis Ferdinand, heir apparent to the throne of Austria-Hungary, and his wife. The assassination took place in Sarajevo, the capital of Bosnia-Herzegovina, which had been annexed by Austria-Hungary but whose population was mostly Serbian.

Austrian leaders quickly decided to use the occasion to force a diplomatic confrontation with Serbia. For a long time, Serbians had called fellow Slavs within the Austrian Empire to join them in creating a great Slavic nation, and Austria now suspected the Serbian government of having supported the plot to assassinate Francis Ferdinand.

With the support of Germany, Austria wrote an ultimatum to present to the Serbian government. The ultimatum's demands were quite severe, and Serbia was given only two days to comply. Yet on July 25, Serbia agreed to nearly all the Austrians' demands.

Austria-Hungary and its main ally, Germany, apparently thought that Serbia's ally Russia would avoid becoming involved in the conflict. Though Russia had a long-standing interest in Serbia and the other Balkan states, leaders of Austria and Germany decided that Russia would have too much to lose by joining the dispute. Even if Russia and its ally France intervened, they thought, the other member of the Triple Entente, Great Britain, would remain neutral.

So it was that Austria-Hungary, declaring that Serbia had not met all of its demands, broke off all diplomatic relations with Serbia and prepared for war. The formal declaration of war came on July 28, 1914.

## War Breaks Out

The Austrians had not understood the Russians' position. Russia had been defeated in two important wars—the Crimean War of 1853-1856 and the Russo-Japanese War of 1904-1905—and had suffered a diplomatic embarrassment in the Balkans in 1909. Its leaders wanted to restore their country to a place of world leadership and respect, and they did not want to be humiliated again.

For these reasons, Russia responded to Austria-Hungary with heat. On July 28, the same day that

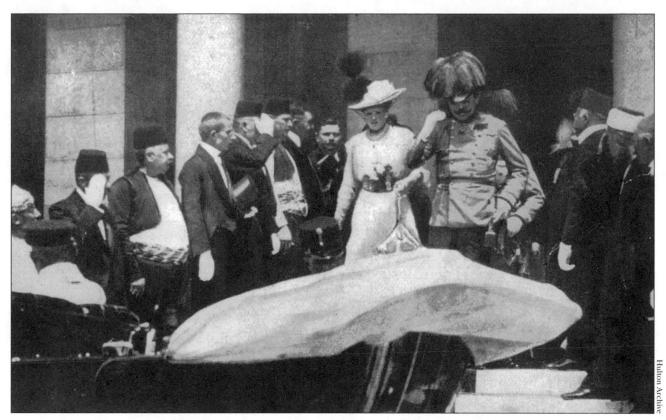

*Austrian archduke Francis Ferdinand and his wife moments before both were assassinated.*

Austria declared war on Serbia, Czar Nicholas II and his council of ministers ordered the Russian army to begin mobilizing—gearing up for the possibility of war. Two days later, Nicholas announced an all-out mobilization of Russia's troops.

Within Russia, there was widespread support for the war effort. Though the Bolsheviks opposed war, other parties, both conservative and liberal, agreed that Russia should be willing to fight if necessary.

Great Britain and Germany had both made efforts to end the crisis through some kind of international conference, but their attempts led nowhere. Germany's plan of defense was based on its ability to mobilize quickly and to act more swiftly than its enemies. Now German officials believed their country to be in serious danger from Russia. Having sent a "war warning" to Russia, Germany began mobilizing on August 1. France started mobilizing its army the same day.

Germany declared war on Russia on August 1, and two days later it declared war on France. Russia was deep in its own preparations for war. On August 4, Germany began an invasion of Belgium. In response, that same day Great Britain joined the fray. World War I had begun.

## Consequences

The Great War, as it has come to be called, was the product of tensions that had been building throughout Europe for many years. One key element in these tensions was the drive to build empires—to extend national influence into weaker countries, not only in Africa, Asia, and South America but also in the Balkan states of southeastern Europe.

World War I was a long, bloody, expensive conflict. One of its effects was the breaking up of the European empires. Throughout the twentieth century, colony after colony gained independence, and the European powers eventually retreated, for the most part, from ambitions of vast political empires.

*John H. Morrow, Jr.*

# France and Germany Fight First Battle of the Marne

> *With a French and British victory in a battle in the Marne Valley, German attempts to gain control of the European continent were frustrated.*

**What:** War
**When:** September 6-9, 1914
**Where:** Marne Valley, France
**Who:**
ALFRED VON SCHLIEFFEN (1833-1913), chief of staff of the German army from 1891 to 1906
HELMUTH VON MOLTKE (1848-1916), chief of staff of the German army from 1906 to 1914
ALEXANDER VON KLUCK (1846-1934), commander of the first army of Germany
KARL VON BÜLOW (1846-1921), commander of the second army of Germany
JOSEPH-JACQUES-CÉSAIRE JOFFRE (1852-1931), commander in chief of the French armies from 1914 to 1916
JOSEPH-SIMON GALLIENI (1849-1916), military governor of Paris from 1914 to 1915
LOUIS-FÉLIX-MARIE-FRANÇOIS FRANCHET D'ESPEREY (1856-1942), commander of the Fifth Army of France
SIR JOHN DENTON FRENCH (1852-1925), field marshal in command of the British Expeditionary Force

## Strategies

Before the outbreak of World War I, General Alfred von Schlieffen, chief of staff of the German army from 1891 to 1906, developed a plan to be followed in case of war with France and Russia. He knew that the vast size of Russia and its lack of a good railroad system meant that the Russian reserve divisions and the Siberian armies could not be made ready for war on the German frontier with less than two months' preparation time. France, on the other hand, was a much smaller country and had a dense network of railroads; it could bring up all of its troops against Germany in three weeks.

In the light of these realities, Schlieffen planned that seven of the eight German armies would be sent up against the French at first. His proposal was to send two strong armies through Belgium toward the west and then wheel them southward, with the German First Army, on the right of the advance, passing west of Paris. The French armies would be trapped in a pocket southeast of Paris, and they would be forced to surrender. German forces could then be sent to the Eastern Front to fight the Russians, who were expected to have gotten as far as East Prussia by then, but no farther than the Vistula River in Poland. If this plan were followed, Schlieffen estimated that the war would be over in about four months.

General Helmuth von Moltke succeeded General Schlieffen as chief of staff of the German army in 1906. Fearful that the French might invade German Alsace while the right wing of the German forces was still advancing, Moltke changed Schlieffen's plan somewhat to strengthen the army's left flank with new divisions. These new divisions were formed before 1913.

## Battle in the Marne

With the outbreak of World War I in 1914, Moltke's version of Schlieffen's plan was put into action. The German First and Second armies, with thirty-two of the seventy-eight German infantry divisions in the West, began to move through Belgium. Meanwhile, the French began to fight the advancing Germans directly. Between August 20 and 24, 1914, a series of small but bloody battles took place along the border between France and Belgium.

Although the Germans were attacking, they were often in the defensive position, having to protect themselves against French onslaughts. Yet the German artillery and machine guns did turn back the French attacks. On August 25, General Joseph-Jacques-Césaire Joffre, commander in chief of the French Armies, was forced to order a general retreat of all the French armies. German troops followed close behind.

At the same time, France's allies were mobilizing and sending troops. The British Expeditionary Force, commanded by Field Marshal Sir John Denton French, entered the fight on August 23. Three days later, Joffre ordered the formation of a new French Sixth Army in Paris, under the orders of the capital's military governor, General Joseph-Simon Gallieni.

Moltke was still determined to get past the French Fifth Army at the Franco-Belgian border. On August 30, he decided not to follow Schlieffen's plan of an advance to the west of Paris. Instead, he directed the German First Army to advance east of Paris, one day's march behind the German Second Army.

The proud commander of the German First Army, General Alexander von Kluck, disobeyed the order and continued to move his army forward in line with the German Second Army, commanded by General Karl von Bülow, on his left. Kluck did move one army corps from his left to his right as a flanking guard against forces that might come from Paris, but this maneuver opened a small gap between his forces and Bülow's.

On September 3, aware that Kluck's and Bülow's forces were slightly separated, Gallieni suggested attacking Kluck's army. Joffre re-

*French soldiers shoot over the head of their fallen comrade during the First Battle of the Marne.*

**303**

sponded on September 4, by ordering an "about-face" of the Allied armies, and a general offensive against the Germans began on September 6.

When Gallieni's troops came up to Kluck's flanking guard, Kluck transferred more men from his left to his right in order to avoid being overcome from the rear. As a result, the gap between his army and Bülow's widened to about twenty miles. If their armies had had the extra troops that Moltke had assigned to the left flank at the beginning of the war, there might have been no gap, and the Battle of the Marne might have turned out differently.

On September 6, the French Fifth Army, under the command of General Louis-Félix-Marie-François Franchet d'Esperey, and the British Expeditionary Force, sent forward by Field Marshal French, moved slowly into the gap to threaten the flanks of the German First and Second armies. There was little serious fighting. The Germans were alarmed and fell back to regroup and prepare a new move to try to outflank the French. Their retreat ended the First Battle of the Marne.

## Consequences

The German plan to overwhelm France quickly and then crush Russia with Austria-Hungary's help was frustrated by the French victory at Marne. The Germans would now have to face a long conflict on the Western Front, with Allied troops established in position. Joffre's decision to stop the withdrawal of the French armies and to counterattack stopped the German invasion, saved the French from defeat, and kept the Germans from winning quick control over Europe.

*Samuel K. Eddy*

# Submarine Warfare Begins

With the Germans' use of submarines to attack British ships in World War I, a new era of war at sea was launched.

**What:** War; Weapons technology
**When:** September 22, 1914
**Where:** North Sea
**Who:**
LOUIS ALEXANDER OF BATTENBERG (1854-1921), British prince and first sea lord of the British Admiralty from 1912 to 1914
HENRY H. CAMPBELL (1865-1933), commander of the British Seventh Cruiser Squadron
OTTO WEDDIGEN (1882-1915), captain of the German submarine *U-9*

## The First Attacks

At the beginning of World War I, the First Sea Lord of the British Admiralty, Vice Admiral Prince Louis Alexander of Battenberg, set up patrols in the southern North Sea to protect the eastern entrance of the English Channel against raids by German destroyers and minelayers. The force that was assigned to this task included two flotillas of destroyers and a squadron of five old armored cruisers. Copies of the orders given at the time show that British naval officers were not taking seriously the possibility of an attack from German submarines.

On September 5, 1914, the British light cruiser HMS *Pathfinder* was sunk by the German submarine *U-21* off the Firth of Forth. On September 13, the British submarine *E-9* was able to destroy the German light cruiser *Hela* off Heligoland. These events should have made it clear that the submarine was making its appearance in war and would have to be seen as a serious danger to surface ships. Yet the First Sea Lord did nothing to prepare the southern patrol for further submarine attacks.

On September 22, three ships of the British Seventh Cruiser Squadron, HMS *Aboukir, Cressy,* and *Hogue,* were patrolling west of the Dutch coast. The squadron commander, Admiral Henry H. Campbell, was on his flagship, which was refueling in port. When bad weather drove the escorting destroyers into port, the three cruisers were left unprotected. Following the admiral's orders, they were steaming on a straight course at less than ten knots per hour.

Soon after sunrise, they were sighted by Lieutenant Otto Weddigen, captain of the German submarine *U-9*. At 6:30 A.M. he hit the *Aboukir* with a single, well-directed torpedo. The cruiser quickly sloped dangerously to one side.

The captain of the *Aboukir,* Captain John E. Drummond, believed that he had hit a mine, and, being the senior officer, he ordered the other two British cruisers to come close to his ship. The *Hogue* neared the sinking ship, came to a stop, and launched its boats to rescue the crew of the *Aboukir. U-9* then took aim at the *Hogue,* which was sitting still, and fired at close range. Two torpedoes hit their mark.

The *Aboukir,* meanwhile, capsized and sank. Ten minutes later the *Hogue* sank as well, and its survivors joined those from the *Aboukir* in the sea. Unwilling to abandon the struggling men, Captain R. W. Johnson of the *Cressy* brought his vessel to a dead stop and thus provided another helpless target for the *U-9.*

Weddigen did not miss the opportunity. He reloaded his torpedo tubes and fired three missiles at the *Cressy*. Two torpedoes exploded against it, ripping out its side. The cruiser rolled over on its beam-ends, hung there while it filled, and then sank in fifteen minutes.

In just over an hour, Weddigen had sunk three 12,000-ton armored cruisers with his small 493-ton submarine, manned by twenty-nine men. The British loss of life was very heavy; fourteen hundred men out of the twenty-two hundred in the three ships' crews were either killed by the torpedoes or drowned.

## The Attacks Spread

Soon after this disaster, German submarines began to attack merchant vessels, and they proved to be very effective destroyers of naval commerce. On October 20, 1914, *U-17* sank the British *Glitra*; this was the first of more than thirteen thousand merchant ships, thirty-six million tons in all, to be sunk by submarines of all navies in World Wars I and II.

In 1917, the Germans began a campaign of unrestricted submarine warfare. "Unrestricted" meant that the U-boats did not follow the rules of international law, which said that warship crews must search merchant ships suspected of carrying contraband before destroying them. According to international law, merchant ships were not to be sunk at all if the crew could not first be put in a safe place.

Through unrestricted torpedo attacks and minelaying in the first four months of 1917, German submarines sank 1,147 ships, totaling 2,224,000 tons. As a result, Great Britain faced the possibility of starvation and defeat. The British responded by collecting their merchant ships into convoys, so that they could be directly protected by an escort of warships.

## Consequences

The Germans' attack against commercial shipping was considered cowardly and uncivilized by the standards of those days. This was the development that brought the United States into the war against Germany. Clearly, submarine warfare was one of Germany's most important strategies throughout World War I.

In World War II, the British and their allies used the convoy system to protect merchant ships from the beginning. The Germans retaliated by grouping their U-boats into packs. Packs of five or ten submarines and the convoy escorts fought one another fiercely during the grim Battle of the Atlantic. In 1942 alone, German submarines sank 1,054 ships totaling 5,764,000 tons—a very serious set of losses for the Allies. In the Mediterranean, British submarines were successful in sinking a quarter of the entire Italian merchant marine. In this way the British helped to cut off the Italian and German armies in Libya from their sources of food, fuel, and ammunition.

*Samuel K. Eddy*

Library of Congress

*German officers and crew members pose on top of their submarine in 1916.*

# Spain Declares Neutrality in World War I

*Though Spain's leaders decided to keep the country out of World War I because of internal divisions, the war served to make the divisions more serious, so that Spain's constitutional government fell.*

**What:** International relations; War
**When:** October 30, 1914
**Where:** Madrid
**Who:**

EDUARDO DATO IRADIER (1856-1921), Conservative prime minister of Spain from 1913 to 1915 and in 1917

ALEJANDRO LERROUX GARCÍA (1864-1949), a radical Republican leader

ANTONIO MAURA Y MONTANER (1853-1925), a Conservative leader, and prime minister in 1918 and 1919

ÁLVARO DE FIGUEROA Y DE TORRES, CONDE DE ROMANONES (1863-1950), Liberal prime minister in 1915

FRANCISCO CAMBÓ, leader of the Catalan Lliga

MANUEL GARCÍA PRIETO, MARQUÉS DE ALHUEMAS, Liberal prime minister from 1917 to 1918

## The Necessary Neutrality

When World War I broke out in August, 1914, Spain was a deeply divided nation. Some powerful groups supported the Allies, while other equally powerful ones supported the Central Powers. All agreed, however, that Spain would be better off by remaining neutral in the conflict. There was a general sense that Spain's internal problems were more important at the time than foreign issues.

After Conservative prime minister Eduardo Dato Iradier declared the neutrality of Spain, his declaration was confirmed by the Cortes (the Spanish parliament), when it met on October 30, 1914.

For some time, Spain's opposing parties had taken turns governing the country for agreed-upon periods of time. This system had broken down, however, by 1914. Demands for reform were fought bitterly in the Cortes and aroused violence across the country. The republicans, led by Radical Republican Alejandro Lerroux García, had been growing into a sizable force, and so had the socialists and the anarchists. Even the Conservatives had been preaching the cause of reform—especially Antonio Maura y Montaner. He demanded that the government begin "reform from above" so that the people would not rebel and start a revolution.

## Weathering the War

At first, World War I had a positive effect on Spain. Trade increased enormously, as orders poured in from nations involved in fighting. Investment funds from abroad increased as well. Spain suddenly found itself exporting more goods than it imported. An industrial boom led factories to operate overtime, and there was no unemployment in industrial areas.

Yet the boom also brought inflation. Wages kept up with prices in only a few industries, so that most of the Spanish people found that their money covered fewer of their expenses. Soon the working classes threatened to strike unless the government took action to stop rising prices.

The Conservative government of Dato Iradier could not control this unrest, and late in 1915 it was replaced by the Liberal government of Conde de Romanones. Romanones faced two big problems. First, the younger army officers had responded to inflation by organizing their own association, similar to a union, to help them get higher pay and to demand that political promotions within the army come to an end. Most working-class organizations, along with the country's intellectuals, sympathized with the goals of this group, seeing it as a reforming trade-union movement within the army. The Catalans were also supportive.

**307**

The people of Catalonia, in response to the Industrial Revolution and a cultural renaissance in the late nineteenth century, had begun a strong movement for their region's independence. They complained that the Madrid government stood in the way of their well-being and that their taxes were being used to support the rest of Spain. Catalan leaders in business and industry formed the Lliga, an organization to protest Madrid's policies and to demand that Catalonia be given greater control over its own affairs.

The coming of World War I made the Catalan problem more intense. Because Barcelona, the principal city in Catalonia, was the nation's largest center of industry and trade, inflation and other economic problems became worst there.

The Lliga, led by Francisco Cambó, demanded that Barcelona be made a free port. Romanones refused, and his government fell in 1915.

Romanones's successor was Manuel García Prieto, Marqués de Alhuemas, another Liberal. He also failed to solve the nation's problems, and the Conservative Dato Iradier was returned to leadership in 1917.

By then, the crisis in Spain had grown out of hand. The Catalans, the young army officers, the socialists, Lerroux, and Maura all joined in demanding that a special Cortes be gathered to draft a new constitution for the country. Dato Iradier refused and, saying that the situation demanded emergency measures, took away the freedoms promised by the Spanish constitutions.

The response was a general strike, and the army was called to put it to an end. With this act, parliamentary government in Spain ended; the army was now the ruling force in the country. The army took charge completely in 1923.

## Consequences

Spain was in no condition to send an army into World War I, and its leaders were only displaying common sense when they decided to declare neutrality. Yet the war still affected the country deeply. With prices rising faster than incomes, the people of Spain found it more and more difficult to meet their daily expenses.

As their discontent increased, they continued to divide into hostile factions. All the groups became less and less willing to accept the inadequate solutions their government offered. The end result was the collapse of parliamentary government in Spain.

*José M. Sánchez*

# Morgan Develops Gene-Chromosome Theory

> *Thomas Hunt Morgan's experiments led to the discovery of the principles of the gene-chromosome theory of hereditary transmission.*

**What:** Biology; Genetics
**When:** 1915
**Where:** New York, New York
**Who:**

THOMAS HUNT MORGAN (1866-1945), a professor of experimental zoology who won the 1933 Nobel Prize in Physiology or Medicine

CALVIN BLACKMAN BRIDGES (1889-1938), a geneticist and one of Morgan's assistants

ALFRED HENRY STURTEVANT (1891-1970), a geneticist and another of Morgan's assistants

HERMANN JOSEPH MULLER (1890-1967), a geneticist and Morgan's student, who won the 1946 Nobel Prize in Physiology or Medicine

## The Fortuitous Fruit Fly

In 1904, Thomas Hunt Morgan, a young professor of biology, began work at Columbia University as professor of experimental zoology. That same year, through a friend, Morgan met Hugo de Vries, the Dutch biologist who had been one of the trio of scientists who in 1900 had rediscovered the work of the Austrian botanist Gregor Johann Mendel. (Mendel's work had been published first in 1866, but it was promptly ignored and forgotten.) De Vries had a theory that new species originated through mutation. (A mutation is defined as a change occurring within a gene or chromosome that produces a new and inheritable feature or characteristic.)

The rediscovery of Mendel and the contact with de Vries influenced Morgan to initiate experiments to try to discover mutations and to test the Mendelian laws. He began to experiment with *Drosophila melanogaster*, the fruit fly. It bred rapidly and ate little, and Morgan and his students found that thousands of these tiny insects could be contained in a small collection of milk bottles that they "borrowed" from the Columbia cafeteria.

In 1908, Morgan had one of his graduate students perform an experiment to breed generations of *Drosophila* in the dark, in an attempt to produce flies whose eyes would waste away and become useless. After sixty-nine generations, however, the experiment was abandoned; the sixty-ninth generation could see as well as the first.

Morgan's experiments did not reveal mutations. In fact, they turned up enough exceptions to Mendel's laws that Morgan began to doubt the validity of the laws.

One day in 1910, however, he found a single male fly with white eyes rather than the standard red. Morgan bred the white-eyed male with a red-eyed female and the first-generation offspring were all red-eyed, suggesting that white-eyed was a Mendelian recessive factor. He then bred the first-generation flies among themselves, and the second generation were red-eyed and white-eyed in a 3:1 ratio, appearing to confirm that white eyes were Mendelian recessive.

Morgan noted that all the white-eyed flies were males. He had discovered "sex-limited" heredity (in which inheritable characteristics may be expressed in one sex but not in the other). The Mendelian factor, or gene, that determined white eyes was located on the same chromosome as the gene that determined male sex—on the male chromosome, that is.

## Linking and Crossing Over

Chromosomes—which appear as stringlike structures within cells—had been discovered by cell researchers in the 1850's. It had been theorized, without much hard evidence, that they were involved in heredity. Morgan had considered such theories and rejected them. Meanwhile, researchers had discovered an odd-shaped chromosome (all other chromosomes occurred in similarly shaped pairs) that seemed to be related to male sex (now called the *Y* chromosome). The discovery of sex-limited heredity revealed the association of Mendelian genes with chromosomes and the function of chromosomes in heredity.

Following this discovery, Morgan saw that a concerted effort was required to expound fully the Mendelian-chromosome theory and therefore enlisted a group of exceptional students to share the work in his so-called fly room. The nucleus of the group consisted of Calvin Blackman Bridges, Alfred Henry Sturtevant, and Hermann Joseph Muller.

Between 1910 and 1915, Morgan and his team developed and perfected the concepts of linkage, in which various genes are located on the same chromosome and are transmitted together, and crossing-over, in which these paired (linked) chromosomes break apart and recombine during meiosis. This creates new gene combinations. Crossing-over and linkage therefore produce two opposite results. Based on an understanding of linkage and crossing-over, the team was also able to create chromosome maps, plotting the relative locations and distances of the genes on the chromosomes.

The culmination of the work of Morgan's team was the publication in 1915 of *The Mechanism of Mendelian Heredity*, coauthored by Morgan, Sturtevant, Muller, and Bridges. For the next twelve years, strictly genetics studies were performed mainly by Sturtevant and Bridges and other team members. Morgan returned to his previous areas of interest of embryology and evolution, pursuing connections between those areas and the new discoveries in genetics. He also was occupied in publicizing the new views of heredity and their ramifications through publications and lectures.

## Consequences

The discovery and demonstration that genes reside on chromosomes provided the key to all further work in the area of genetics. Mendel was extremely fortunate in the design of his experiments in that each of the characteristics he investigated in his pea plants happened to reside on a separate chromosome. This facilitated discovery of the Mendelian hereditary principles but made for experimental results that were rather predictable. When a larger number of characteristics is investigated, because of linkage of genes located on the same chromosomes and crossing-over of chromosomes, the results will be more complicated and will not reveal so clearly the Mendelian pattern. This is what happened in Morgan's early experiments. It was only by carefully examining—using tweezers and a magnifying glass—generations upon generations of the tiny *Drosophila* that the mechanism of inheritance began to emerge.

Although Morgan did not pursue medical studies, his studies laid the groundwork for all genetically based medical research. As was stated when the Nobel Prize was presented to Morgan, without Morgan's work "modern human genetics and also human eugenics would be impractical." It is generally accepted that Mendel's and Morgan's discoveries were responsible for the investigation and understanding of hereditary diseases.

# Turks Massacre Armenians

*A Turkish assault on Armenians within the Ottoman Empire killed many and drove survivors to new homes in the United States, the Soviet Union, and other parts of the world.*

**What:** Civil strife; Ethnic conflict
**When:** 1915-1916
**Where:** Ottoman Empire (modern Turkey and Syria)
**Who:**
MEHMET TALAAT PASHA (1872-1921),
ENVER PASHA (1881?-1922), and
AHMED DJEMAL PASHA (1872-1922),
   members of the Young Turk
   triumvirate that ruled the Ottoman
   Empire from 1913 to 1918
DJEVDET BEY, Turkish governor of Van in
   1915
KEMAL ATATÜRK (1881-1938), a Turkish
   nationalist who founded the Turkish
   Republic in 1923

## Armenian-Turkish Conflict

The Armenian people, who had been converted to Christianity in the fourth century C.E., had once enjoyed political independence. By the beginning of the twentieth century, however, the lands they inhabited were divided among three states: Persia, the Ottoman Empire, and the empire of the Russian czars. In each of these states, the Armenians were an ethnic minority.

The Ottoman Armenians lived throughout the eastern half of what is now Turkey in Asia Minor. There were also large Armenian populations in Smyrna (an Aegean seaport), the Syrian cities of Damascus and Aleppo, and the capital city of Constantinople (modern Istanbul). The Armenians looked much like their Turkish neighbors, but there were important cultural differences. The Armenians were Christian, while the Turks were Muslim. The Turks were composed of an illiterate peasant majority ruled by a small elite, while the Armenians had a healthy middle class.

In the second half of the nineteenth century, the Ottoman Armenians became increasingly nationalistic, and many of them came to believe that they should have a state of their own. The Treaty of Berlin, which ended the Russo-Turkish War of 1877-1878, forced the Turks to promise fair treatment of the Armenian Christians within their borders, but set up no practical way of enforcing the promise. The treaty encouraged the Armenians to become bolder and bolder in demanding their rights, while making the Ottoman authorities more and more anxious and resentful.

In 1894-1896, Sultan Abdul-Hamid II, obsessed with the threat of revolution among the Armenians, had 300,000 of them massacred. Although the killings aroused a cry of indignation throughout the United States and Western Europe, disagreements among the Great Powers prevented anything from being done to help the Armenians.

In July, 1908, a military uprising by officers loyal to the underground Committee of Union and Progress (the so-called Young Turks) forced Abdul-Hamid II to give the empire a liberal constitution, with a representative parliament. Yet the Balkan Wars of 1912-1913 put an end to the Young Turks' dream of reviving the Ottoman Empire through liberal, constitutional government. Under the treaty that ended the war, the Ottoman Empire lost Albania, Macedonia, and Western Thrace. By the thousands, Turkish Muslim refugees from these lost territories poured into Constantinople.

The shock of defeat enraged many Young Turks and destroyed their liberal idealism. There was another coup in January, 1913, and the Turkish cabinet gradually came to be dominated by three Young Turks: War Minister Enver Pasha, Navy Minister Ahmed Djemal Pasha, and Interior Minister Mehmet Talaat Pasha. In the fall of 1914, this group committed the Ottoman Empire to war against Russia, Great Britain, and France on the side of Germany.

The war did not go well for the Turks. They suffered some early defeats, and though they managed to turn back the Allied invasion of Gallipoli, their losses in that campaign were heavy.

## Attack on Armenians

Fearing military defeat, the Young Turk triumvirate decided to strike out against the group they saw as the enemy within the gates: the Armenian minority. The Armenians were seen as allied with Russia. An Armenian revolt in the town of Van was seen as proof: Enraged by the brutality of Governor Djevdet Bey, the Armenians in Van rose up against the Turkish authorities on April 20, 1915. On May 16, after much hard fighting, the Armenians were rescued when the Russian army captured the town from the Turks.

The Ottoman rulers' response was swift. Under the direction of Talaat Pasha, Armenians began to be forced to leave the Ottoman lands. Throughout eastern Anatolia, Armenian men of military age were rounded up, marched off for several miles, and shot. Armenian women, children, and old men were ordered, at bayonet point, to leave their home villages and move to relocation centers in the Syrian desert.

No effort was made to provide these forced emigrants with food, water, or shelter, and thousands of them dropped dead of hunger, thirst, exhaustion, or disease during the long march to Syria. Many of them were murdered. Survivors were sometimes raped or forced to convert to Islam.

The deportations began in April, 1915, in Cilicia, a Mediterranean coastal province, and spread into other provinces through October. On August 4, Van was recaptured from the Russians. Only in Smyrna and in Constantinople

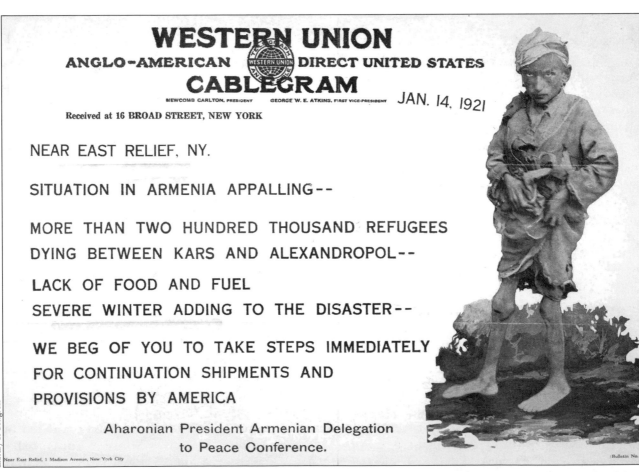

*Threats to Armenian survival in Turkey continued after 1918, and residents of the neighboring Armenian homeland faced new challenges when the Soviet Union was formed in the early 1920's, as this 1921 appeal for American help shows.*

were most of the Armenians spared. On a few occasions groups of Armenians were able to mount an armed resistance, but generally the Turkish army and police had superior power.

By the beginning of 1916, the deportations had been mostly completed, but occasional outbreaks of violence against Armenians continued until the Turks signed an armistice with the Allies on October 30, 1918. The exact number of Armenians killed through deportations and massacres will never be known; it seems likely that about one million perished.

## Consequences

Embroiled in its great war against the Allies, Germany had been unwilling to risk losing Turkish support by pressuring for an end to the persecution of Armenians. Within the United States, however, reports from American diplomats and Protestant missionaries in the Ottoman Empire led to an outpouring of support for the Armenian cause. Money was raised to aid the survivors, and in 1918 former missionary James Barton helped organize the American Committee

for the Independence of Armenia, a lobbying group. Its goal was that a state of Greater Armenia be created out of the ruins of the Ottoman Empire.

The Young Turk triumvirate was forced to flee to exile in Germany, and Russian Armenians had organized a small but independent republic of their own. Yet the Soviet Republic of Armenia's hopes for annexing additional territory eventually came to nothing. Military hero Mustafa Kemal came to power over the Turks, and he refused to give up eastern Anatolian territory to the Armenians. Now that World War I was over, the Allied powers were reluctant to become involved in further fighting. In 1919, Kemal's forces entered Smyrna—which had been occupied by the Greek army—and slaughtered many Armenians in that city.

After 1921, many Armenian survivors of the 1915-1916 massacre migrated to Syria and Lebanon, which had become protectorates of France. Others moved to Soviet Armenia and to the United States.

*Claude Hargrove*

# Japan Presents China with Twenty-one Demands

> *With a list of demands for increased political, military, and economic rights, Japan tried to become the controlling foreign power in China.*

**What:** International relations
**When:** January 18, 1915
**Where:** Japan and China
**Who:**

YUAN SHIKAI (YÜAN SHIH-K'AI; 1859-1916), president of China from 1912 to 1916

BARON TAKAAKIRA KATO (1859-1926), foreign minister of Japan from 1914 to 1915

ARIGA NAGAO (1860-1921), Japanese adviser to Yüan Shih-k'ai

WOODROW WILSON (1856-1924), president of the United States from 1913 to 1921

PAUL SAMUEL REINSCH (1869-1923), United States ambassador to China from 1913 to 1919

SIR EDWARD GREY (1862-1933), foreign secretary of Great Britain from 1905 to 1916

JOHN NEWELL JORDAN (1852-1925), British minister to China from 1906 to 1920

## Reaching for Power

The island nation of Japan borrowed many of its traditions from China, and there have always been similarities of culture between the two countries. The Buddhist religion first came to Japan from China, and the Japanese system of writing is based on that of China.

In the early twentieth century, however, the international status of the two countries was quite different. Under the Manchu emperors, China had remained wrapped up in the traditions of the past, while greedy Western states took more and more special privileges for their business interests within the country. The Republican Revolution of 1911 did not stop China's decline.

Japan, on the other hand, had successfully reformed its economy, army, and school system, though it had kept the traditional imperial form of government. In the Sino-Japanese War of 1895, Japan was strong enough to defeat the much larger China on the field of battle. As a result, Japan annexed the island of Taiwan and gained great influence over the kingdom of Korea. Japan also joined the soldiers of the Western nations to defeat the Chinese who rose up against foreign domination in the Boxer Rebellion of 1900. Clearly, Japan was moving toward building its own empire.

In August, 1914, Takaakira Kato, who had become foreign minister of Japan earlier that year, led his country into World War I on the side of Great Britain, which had been Japan's ally since 1902. Japanese forces soon took over the land Germany had been leasing in the Chinese province of Shantung.

Kato believed in government by political parties, and he wanted to get rid of the traditional power of the elder statesmen, or genro, in Japanese politics. Now he thought he saw an opportunity. If he could gain new power for Japan within China, perhaps he could win the political struggle at home. With the European powers tied up in World War I, China would be quite helpless and Japan could gain valuable advantages.

So it was that on January 18, 1915, Hioki Eki, Japanese minister to China, presented to Chinese president Yuan Shikai a series of proposals that came to be known as the Twenty-one Demands. These demands were divided into five groups. In groups one through four, Japan demanded that China's government approve Japan's recent gains in Shandong (Shantung) and its privileges in overseeing certain territories, trade, and railroads in Manchuria. China was also asked to promise never to let any other power lease or gain control over any harbor, bay, or island along its coast.

It was the group five demands, however, that were most serious. China was to consult with Japan before allowing any foreign investments in the Fujian (Fukien) Province. Japan also demanded new rights over railroads in the Yangtze River Valley and said that China must buy at least half of its arms from Japan. What is more, the Chinese government would be required to hire Japanese men as political, financial, and military advisers, and to share with Japan the control of the police forces in certain parts of China. If China had agreed to all these terms, it would have become a protectorate of Japan.

## Responding to the Demands

President Yuan knew that because of his country's military weakness, it could not resist Japan on its own. He would have to make sure other nations became involved. So when negotiations began, he stalled for time, dragging the talks out as long as possible. The talks were supposed to be secret, but Yuan made sure that United States ambassador Paul Samuel Reinsch and British ambassador John Newell Jordan found out what was going on.

At first, this strategy seemed to be working. Reinsch sent numerous telegrams to U.S. president Woodrow Wilson, urging him to protect China. Wilson was quite disturbed by Japan's bullying of China, for the American policy had been to keep any single power from dominating that vast country. On March 13, 1915, the U.S. government sent a note to Japan, expressing its objections to various parts of the Twenty-one Demands.

When British foreign secretary Sir Edward Grey learned from Jordan about Japan's demands, he, too, became alarmed. He wanted to keep Japan as an ally of Great Britain, yet he did not want Japan to elbow Great Britain out of its rights in China. So on May 3, 1915, the British government followed the United States' lead and sent a note to Japan, objecting to the Twenty-one Demands.

Yuan, meanwhile, sent Ariga Nagao, a Japanese adviser who was loyal to China, to the Japanese capital of Tokyo. There Ariga warned various Japanese politicians that Kato's policy might lead Great Britain to abandon Japan as an ally. Many of the Japanese genro, who had great influence over Japan's cabinet, were persuaded by his arguments.

On May 4, 1915, just after Japan had received Great Britain's warning, the Japanese cabinet held an emergency meeting. Its decision was to alter Kato's China policy. Two days later, Japan would insist that China meet certain of its demands, but none of the demands of group five would be included. China was given until May 9, to accept the first four groups of the original Twenty-one Demands. If it refused, Japan would declare war.

British ambassador Jordan urged Yuan to accept this reduced set of demands as soon as possible. The Americans had become preoccupied by their deteriorating relations with Germany, so that they had lost interest in backing China. Yuan felt compelled to sign a treaty with Japan, granting all the first four groups of the Twenty-one Demands. He did so on May 25, 1915.

## Consequences

Within Japan, many nationalists were not satisfied with the terms of the treaty. Many blamed Great Britain for limiting Japan's gains. As a result, in August, 1915, Kato resigned from the Foreign Ministry; he would never hold office again.

Many Chinese nationalists were similarly upset. They saw the treaty as a national humiliation, and a new wave of anger against foreigners—Japanese and Westerners alike—swept across China. When the 1919 Treaty of Versailles confirmed Japan's gains in the province of Shandong, the Chinese became even more enraged.

The United States continued to be concerned about Japanese expansion in China. On May 11, 1915, U.S. secretary of state William Jennings Bryan publicly warned Japan that the United States would not recognize any agreement that damaged China as a sovereign state. Great Britain eventually followed suit: In 1923, wishing to please the United States, it ended its alliance with Japan. The end result of the break between Japan and its former Western allies was the Japanese attack on the U.S. fleet at Pearl Harbor in 1941.

# Allies Fight Turkey at Gallipoli

*The long Allied effort to gain control of the Gallipoli peninsula—and use of the important Dardanelles Straits—ended in a costly defeat.*

**What:** War
**When:** February, 1915-January, 1916
**Where:** Gallipoli peninsula of Turkey
**Who:**

SIR WINSTON CHURCHILL (1874-1965), British first lord of the admiralty from 1911 to 1915

SIR JOHN ARBUTHNOT FISHER (1841-1920), British first sea lord

SIR SACKVILLE HAMILTON CARDEN (1857-1930), commander of the British expeditionary fleet

JOHN MICHAEL DE ROBECK (1862-1928), Carden's replacement

SIR IAN STANDISH HAMILTON (1853-1947), commander of the Allied expeditionary army

SIR CHARLES CARMICHAEL MONRO (1860-1929), Hamilton's replacement

HORATIO HERBERT KITCHENER, EARL KITCHENER OF KHARTOUM AND OF BROOME (1850-1916), British secretary of state for war in 1914

ENVER PASHA (1881?-1922), the head of the ruling junta in Turkey

MUSTAFA KEMAL (later KEMAL ATATÜRK) (1881-1938), a Turkish commander

OTTO LIMAN VON SANDERS (1855-1929), German adviser to the Turkish army and commander of Turkish forces in the Dardanelles

## The Offensive Begins

The Gallipoli peninsula is a finger of land about sixty miles long and fifteen miles wide, jutting southwestward from European Turkey into the Aegean Sea. The Dardanelles Strait separates this rocky, scrubby peninsula from Asiatic Turkey and connects the Mediterranean to the entrances to the Black Sea.

When Turkey entered World War I on the side of the Central Powers in late 1914, one of its first moves was to close the Dardanelles to Allied traffic. This was done to keep Russia from receiving arms from its Western allies, Great Britain and France, and to prevent Russia from exporting wheat to the West.

Soon military leaders in Great Britain were debating whether to send in battleships to reopen the Dardanelles. Sir John Fisher warned that the Dardanelles would be strongly defended and that a large number of ships would probably be lost in the effort to force a passage. Sir Winston Churchill and others, however, overcame his arguments. Forcing the Dardanelles would secure the supply line to Russia, an Allied front in the Balkans could be established, Bulgaria would be won to the Allied side, Serbia would be saved, and Constantinople could be attacked, which might knock Turkey out of the war.

Vice Admiral Sir Sackville Hamilton Carden would command the proposed attack. He drew up a plan for destroying the Turkish defenses at the Dardanelles' western entrance, then moving up the forty-mile channel. A major challenge would be to break through the Narrows, a tight neck of the channel that was heavily protected by land artillery. If the Allied ships were successful, they could move on to bomb Constantinople.

The French government agreed to send a squadron of ships to join the British fleet. Vice Admiral Carden would command fifteen British battleships, five French battleships, seven Allied cruisers, and various smaller vessels. It would be one of the greatest Allied naval efforts of World War I.

The Turks and their German advisers, meanwhile, had been busily improving the land defenses on both sides of the straits. Mines were placed along the straits, along with antisubmarine nets, searchlights, and torpedoes.

Allied ships began bombarding Turkish forts on the western approach in the early morning of February 19, 1915. After several hours, the Turkish defenders withdrew from the outer forts, and marines were sent in to destroy the abandoned guns.

About a week later, the Allied fleet moved forward again. Minesweeping trawlers were sent ahead to clear a channel through the mine fields. Meanwhile, the Turks strengthened the forts at the Narrows. Ill and exhausted, Vice Admiral Carden returned home; he was replaced by Vice Admiral John Michael de Robeck.

On March 5, the Allies attacked the Narrows' forts and succeeded in smashing Fort Chanak on the Asiatic side. Yet the Allies did suffer serious losses. Of the fifty-eight Allied ships in the Narrows engagement, about a dozen were lost on the afternoon of March 5, and some two thousand men died.

Still, Lord Herbert Kitchener, British secretary of state for war, joined Churchill in urging Robeck to push forward through the Narrows. The Turks and Germans expected the British to continue trying, but Robeck decided that naval action would not be enough to gain control of the Dardanelles; the forts would have to be defeated by the Allied army. He withdrew the Allied fleet from the Narrows.

## The Land Battles

General Ian Hamilton had been appointed commander of an Allied expeditionary force consisting of British and French soldiers. It took some time for all the forces to arrive and for landing craft to be assembled and transports to be repacked. The Turks used the delay to improve the Gallipoli defenses. Turkish leader Enver Pasha and Otto Liman von Sanders, German com-

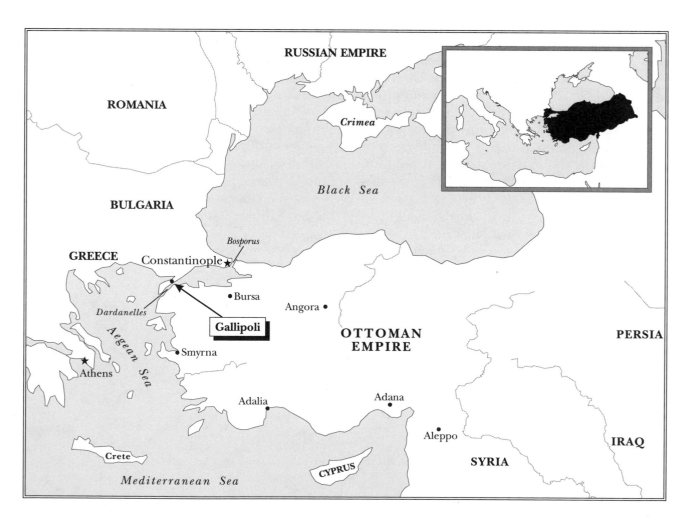

mander of the Turkish forces in the Dardanelles, raised six divisions of soldiers and put them at strategic points on the peninsula.

The first British landings on Gallipoli were made from Lemnos on April 25, at Cape Helles and Ari Burnu. The French landed at Kum-Kale, on the Asiatic side of the entrance to the Dardanelles. They soon joined the British forces on the peninsula. Turkish artillery on the Gallipoli heights kept up a heavy fire against the invaders. Several of the landing parties were driven back off the beaches, but other groups were able to establish their position on beachheads along Gallipoli's southwestern edge.

By early May, 1915, the Allies had about seventy thousand troops on the western end of Gallipoli. They moved eastward toward the village of Krithia, which controlled the only road through the peninsula. The fight for Krithia was very intense, and the Allies finally had to pull back when their ammunition ran out. A week later, they tried again. The British took about half of the village, but then a little-known commander of a Turkish reserve brigade, Mustafa Kemal, led his men to drive the British back again.

Both sides then settled into trench warfare, bombarding each other across the width of the peninsula. The slaughter and the suffering were terrible. The British suffered from a scarcity of fresh water, which had to be brought ashore by ships. Malaria and dysentery struck down hundreds of soldiers, and the battle was in a stalemate.

Hamilton decided to break the deadlock by attempting a landing at Suvla Bay, on the peninsula's north shore. More Allied troops were brought in, along with stocks of ammunition.

At Suvla Bay, Australian and New Zealand troops and Indian Ghurkas went in first. They got ashore on August 6 and headed for the twelve-hundred-foot ridges called Sari Bair. On the way, they had to cross a wide, flat beach protected by Turkish guns on the cliffs above. The Allied troops were hit hard as they advanced, but some of them reached the Sari Bair crests.

Then the Turks, led by Kemal, counterattacked to drive the Allied forces off the crests. Again, the losses on both sides were heavy. Finally, the Allied attackers were forced to withdraw to the beaches. The Allied offensive had failed.

General Hamilton was relieved of his command. Allied armies still held points around Gallipoli's perimeter, but they were not strong enough to return to the offensive. In November, 1915, the new commander, General Charles Monro, recommended that the troops be withdrawn. The withdrawal began the following month and finished in January, 1916.

## Consequences

Gallipoli was a serious defeat for the British. The human costs were enormous: Each side lost more than 250,000 men. There were consequences, too, for various British leaders. Churchill was forced out of office, Fisher resigned, and Hamilton left the army.

On the Turkish side, Mustafa Kemal was hailed as the "Savior of Gallipoli." He went on to become the first president of the Turkish Republic.

# First Transcontinental Telephone Call Is Made

---

*Bell and Watson, the inventors of the telephone, officially inaugurated long-distance service.*

---

**What:** Communications
**When:** April, 1915
**Where:** Jekyll Island, Georgia; New York City; San Francisco, California
**Who:**
ALEXANDER GRAHAM BELL (1847-1922), an American inventor
THOMAS A. WATSON (1854-1934), an American electrical engineer

## The Problem of Distance

The telephone may be the most important invention of the nineteenth century. The device developed by Alexander Graham Bell and Thomas A. Watson opened a new era in communication and made it possible for people to converse over long distances for the first time. During the last two decades of the nineteenth century and the first decade of the twentieth century, the American Telephone and Telegraph (AT&T) Company continued to refine and upgrade telephone facilities, introducing such innovations as automatic dialing and long-distance service.

One of the greatest challenges faced by Bell engineers was to develop a way of maintaining signal quality over long distances. Telephone wires were susceptible to interference from electrical storms and other natural phenomena, and electrical resistance and radiation caused a fairly rapid drop-off in signal strength, which made long-distance conversations barely audible or unintelligible.

By 1900, Bell engineers had discovered that signal strength could be improved somewhat by wrapping the main wire conductor with thinner wires called "loading coils" at prescribed intervals along the length of the cable. Using this procedure, Bell extended long-distance service from New York to Denver, Colorado, which was then considered the farthest point that could be reached with acceptable quality. The result, however, was still unsatisfactory, and Bell engineers realized that some form of signal amplification would be necessary to improve the quality of the signal.

A breakthrough came in 1906, when Lee de Forest invented the "audion tube," which could send and amplify radio waves. Bell scientists immediately recognized the potential of the new device for long-distance telephony and began building amplifiers that would be placed strategically along the long-distance wire network.

Work progressed so quickly that by 1909, Bell officials were predicting that the first transcontinental long-distance telephone service, between New York and San Francisco, was imminent. In that year, Bell president Theodore N. Vail went so far as to promise the organizers of the Panama-Pacific Exposition, scheduled to open in San Francisco in 1914, that Bell would offer a demonstration at the exposition. The promise was risky, because certain technical problems associated with sending a telephone signal over a 4,800-kilometer wire had not yet been solved. De Forest's audion tube was a crude device, but progress was being made.

Two more breakthroughs came in 1912, when de Forest improved on his original concept and Bell engineer Harold D. Arnold improved it further. Bell bought the rights to de Forest's vacuum-tube patents in 1913 and completed the construction of the New York-San Francisco circuit. The last connection was made at the Utah-Nevada border on June 17, 1914.

**319**

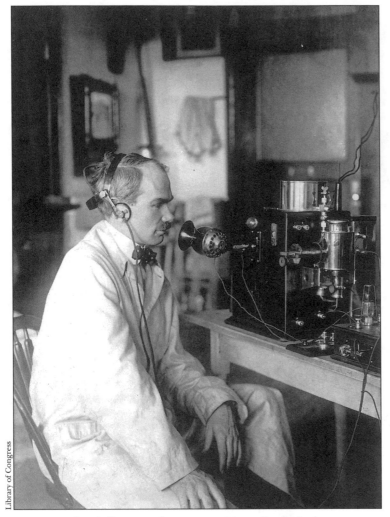

*Lee de Forest sends a message via radiotelephony.*

Library of Congress

## Success Leads to Further Improvements

Bell's long-distance network was tested successfully on June 29, 1914, but the official demonstration was postponed until January 25, 1915, to accommodate the Panama-Pacific Exposition, which had also been postponed. On that date, a connection was established between Jekyll Island, Georgia, where Theodore Vail was recuperating from an illness, and New York City, where Alexander Graham Bell was standing by to talk to his former associate Thomas Watson, who was in San Francisco. When everything was in place, the following conversation took place. Bell: "Hoy! Hoy! Mr. Watson? Are you there? Do you hear me?" Watson: "Yes, Dr. Bell, I hear you perfectly. Do you hear me well?" Bell: "Yes, your voice is perfectly distinct. It is as clear as if you were here in New York."

The first transcontinental telephone conversation transmitted by wire was followed quickly by another that was transmitted via radio. Although the Bell company was slow to recognize the potential of radio wave amplification for the "wireless" transmission of telephone conversations, by 1909 the company had made a significant commitment to conduct research in radio telephony. On April 4, 1915, a wireless signal was transmitted by Bell technicians from Montauk Point on Long Island, New York, to Wilmington, Delaware, a distance of more than 320 kilometers. Shortly thereafter, a similar test was conducted between New York City and Brunswick, Georgia, via a relay station at Montauk Point. The total distance of the transmission was more than 1,600 kilometers. Finally, in September, 1915, Vail placed a successful transcontinental radio-telephone call from his office in New York to Bell engineering chief J. J. Carty in San Francisco.

Only a month later, the first telephone transmission across the Atlantic Ocean was accomplished via radio from Arlington, Virginia, to the Eiffel Tower in Paris, France. The signal was detectable, although its quality was poor. It would be ten years before true transatlantic radio-telephone service would begin.

The Bell company recognized that creating a nationwide long-distance network would increase the volume of telephone calls simply by increasing the number of destinations that could be reached from any single telephone station. As the network expanded, each subscriber would have more reason to use the telephone more often, thereby increasing Bell's revenues. Thus, the company's strategy became one of tying local and regional networks together to create one large system.

## Consequences

Just as the railroads had interconnected centers of commerce, industry, and agriculture all across the continental United States in the nineteenth century, the telephone promised to bring a new kind of interconnection to the country in the twentieth century: instantaneous voice communication. During the first quarter century after the invention of the telephone and during its subsequent commercialization, the emphasis of telephone companies was to set up central office switches that would provide interconnections among subscribers within a fairly limited geographical area. Large cities were wired quickly, and by the beginning of the twentieth century most were served by telephone switches that could accommodate thousands of subscribers.

The development of intercontinental telephone service was a milestone in the history of telephony for two reasons. First, it was a practical demonstration of the almost limitless applications of this innovative technology. Second, for the first time in its brief history, the telephone network took on a national character. It became clear that large central office networks, even in large cities such as New York, Chicago, and Baltimore, were merely small parts of a much larger, universally accessible communication network that spanned a continent. The next step would be to look abroad, to Europe and beyond.

*Andre Millard*

# Italy Enters World War I

> *Though Italy had at first declared neutrality in World War I, British and French promises of new territories helped to propel it into the conflict on the Allied side.*

**What:** War
**When:** May, 1915
**Where:** Rome
**Who:**
ANTONIO SALANDRA (1853-1931), premier of Italy from 1914 to 1916
GIOVANNI GIOLITTI (1842-1928), a leading Liberal-Democrat in Italy's Chamber of Deputies
GABRIELE D'ANNUNZIO (1863-1938), a poet and Italian nationalist
BENITO MUSSOLINI (1883-1945), a socialist editor

### Objections to Neutrality

When World War I broke out in the Balkans, Italy announced its decision to remain neutral. At this time, it was a member of the Central Powers (which had also been known as the Triple Alliance) and might have been expected to join forces with Germany and Austria-Hungary. Yet Italian leaders argued that when Austria declared war on Serbia, it was not acting defensively. Therefore Italy was not under obligation to support Austria in the conflict.

Italy's position, announced by the conservative government of Antonio Salandra in Rome, was at first favored by many Italians. Those who supported neutrality included most of the Socialist Party, the liberal-democrat Giovanni Giolitti and a number of other parliamentarians, and the Roman Catholic clergy.

Yet voices favoring involvement on the side of the Allies (the Triple Entente) began to be heard more and more. Older liberals in Italy tended to see France and Great Britain as leaders of civilized world development, and they believed that Italy would be wise to ally itself with them. The Italian army and the press favored intervention; their spokespersons included a socialist editor named Benito Mussolini and a well-known nationalist poet named Gabriele D'Annunzio.

D'Annunzio and other Italian nationalists wanted to see their nation restored to unity and to a position of international strength and respect. D'Annunzio's efforts to get Italy involved in the war were especially effective. His arguments persuaded many Italians who had originally been undecided, and the result was mass demonstrations in the streets of Rome, calling for an Italian war commitment.

### Negotiations with the Allies

Diplomats representing both sides of the war had been busy in Rome, each group trying to win the Italian government to its side. The Allies were pleased at first with Italy's neutrality—and became even more pleased when the Italian people began calling openly for involvement on the Allied side.

Salandra's government, however, insisted that if Italy were to join the war effort, it should be promised generous rewards. Talks between the Italian and the Allied governments focused on this issue. The British and the French were willing to guarantee an array of new territories for Italy.

The British and French offers included many areas in southeastern Europe and the eastern Mediterranean. The Allies promised that Italy would receive the Austrian Tyrol as far north as the Brenner Pass; Trieste; a number of Adriatic regions, including northern Albania; and, if Turkey were to be divided up, a piece of Asia Minor. In addition, a share of the war-reparations funds from the defeated Central Powers would come to Italy.

With such lavish promises, Italy's commitment to neutrality began to weaken, and the government became more open to intervening in the war. Italian leaders were especially interested in acquiring territories inhabited by people of Italian origin. Control over wider regions to its north and the Adriatic Sea communities to the east would put Italy in the position of influencing Mediterranean affairs.

By the spring of 1915, the only remaining obstacles were the Church, the Socialist Party, and the supporters of Giolitti in the Chamber of Deputies. Giolitti, the leading Italian liberal-democrat since the turn of the century, remained determined to oppose Italian involvement in the war, but most of his allies were not as determined. Street mobs in Rome were already shouting their objections to Giolitti's position, and both the Church and the socialists were intimidated by public opinion. Without strong backing for his position, Giolitti's efforts to preserve neutrality were doomed.

On May 18, 1915, Salandra's government offered the Chamber of Deputies a war resolution. On April 26, 1915, a secret treaty had already been signed in London with Great Britain and France. This treaty promised Italy many territo-rial rewards and part of the war-reparations money. With these guarantees, the war resolution passed easily; only a few Giolitti loyalists voted against it. In May, Italy formally entered the war by declaring war on Austria-Hungary.

## Consequences

Italy's military contribution to the Allied war effort proved to be fairly small. The country's armed forces were not prepared to create any major threat to Germany and Austria-Hungary, and the Italians were unable to keep the Central Powers from overrunning Romania.

At the end of World War I, United States president Woodrow Wilson's Fourteen Points were influential in the drafting of the Treaty of Versailles. Wilson stressed that the war should be ended in a spirit of restoration, not punishment.

As a result, Italy failed to receive all the territories that had been promised in the Treaty of London. The Italian people's resentment over the conclusion of the war combined with turmoil inside the country to encourage the rise of fascism under Benito Mussolini.

# Corning Trademarks Pyrex Glass

*After developing heat-resistant borosilicate glass in 1915, Corning Glass Works named the material and began marketing it for bakeware and laboratory applications.*

**What:** Materials
**When:** May, 1915
**Where:** Corning, New York
**Who:**
JESSE T. LITTLETON (1888-1966), the chief physicist of Corning Glass Works' research department
EUGENE G. SULLIVAN (1872-1962), the founder of Corning's research laboratories
WILLIAM C. TAYLOR (1886-1958), an assistant to Sullivan

## Cooperating with Science

By the twentieth century, Corning Glass Works had a reputation as a corporation that cooperated with the world of science to improve existing products and develop new ones. In the 1870's, the company had hired university scientists to advise on improving the optical quality of glasses, an early example of the common modern practice of academics consulting for industry.

When Eugene G. Sullivan established Corning's research laboratory in 1908 (the first of its kind devoted to glass research), the task that he undertook with William C. Taylor was that of making a heat-resistant glass for railroad lantern lenses. The problem was that ordinary flint glass (the kind in bottles and windows, made by melting together silica sand, soda, and lime) has a fairly high thermal expansion, but a poor heat conductivity. The glass thus expands unevenly when exposed to heat. This condition can cause the glass to break, sometimes violently. Colored lenses for oil or gas railroad signal lanterns sometimes shattered if they were heated too much by the flame that produced the light and were then sprayed by rain or wet snow. This changed a red

"stop" light to a clear "proceed" signal and caused many accidents or near misses in railroading in the late nineteenth century.

Two solutions were possible: to improve the thermal conductivity or reduce the thermal expansion. The first is what metals do: When exposed to heat, most metals have an expansion much greater than that of glass, but they conduct heat so quickly that they expand nearly equally throughout and seldom lose structural integrity from uneven expansion. Glass, however, is an inherently poor heat conductor, so this approach was not possible.

Therefore, a formulation had to be found that had little or no thermal expansivity. Pure silica (one example is quartz) fits this description, but it is expensive and, with its high melting point, very difficult to work.

The formulation that Sullivan and Taylor devised was a borosilicate glass—essentially a soda-lime glass with the lime replaced by borax, with a small amount of alumina added. This gave the low thermal expansion needed for signal lenses. It also turned out to have good acid-resistance, which led to its being used for the battery jars required for railway telegraph systems and other applications. The glass was marketed as "Nonex" (for "nonexpansion glass").

## From the Railroad to the Kitchen

Jesse T. Littleton joined Corning's research laboratory in 1913. The company had a very successful lens and battery jar material, but no one had even considered it for cooking or other heat-transfer applications, because the prevailing opinion was that glass absorbed and conducted heat poorly. This meant that, in glass pans, cakes, pies, and the like would cook on the top, where they were exposed to hot air, but would remain cold and wet (or at least undercooked) next to the glass surface. As a physicist, Littleton knew that glass absorbed radiant energy very well. He thought that the heat-conduction problem could be solved by using the glass vessel itself to absorb and distribute heat. Glass also had a significant advantage over metal in baking. Metal bakeware mostly reflects radiant energy to the walls of the oven, where it is lost ultimately to the surroundings. Glass would absorb this radiation energy and conduct it evenly to the cake or pie, giving a better result than that of the metal bakeware. Moreover, glass would not absorb and carry over flavors from one baking effort to the next, as some metals do.

Littleton took a cut-off battery jar home and asked his wife to bake a cake in it. He took it to the laboratory the next day, handing pieces around and not disclosing the method of baking until all had agreed that the results were excellent. With this agreement, he was able to commit laboratory time to developing variations on the Nonex formula that were more suitable for cooking. The result was Pyrex, patented and trademarked in May of 1915.

## Consequences

In the 1930's, Pyrex "Flameware" was introduced, with a new glass formulation that could resist the increased heat of stovetop cooking. In the half century after Flameware was introduced, Corning went on to produce a variety of other products and materials: tableware in tempered opal glass; cookware in Pyroceram, a glass product that during heat treatment gained such mechanical strength as to be virtually unbreakable;

even hot plates and stoves topped with Pyroceram.

In the same year that Pyrex was marketed for cooking, it was also introduced for laboratory apparatus. Laboratory glassware had been coming from Germany at the beginning of the twentieth century; World War I cut off the supply. Corning filled the gap with Pyrex beakers, flasks, and other items. The delicate blown-glass equipment that came from Germany was completely displaced by the more rugged and heat-resistant machine-made Pyrex ware.

Any number of operations are possible with Pyrex that cannot be performed safely in flint glass: Test tubes can be thrust directly into burner flames, with no preliminary warming; beakers and flasks can be heated on hot plates; and materials that dissolve when exposed to heat can be made into solutions directly in Pyrex storage bottles, a process that cannot be performed in regular glass. The list of such applications is almost endless.

Pyrex has also proved to be the material of choice for lenses in the great reflector telescopes, beginning in 1934 with that at Mount Palomar. By its nature, astronomical observation must be done with the scope open to the weather. This means that the mirror must not change shape with temperature variations, which rules out metal mirrors. Silvered (or aluminized) Pyrex serves very well, and Corning has developed great expertise in casting and machining Pyrex blanks for mirrors of all sizes.

*Robert M. Hawthorne, Jr.*

# Fokker Designs First Propeller-Coordinated Machine Guns

> *Anthony Herman Gerard Fokker designed a mechanism that permitted a machine gun to be fired through the moving propeller of an aircraft.*

**What:** Weapons technology
**When:** May, 1915
**Where:** Schwerin, Germany
**Who:**
ANTHONY HERMAN GERARD FOKKER (1890-1939), a Dutch-born American entrepreneur, pilot, aircraft designer, and manufacturer
ROLAND GARROS (1888-1918), a French aviator
MAX IMMELMANN (1890-1916), a German aviator
RAYMOND SAULNIER (1881-1964), a French aircraft designer and manufacturer

## French Innovation

The first true aerial combat of World War I took place in 1915. Before then, weapons attached to airplanes were inadequate for any real combat work. Handheld weapons and clumsily mounted machine guns were used by pilots and crew members in attempts to convert their observation planes into fighters. On April 1, 1915, this situation changed. From an airfield near Dunkerque, France, a French airman, Lieutenant Roland Garros, took off in an airplane equipped with a device that would make his plane the most feared weapon in the air at that time.

During a visit to Paris, Garros met with Raymond Saulnier, a French aircraft designer. In April of 1914, Saulnier had applied for a patent on a device that mechanically linked the trigger of a machine gun to a cam on the engine shaft. Theoretically, such an assembly would allow the machine gun to fire between the moving blades of the propeller. Unfortunately, the available machine gun Saulnier used to test his device was a Hotchkiss gun, which tended to fire at an uneven rate. On Garros's arrival, Saulnier showed him a new invention: a steel deflector shield that, when fastened to the propeller, would deflect the small percentage of mistimed bullets that would otherwise destroy the blade.

The first test-firing was a disaster, shooting the propeller off and destroying the fuselage. Modifications were made to the deflector braces, streamlining its form into a wedge shape with gutter-channels for deflected bullets. The invention was attached to a Morane-Saulnier monoplane, and on April 1, Garros took off alone toward the German lines. Success was immediate. Garros shot down a German observation plane that morning. During the next two weeks, Garros shot down five more German aircraft.

## German Luck

The German high command, frantic over the effectiveness of the French "secret weapon," sent out spies to try to steal the secret and also ordered engineers to develop a similar weapon. Luck was with them. On April 18, 1915, despite warnings by his superiors not to fly over enemy-held territory, Garros was forced to crash-land behind German lines with engine trouble. Before he could destroy his aircraft, Garros and his plane were captured by German troops. The secret weapon was revealed.

The Germans were ecstatic about the opportunity to examine the new French weapon. Unlike the French, the Germans had the first air-cooled machine gun, the Parabellum, which shot continuous bands of one hundred bullets and was reliable enough to be adapted to a timing mechanism.

In May of 1915, Anthony Herman Gerard Fokker was shown Garros's captured plane and was ordered to copy the idea. Instead, Fokker and his assistant designed a new firing system. It is unclear whether Fokker and his team were already working on a synchronizer or to what extent they knew of Saulnier's previous work in France. Within several days, however, they had constructed a working prototype and attached it to a Fokker *Eindecker 1* airplane. The design consisted of a simple linkage of cams and push-rods connected to the oil-pump drive of an Oberursel engine and the trigger of a Parabellum machine gun. The firing of the gun had to be timed precisely to fire its six hundred rounds per minute between the twelve-hundred-revolutions-per-minute propeller blades.

Fokker took his invention to Doberitz air base, and after a series of exhausting trials before the German high command, both on the ground and in the air, he was allowed to take two prototypes of the machine-gun-mounted airplanes to Douai in German-held France. At Douai, two German pilots crowded into the cockpit with Fokker and were given demonstrations of the plane's capabilities. The airmen were Oswald Boelcke, a test pilot and veteran of forty reconnaissance missions, and Max Immelmann, a young, skillful aviator who was assigned to the front.

When the first combat-ready versions of Fokker's *Eindecker 1* were delivered to the front lines, one was assigned to Boelcke, the other to Immelmann. On August 1, 1915, with their aerodrome under attack from nine English bombers, Boelcke and Immelmann manned their aircraft and attacked. Boelcke's gun jammed, and he was forced to cut off his attack and return to the aerodrome. Immelmann, however, succeeded in shooting down one of the bombers with his synchronized machine gun. It was the first victory credited to the Fokker-designed weapon system.

## Consequences

At the outbreak of World War I, military strategists and commanders on both sides saw the wartime function of airplanes as a means to supply intelligence information behind enemy lines or as airborne artillery spotting platforms. As the war progressed and aircraft flew more or less freely across the trenches, providing vital information to both armies, it became apparent to ground commanders that while it was important to obtain intelligence on enemy movements, it was important also to deny the enemy similar information.

Early in the war, the French used airplanes as strategic bombing platforms. As both armies began to use their air forces for strategic bombing of troops, railways, ports, and airfields, it became evident aircraft would have to be employed against enemy aircraft to prevent reconnaissance and bombing raids.

With the invention of the synchronized forward-firing machine gun, pilots could use their aircraft as attack weapons. A pilot finally could coordinate control of his aircraft and his armaments with maximum efficiency. This conversion of aircraft from nearly passive observation platforms to attack fighters is the single greatest innovation in the history of aerial warfare. The development of fighter aircraft forced a change in military strategy, tactics, and logistics and ushered in the era of modern warfare. Fighter planes are responsible for the battle-tested military adage: Whoever controls the sky controls the battlefield.

*Randall L. Milstein*

*Hulton Archive*

*Anthony Herman Gerard Fokker.*

# McLean Discovers Anticoagulant Heparin

*A medical student discovered a natural chemical that prevents blood from clotting, a finding that had important medical consequences.*

**What:** Medicine
**When:** September, 1915-February, 1916
**Where:** Baltimore, Maryland
**Who:**
JAY MCLEAN (1890-1957), an American physician
WILLIAM HOWELL (1860-1945), an American physiologist who directed research studies on coagulation
CHARLES H. BEST (1899-1978), a Canadian physiologist

## The Importance of Learning German

Jay McLean, a young medical student at the Johns Hopkins University, accidentally discovered the natural anticoagulant heparin in 1916 while working on a physiology research project for William Howell. For more than thirty years, Howell had been studying the various aspects of the coagulation of blood. At the time, he was performing blood-clotting experiments involving a substance known as cephalin. Howell considered cephalin to be the thromboplastic substance of the body, meaning that cephalin caused the blood to clot. The problem Howell assigned to McLean was to develop a method to prepare cephalin in a highly purified form and to determine if it enhanced the clotting process.

The method of obtaining cephalin previously employed by Howell used brain tissue, which contains large amounts of cephalin. The brain tissue was dried on glass plates and then taken through a series of complex, time-consuming steps, using ether and alcohol precipitation in order to separate out the cephalin.

As McLean began work on his project, he began taking an advanced course in German. While reading in German chemical literature, he learned of work in Germany in which substances similar to cephalin were being extracted from heart and liver tissue. McLean suggested to Howell that it might be advantageous to use other organs in his experiments. Subsequently, he found that extracts from heart and liver tissue did indeed possess thromboplastic activity, but apparently the extracts were not pure cephalin. This was demonstrated by the fact that the substances extracted from the heart and liver tissue had different appearances from the cephalin extracted from brain tissue. McLean referred to the extract from the heart as "cuorin" and the extract from the liver as "heparphosphatid," the terms used in the German literature.

In the course of his experiments, McLean also found that, over time, the thromboplastic activity of cephalin deteriorated in these tissues. While he was attempting to determine how long the thromboplastic activity lasted, McLean discovered that, at a certain point, the liver extract not only failed to cause blood to clot but also actually prevented the blood from clotting. McLean concluded that this inhibition of coagulation was apparently caused by substances contained in the liver extract.

## The Proof Is in the Cat

Since this discovery was not a part of the assigned problem, McLean did not inform Howell immediately. He instead repeated the experiment several times to prove that the conclusion was correct. When McLean informed Howell of his finding, Howell was understandably skeptical that McLean had discovered a naturally occurring anticoagulant. In order to convince Howell, McLean added some of his extract to the freshly drawn blood of a laboratory cat and, as McLean predicted, the blood failed to clot. Even though Howell still was not totally convinced, he realized the importance of McLean's work and decided to continue research on the project.

**328**

The first presentation of McLean's results was made on February 19, 1916, at a meeting of the Society of the Normal and Pathological Physiology at the University of Pennsylvania. In June, 1916, McLean submitted his work on the thromboplastic action of cephalin for publication in *The American Journal of Physiology.*

After McLean left Johns Hopkins University, Howell, working with Emmett Holt, continued McLean's work. Howell and Holt, after evaluating many variations of McLean's methods, developed a technique that would yield a reliable preparation of the anticoagulant. In their publication in *The American Journal of Physiology* submitted in October, 1918, Howell and Holt first introduced the word "heparin" for the heparphosphatid which McLean had discovered.

## Consequences

The first clinical trials with heparin were done by E. C. Mason in 1924 and Howell in 1928 in blood-transfusion studies. These trials were largely unsuccessful because of the undesirable reactions in the patients. As soon as a purified, concentrated form of heparin became available in 1935, Gordon Murray began clinical trials in Toronto. His results, published in 1937, clearly indicated that treatment with heparin could prevent some types of clinical thrombosis (clotting within blood vessels). The availability of heparin also made possible a large number of experimental studies. These resulted in the first exchange

transfusion in 1938 and the development of the artificial kidney in 1944. In the 1940's, large-scale clinical experiments with anticoagulant therapy in the United States, Sweden, and Switzerland were begun. These experiments involved heparin treatment of medical conditions such as leg thrombosis, thrombophlebitis of deep veins, and pulmonary embolism. The success of these studies firmly established the use of heparin in medical practice. It was not until 1966, however, that studies were done to demonstrate the use of heparin in the prevention of postoperative thrombosis, a frequent cause of complications and mortality following surgery.

Even though heparin is tremendously valuable in reducing sickness and death, it is not without its problems. Since clotting is a natural protective mechanism of the body, its inhibition by heparin can lead to excessive bleeding. A 1977 study indicated that heparin was the most common cause of drug-related deaths of hospitalized patients. Heparin also has uses in medical science beyond treatment and prevention of thrombosis. It makes possible the maintenance of blood circulation outside the body, a process that is necessary in heart surgery and in kidney dialysis. Heparin is also used to prevent clotting in devices implanted in a vein for the intermittent injection of medication or the withdrawing of blood for laboratory testing.

*Deborah L. Rogers*

# Transatlantic Radio Communication Begins

> *The first radio messages from the United States to Europe were hailed as marking a new era in long-distance telecommunications.*

**What:** Communications
**When:** October 21, 1915
**Where:** Arlington, Virginia, and Paris, France
**Who:**

Reginald Aubrey Fessenden (1866-1932), an American radio engineer

Lee de Forest (1873-1961), an American inventor

Harold D. Arnold (1883-1933), an American physicist

John J. Carty (1861-1932), an American electrical engineer

## An Accidental Broadcast

The idea of commercial transatlantic communication was first conceived by Italian physicist and inventor Guglielmo Marconi, the pioneer of wireless telegraphy. Marconi used a spark transmitter to generate radio waves that were interrupted, or modulated, to form the dots and dashes of Morse code. The rapid generation of sparks created an electromagnetic disturbance that sent radio waves of different frequencies into the air—a broad, noisy transmission that was difficult to tune and detect.

The inventor Reginald Aubrey Fessenden produced an alternative method that became the basis of radio technology in the twentieth century. His continuous radio waves kept to one frequency, making them much easier to detect at long distances. Furthermore, the continuous waves could be modulated by an audio signal, making it possible to transmit the sound of speech.

Fessenden used an alternator to generate electromagnetic waves at the high frequencies required in radio transmission. It was specially constructed at the laboratories of the General Electric Company. The machine was shipped to Brant Rock, Massachusetts, in 1906 for testing. Radio messages were sent to a boat cruising offshore, and the feasibility of radiotelephony was thus demonstrated. Fessenden followed this success with a broadcast of messages and music between Brant Rock and a receiving station constructed at Plymouth, Massachusetts.

The equipment installed at Brant Rock had a range of about 160 kilometers. The transmission distance was determined by the strength of the electric power delivered by the alternator, which was measured in watts. Fessenden's alternator was rated at 500 watts, but it usually delivered much less power.

Yet this was sufficient to send a radio message across the Atlantic. Fessenden had built a receiving station at Machrihanish, Scotland, to test the operation of a large rotary spark transmitter that he had constructed. An operator at this station picked up the voice of an engineer at Brant Rock who was sending instructions to Plymouth. Thus, the first radiotelephone message had been sent across the Atlantic by accident. Fessenden, however, decided not to make this startling development public. The station at Machrihanish was destroyed in a storm, making it impossible to carry out further tests. The successful transmission undoubtedly had been the result of exceptionally clear atmospheric conditions that might never again favor the inventor.

One of the parties following the development of the experiments in radio telephony was the American Telephone and Telegraph (AT&T) Company. Fessenden entered into negotiations to sell his system to the telephone company, but, because of the financial panic of 1907, the sale was never made.

## Virginia to Paris and Hawaii

The English physicist John Ambrose Fleming had invented a two-element (diode) vacuum tube in 1904 that could be used to generate and detect radio waves. Two years later, the American inventor Lee de Forest added a third element to the diode to produce his "audion" (triode), which was a more sensitive detector. John J. Carty, head of a research and development effort at AT&T, examined these new devices carefully. He became convinced that an electronic amplifier, incorporating the triode into its design, could be used to increase the strength of telephone signals and to generate continuous radio waves that could travel long distances.

On Carty's advice, AT&T purchased the rights to de Forest's audion. A team of about twenty-five researchers, under the leadership of physicist Harold D. Arnold, were assigned the job of perfecting the triode and turning it into a reliable amplifier. The improved triode was responsible for the success of transcontinental cable telephone service, which was introduced in January, 1915. The triode was also the basis of AT&T's foray into radio telephony.

Carty's research plan called for a system with three components: an oscillator to generate the radio waves, a modulator to add the audio signals to the waves, and an amplifier to transmit the radio waves. The total power output of the system was 7,500 watts, enough to send the radio waves over thousands of kilometers.

The apparatus was installed in the U.S. Navy's radio tower in Arlington, Virginia, in 1915. Radio messages from Arlington were picked up at a receiving station in California, a distance of 4,000 kilometers, then at a station in Pearl Harbor, Hawaii, which was 7,200 kilometers from Arlington. AT&T's engineers had succeeded in joining the company telephone lines with the radio transmitter at Arlington; therefore, the president of AT&T, Theodore Vail, could pick up his telephone and talk directly with someone in California.

The next experiment was to send a radio message from Arlington to a receiving station set up in the Eiffel Tower in Paris. After several unsuccessful attempts, the telephone engineers in the Eiffel Tower finally heard Arlington's messages on October 21, 1915. The AT&T receiving sta-

tion in Hawaii also picked up the messages. The two receiving stations had to send their reply by telegraph to the United States because both stations were set up to receive only. Two-way radio communication was still years in the future.

## Consequences

The announcement that messages had been received in Paris was front-page news and brought about an outburst of national pride in the United States. The demonstration of transatlantic radio telephony was more important as publicity for AT&T than as a scientific advance. All the credit went to AT&T and to Carty's laboratory. Both Fessenden and de Forest attempted to draw attention to their contributions to long-distance radio telephony, but to no avail. The Arlington-to-Paris transmission was a triumph for corporate public relations and corporate research.

The development of the triode had been achieved with large teams of highly trained scientists—in contrast to the small-scale efforts of Fessenden and de Forest, who had little formal scientific training. Carty's laboratory was an example of the new type of industrial research that was to dominate the twentieth century. The golden days of the lone inventor, in the mold of Thomas Edison or Alexander Graham Bell, were gone.

In the years that followed the first transatlantic radio telephone messages, little was done by AT&T to advance the technology or to develop a commercial service. The equipment used in the 1915 demonstration was more a makeshift laboratory apparatus than a prototype for a new radio technology. The messages sent were short and faint. There was a great gulf between hearing "hello" and "good-bye" amid the static. The many predictions of a direct telephone connection between New York and other major cities overseas were premature. It was not until 1927 that a transatlantic radio circuit was opened for public use. By that time, a new technological direction had been taken, and the method used in 1915 had been superseded by shortwave radio communication.

*Andre Millard*

# Garvey Founds Universal Negro Improvement Association

> *Marcus Garvey's emphasis on black pride and self-help was a major contribution to race consciousness in the United States.*

**What:** Civil rights and liberties; Social reform
**When:** 1916
**Where:** Harlem, New York
**Who:**
MARCUS MOZIAH GARVEY (1887-1940), the founder of the Universal Negro Improvement Association
BOOKER TALIAFERRO WASHINGTON (1856-1915), an African American educator
DUSE MOHAMMED, an African scholar in London

## Garvey's Work Begins

Marcus Garvey was born in St. Ann's Bay, Jamaica, in 1887. He later claimed to be of pure African descent. His father was a descendant of the Maroons, or Jamaican slaves, who successfully revolted against their British masters in 1739.

During his early years, Garvey gradually realized that his color was a badge of inferiority. In Jamaica, the mulattoes (mixed-race blacks with light skin) had higher status than dark-skinned blacks. Though Garvey grew up with a sense of isolation, he also came to take pride in his color.

By his twentieth birthday, Garvey had started a program to change the lives of blacks in Jamaica. While working as a foreman in a printing shop in 1907, he became a leader in a labor strike. The strike was quickly broken by the shop owners, however, and as a result Garvey lost faith in labor unions as a way to bring about social reform.

In 1910, he began publishing a newspaper, the *Watchman*, and helped form a political organiza-tion called the National Club. These efforts were not especially successful either, but they did inspire Garvey to visit Central America, where he was able to observe the miserable conditions of blacks in Costa Rica and Panama.

Garvey's travels eventually led him to London, the center of the British Empire. There he met Duse Mohammed, an African scholar who increased the young Jamaican's knowledge and awareness of Africa. During this time, Garvey also read Booker T. Washington's *Up from Slavery* (1901), which acquainted him with the problems of African Americans. Reading Washington's autobiography raised questions in Garvey's mind: "I asked, where is the black man's Government? Where is his President, his country and his ambassador, his army, his navy, his men of big affairs? I could not find them, and then I declared, I will help to make them."

Returning to Jamaica in 1914, Garvey started a self-help organization for black people and called it the Universal Negro Improvement and Conservation Association and African Communities League. This new organization, whose name was soon shortened to the Universal Negro Improvement Association (UNIA), called for the union of "all people of Negro or African parentage." The goals of the UNIA were to increase racial pride, aid black people throughout the world, and "establish a central nation for the race."

Garvey, who was elected the first president of the UNIA, realized that black people would have to achieve these goals without help from white people. This self-help concept, which resembled the ideas of Booker T. Washington, led Garvey to propose a trade school for blacks in Kingston, Jamaica, similar to Washington's Tuskegee Insti-

Library of Congress

*Marcus Garvey.*

the South. In New York City, for example, the population of African Americans increased from 91,709 in 1910, to 152,467 in 1920. They were attracted by the promise of jobs and by the possibility of escaping the rigid system of segregation in the South.

Yet they found that they could not escape racism simply by moving north. Northern whites believed in the racial inferiority of blacks just as Southern whites did, and they resented having to compete with African Americans for jobs. Like immigrants from abroad, the new African American residents were crowded in ghettos without proper housing. Racial violence broke out in several Northern cities.

The people of Harlem were not particularly interested in Booker T. Washington's philosophy, which they saw as giving in too much to the white power structure, and the National Association for the Advancement of Colored People seemed too intellectual and middle class. Yet Garvey was able to attract support from the Jamaican immigrants in Harlem, who felt isolated, and he established a branch of the UNIA there in 1916.

tute. Yet he was not able to gather much support for the plan.

### Garvey in America

In 1915, Garvey decided to come to the United States to seek aid for his organization. He exchanged letters with Washington, but Washington died before Garvey arrived in the United States in 1916. Garvey went immediately to Harlem, which in the early twentieth century was gaining a large population of African Americans.

The lives of African Americans were changing rapidly during these years. Cities in the North were receiving mass migrations of blacks from

At first, the organization met with serious problems. Local politicians tried to gain control of it, and Garvey had to struggle hard to keep the UNIA's independence. The original branch of the UNIA was dissolved, and a charter was obtained from the state of New York which prevented other groups from using the organization's name.

By 1918, under Garvey's leadership, the New York chapter of the UNIA had 3,500 members. By 1919, Garvey claimed two million members for his organization across the world and 200,000 subscribers for his publication, *Negro World.*

To help blacks economically, in 1916 Garvey set up two joint-stock companies—the Black Star

Line, an international commercial shipping company, and the Negro Factories Corporation, which was to build and run factories "to manufacture every marketable commodity." Stock in these companies was sold only to blacks. The Black Star Line was to establish trade with Africa and to transport willing blacks to a new community in Africa. Although both companies failed financially, their existence gave many blacks a feeling of dignity.

## Consequences

At the urging of rival African American leaders, the U.S. government had Garvey indicted for fraudulent use of the mails in 1922. He was tried, found guilty, and sent to prison in 1923. The UNIA began to lose its organizational strength.

In 1927, Garvey was released from prison and deported as an "undesirable alien." He returned to Jamaica and then went to London and Paris, where he tried to resurrect the UNIA, but met with little success. He died in poverty in London in 1940.

Although he failed as a businessman, Garvey was an eloquent spokesman for African Americans and other blacks throughout the world. His speeches and writings contributed to the development of racial awareness and pride among African Americans.

*John C. Gardner*

# Schwarzschild Theorizes Existence of Black Holes

*In 1916, Karl Schwarzschild developed a solution to Albert Einstein's equations of general relativity that describes a gravitational black hole.*

**What:** Astronomy
**When:** 1916
**Where:** Potsdam, Germany
**Who:**
KARL SCHWARZSCHILD (1873-1916), a German physicist
ALBERT EINSTEIN (1879-1955), a German American physicist
SIR ISAAC NEWTON (1642-1727), an English scientist and philosopher

## One of the Twelve

In 1915, Albert Einstein published his general theory of relativity, in which he discussed his theory of a "space-time continuum" that included four dimensions: length, width, height, and time. He believed that physical events could be located precisely on this continuum. This revolutionary new concept of space-time, which included the idea that space-time itself is curved, stood in opposition to the universal law of gravitation, which had been developed by Sir Isaac Newton in 1665. According to Newton, gravity is an attractive force acting between all particles of matter in the universe. Einstein, however, believed that gravity is a consequence of the shape of space-time. Local space-time, according to the general theory of relativity, is distorted by the presence of a large mass such as a star or a planet. Objects traveling close to a massive body would therefore travel in a curved path, causing the appearance of a gravitational field.

When the general theory of relativity was first proposed, the mathematics of its equations were thought to be beyond comprehension. In fact, it was frequently stated that only twelve or so scientists in the world completely understood the theory. Even today, many of its implications remain unexplained. The first person to find an exact solution to the equations of the general theory of relativity was the German physicist Karl Schwarzschild. Prior to the work of Schwarzschild, the only solutions to the equations had been approximations.

In 1916, when Schwarzschild was working on his solution, Germany was at war. Because of his patriotism, the forty-year-old Schwarzschild insisted upon serving in the German armed forces. Various campaigns took him to Belgium, France, and finally Russia. While serving in Russia he contracted the fatal disease pemphigus. Although he became too ill to continue in military service, he continued to work on the equations. Shortly after his return to Germany, he completed his work and sent a copy to Einstein. Within a few months, Schwarzschild died.

Schwarzschild had sought to determine what would happen if gravity around a spherical body became infinitely powerful. He also wanted to find the least complex explanation for the phenomenon. The result, the Schwarzschild solution, describes a "black hole," an object so dense that light itself cannot escape from its surface. Difficulties in interpreting the Schwarzschild solution, however, cast some doubt upon its validity until the 1960's, when its true significance was recognized.

## Black Holes

The formation of a black hole is believed to be the final stage in the decay of a massive star, when nuclear fuel is exhausted in the star's core. The exact sequence of events depends entirely on the mass of the star. Toward the end of its life, a massive star will go through the supernova stage, in which much of its outer material is blasted into

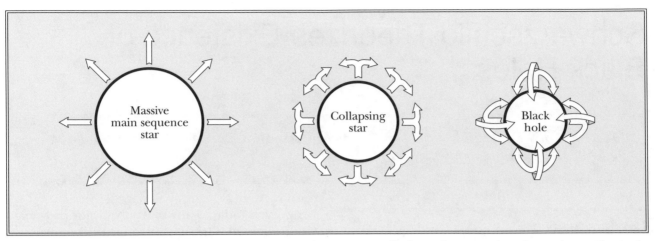

*A massive star may end its life as a black hole: During its main sequence, radiation emits outward; as the core burns, the star begins to collapse in on itself; finally, the increasing mass at the core is so great that gravity is extremely strong, preventing any radiation (including light) from escaping.*

space. The core then begins to collapse. If the star's mass is 1.4 solar masses or smaller, it will end up as a "white dwarf." At this stage, the pressure exerted by the electrons of the atoms is enough to prevent total collapse. This hot carbon mass will eventually cool to become a "black dwarf." If the mass of the decayed star is between 1.4 and 3.1 times the mass of the sun, gravity will cause a much more extensive collapse. At this point, gravity is so intense that electrons and protons combine to form neutrons, resulting in the formation of a "neutron star."

If the mass is greater than 3.1 solar masses, not even neutrons will be able to counteract the force of gravity, and the star will continue to collapse. As the star collapses, its surface gravity will become greater and greater. As a result, the velocity needed to escape this gravitational body increases. After the escape velocity has reached the velocity of light, further collapse results in the formation of a black hole. The distance at which the escape velocity is equal to the velocity of light is the distance calculated by Schwarzschild in his solution to Einstein's equations; it is known as the "Schwarzschild radius" or the "event horizon." Beyond this point, there is no way of determining events. It is an area that is totally disconnected from normal space and time.

In theory, any object could become a black hole if it were compressed enough. If the earth were shrunk to a volume slightly less than a centimeter in radius, it would become a black hole.

If the sun were compressed to a radius of less than three kilometers, it would become a black hole.

The diameter of the Schwarzschild radius of a black hole depends on the mass of the decayed stellar core. For example, a decayed core with a mass five times greater than that of the sun would have an event horizon with a radius of 30 kilometers. A stellar remnant of 20 solar masses would have an event horizon with a 60-kilometer radius. Within this boundary, however, the remains of the star continue to collapse to a point of infinite pressure, infinite density, and infinite curvature of space-time. This point is known as the "Schwarzschild singularity."

### Consequences

The true significance of Schwarzschild's work was not recognized until more study was done on stellar structure and evolution. An important step was taken in 1931 when the astronomer Subrahmanyan Chandrasekhar completed calculations that described the interior of a white dwarf star. At that time, he did not consider the fate of very massive stars, but English astronomer Arthur Stanley Eddington proposed that massive stars in their death stages continue to radiate energy as they become smaller and smaller. At some point, they reach equilibrium. In 1939, the American physicist J. Robert Oppenheimer and his student Hartland Snyder showed that a star that possesses enough mass will collapse indefinitely.

By the end of the twentieth century, it was fully recognized that what the Schwarzschild solution describes is a black hole. However, the type of black hole described by the Schwarzschild equations is essentially static. Static black holes do not rotate and have no electric charges. The only property possessed by such bodies is mass. Later variations on Schwarzschild's work have led to theories about other kinds of objects, such as rotating black holes, black holes with electrical charges, and black holes that both rotate and have electrical charges.

Black holes, by their nature, cannot be seen, so evidence for their existence must necessarily be circumstantial. Nevertheless, proof that they actually exist has grown increasingly strong. Over the last three decades of the twentieth century astrophysicists identified two dozen possible black holes—including one at the center of the Milky Way galaxy. In 2000 alone the discovery of eight supermassive black holes was reported to the American Astronomical Association. With the development of new optical X ray, infrared, and radio telescopes, more discoveries were anticipated, and astrophysicists were hypothesizing that black holes are not rare, but common, throughout the universe.

*David W. Maguire*

# Ricardo Designs Modern Internal Combustion Engine

Harry Ricardo established the standard form of the internal combustion engine.

**What:** Engineering
**When:** 1916-1922
**Where:** London and Shoreham, England
**Who:**
Sir Harry Ralph Ricardo (1885-1974), an English engineer
Oliver Thornycroft (1885-1956), an engineer and works manager
Sir David Randall Pye (1886-1960), an engineer and administrator
Sir Robert Waley Cohen (1877-1952), a scientist and industrialist

## The Internal Combustion Engine: 1900-1916

By the beginning of the twentieth century, internal combustion engines were almost everywhere. City streets in Berlin, London, and New York were filled with automobile and truck traffic; gasoline- and diesel-powered boat engines were replacing sails; stationary steam engines for electrical generation were being edged out by internal combustion engines. Even aircraft use was at hand: To progress from the Wright brothers' first manned flight in 1903 to the fighting planes of World War I took little more than a decade.

The internal combustion engines of the time, however, were primitive in design. They were heavy (10 to 15 pounds per output horsepower, as opposed to 1 to 2 pounds today), slow (typically 1,000 or fewer revolutions per minute or less, as opposed to 2,000 to 5,000 today), and extremely inefficient in extracting the energy content of their fuel. These were not major drawbacks for stationary applications, or even for road traffic that rarely went faster than 30 or 40 miles per hour, but the advent of military aircraft and tanks demanded that engines be made more efficient.

## Engine and Fuel Design

Harry Ricardo, son of an architect and grandson (on his mother's side) of an engineer, was a central figure in the necessary redesign of internal combustion engines. As a schoolboy, Ricardo built a coal-fired steam engine for his bicycle, and at Cambridge University he produced a single-cylinder gasoline motorcycle, incorporating many of his own ideas, which won a fuel-economy competition when it traveled almost 40 miles on a quart of gasoline. He also began development of a two-cycle engine called the "Dolphin," which later was produced for use in fishing boats and automobiles. In fact, in 1911, Ricardo took his new bride on their honeymoon trip in a Dolphin-powered car.

The impetus that led to major engine research came in 1916 when Ricardo was an engineer in his family's firm. The British government asked for newly designed tank engines, which had to operate in the dirt and mud of battle, at a tilt of up to 35 degrees, and could not give off telltale clouds of blue oil smoke. Ricardo solved the problem with a special piston design and with air circulation around the carburetor and within the engine to keep the oil cool.

Design work on the tank engines turned Ricardo into a full-fledged research engineer. In 1917, he founded his own company, and a remarkable series of discoveries quickly followed. He investigated the problem of detonation of the fuel-air mixture in the internal combustion cylinder. The mixture is supposed to be ignited by the spark plug at the top of the compression stroke, with a controlled flame front spreading at a rate about equal to the speed of the piston head as it moves downward in the power stroke. Some fuels, however, detonated (ignited spontaneously throughout the entire fuel-air mixture) as a result of the compression itself, causing loss of fuel efficiency and damage to the engine. With the cooperation of Robert Waley Cohen of Shell Petroleum, Ricardo evaluated chemical mixtures of fuels and found that paraffins (such as *n*-heptane, the current low-octane standard) detonated readily, but aromatics such as toluene were nearly immune to detonation. He established a "toluene number" rating to describe the tendency of various fuels to detonate; this number was replaced in the 1920's by the "octane number" devised by Thomas Midgley at the Delco laboratories in Dayton, Ohio.

The fuel work was carried out in an experimental engine designed by Ricardo that allowed direct observation of the flame front as it spread and permitted changes in compression ratio while the engine was running. Three principles emerged from the investigation: The fuel-air mixture should be admitted with as much turbulence as possible, for thorough mixing and efficient combustion; the spark plug should be centrally located to prevent distant pockets of the mixture from detonating before the flame front reaches them; and the mixture should be kept as cool as possible to prevent detonation.

These principles were then applied in the first truly efficient side-valve ("L-head") engine—that is, an engine with the valves in a chamber at the side of the cylinder, in the engine block, rather than overhead, in the engine head. Ri-

cardo patented this design, and after winning a patent dispute in court in 1932, he received royalties or consulting fees for it from engine manufacturers all over the world.

## Consequences

The side-valve engine was the workhorse design for automobile and marine engines until after World War II. With its valves actuated directly by a camshaft in the crankcase, it is simple, rugged, and easy to manufacture. Overhead valves with overhead camshafts are the standard in automobile engines today, but the side-valve engine is still found in marine applications and in small engines for lawn mowers, home generator systems, and the like. In its widespread use and its decades of employment, the side-valve engine represents a scientific and technological breakthrough in the twentieth century.

Ricardo and his colleagues, Oliver Thornycroft and D. R. Pye, went on to create other engine designs—notably, the sleeve-valve aircraft engine that was the basic pattern for most of the great British planes of World War II and early versions of the aircraft jet engine. For his technical advances and service to the government, Ricardo was elected a Fellow of the Royal Society in 1929, and he was knighted in 1948.

*Robert M. Hawthorne, Jr.*

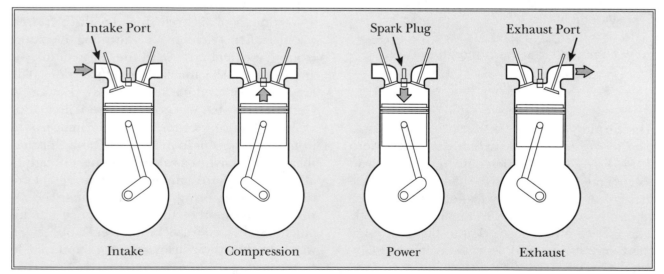

*The four cycles of a standard internal combustion engine (left to right): (1)* intake, *when air enters the cylinder and mixes with gasoline vapor; (2)* compression, *when the cylinder is sealed and the piston moves upward to compress the air-fuel mixture; (3)* power, *when the spark plug ignites the mixture, creating more pressure that propels the piston downward; and (4)* exhaust, *when the burned gases exit the cylinder through the exhaust port.*

# Nearly One Million Die in Battle of Verdun

*Although the French army won a final victory at Verdun, France and Germany both lost hundreds of thousands of men in this "worst battle in history."*

**What:** War
**When:** February 21-December 15, 1916
**Where:** Verdun, France
**Who:**

Erich von Falkenhayn (1861-1922), chief of general staff of the German army from 1914 to 1916

Friedrich Wilhelm Viktor August Ernst, or William (1882-1951), crown prince of Germany from 1888 to 1918, and commander of the Fifth Army of Germany

Friedrich Wilhelm Viktor Albert, or William II (1859-1941), emperor of Germany and king of Prussia from 1888 to 1918

Joseph-Jacques-Césaire Joffre (1852-1931), commander in chief of the French armies from 1914 to 1916

Henri-Philippe Pétain (1856-1951), commander of the French forces at Verdun

Robert-Georges Nivelle (1856-1924), commander of the Second Army of France

## The Strategies

After World War I's First Battle of the Marne in 1914, for a time neither Allied nor German forces could gain a decisive victory on the Western Front. Their military leaders came up with different strategies to try to break the deadlock. The French and British tried to break through the German lines and repeatedly attacked German defenses along the northern portion of the front. In spite of bloody assaults at Ypres and elsewhere, however, the Germans held their ground.

Meanwhile, the Germans concentrated on holding their defensive positions on the Western Front. They realized that offensives in this kind of war would cost many lives. On the Eastern Front, however, a different kind of battle was raging, and there the Germans were able to attack and defeat the Russian armies.

Late in 1915, General Erich von Falkenhayn, chief of general staff of the German army, approached Emperor William II with a plan. He was aware that the Allies were losing many men and beginning to lose morale as a result. To take advantage of this situation, he suggested sending an army against a fortified French position that the French could not surrender without losing prestige. Under pressure to save their fort, the French would allow so many soldiers to go to their deaths that in the end France would be forced to surrender.

The emperor approved this plan, and the fortress town of Verdun was selected as the target. Falkenhayn had no intention of actually capturing Verdun; the point was to force the French to defend it and thus cause them to lose many men.

Verdun, a small town ringed with fortresses, was in a salient—a finger of land along the battlefront that extended toward the German lines—and thus could be attacked from three sides. The French considered it a key defensive position. A German army led by Crown Prince William was to attack the ring of fortresses. German troops would then be sent in to assist the crown prince; they would arrive at a calculated rate, enough to allow the attack to maintain momentum, but not so many as to bring about a German breakthrough. Yet neither the crown prince nor his chief of staff, General Schmidt von Knobelsdorf, was told that their army was not supposed to break through.

## A Long Battle

The German attack began with a massive barrage of artillery on February 21, 1916. The

**340**

French were caught by surprise, so that the German troops made some quick gains and captured some of the important fortresses. General Joseph-Jacques-Césaire Joffre, commander in chief of the French armies, and his chief of staff, General Nöel-Marie-Joseph-Édouard de Curières de Castelnau, reacted as Falkenhayn had predicted. Having decided that Verdun was to be held at all costs, they brought the French Second Army, under General Henri-Philippe Pétain, to the defense of Verdun.

Pétain saw immediately that the main problem was to keep Verdun supplied with troops and equipment. Because it was on a salient, there was only one highway leading into the town. Pétain arranged for large amounts of supplies to be brought in regularly over this road.

The crown prince and his army tried to break through, but they were hampered by the slow arrival of reinforcing troops and by Pétain's defense. The Germans soon found that their losses were equal to and sometimes greater than those of the French. By mid-April, Pétain was attacking the Germans rather than simply defending the Verdun fortresses. The Germans were in a dangerous position, and even Falkenhayn came to believe that to keep their prestige they must break through and capture Verdun.

To complicate matters for the Germans, the British attack on the Somme River in northern France began in midsummer. German troops had to be pulled away from the Verdun area to fight the British. Pétain was made commander of the entire Central Front, and General Robert-Georges Nivelle was placed in command of the French Second Army. Nivelle led a series of French attacks throughout the summer and fall of 1916, and in December, 1916, the Germans gave up their positions at Verdun. The Battle of Verdun was over.

*French troops come under shellfire during the Battle of Verdun.*

## Consequences

Verdun was the longest and most costly battle of World War I. The French lost half a million men, and German casualties came to 400,000. The French soldiers who survived were left with horrible memories of sitting in water-filled shell holes for days on end, with mangled corpses to keep them company. Though the French had won, their army had been so weakened that after this time the burden of the war fell entirely upon the British, until the Americans became involved. Meanwhile, the Germans had been humiliated, and Falkenhayn was replaced by General Paul von Hindenburg.

Verdun had important effects more than two decades later, during World War II. Pétain, who became France's foremost military leader, was convinced through his Verdun experience that only a series of strong forts along the French-German border could enable France to win another war with Germany; this was the concept of the Maginot Line. Moreover, he thought, France should never again take the offensive, for Verdun had proved that an offensive strategy lost more men than a defensive strategy did. These French strategies ended up allowing Germany to overrun France in the Blitzkrieg of 1940.

*José M. Sánchez*

# Pershing Leads U.S. Military Expedition into Mexico

> *Following a raid on a New Mexico town by a band of Mexican rebels, the United States sent troops into Mexico, and the two countries came close to war.*

**What:** Military conflict
**When:** March 15, 1916
**Where:** Mexico
**Who:**
FRANCISCO (PANCHO) VILLA (1877-1923), a
   Mexican revolutionary
VENUSTIANO CARRANZA (1859-1920),
   president of Mexico from 1914 to
   1920
ÁLVARO OBREGÓN (1880-1928), Mexican
   minister of war and marine, and
   commander of military forces in
   northern Mexico
WOODROW WILSON (1856-1924), president
   of the United States from 1913 to 1921
JOHN J. PERSHING (1860-1948),
   commander of the American
   expedition
HUGH L. SCOTT (1853-1934), chief of staff
   of the United States Army

## Raid and Pursuit

After the overthrow of Mexico's dictator Porfirio Díaz in 1911, the new president, Francisco Madero, was in office less than two years before he, in turn, was overthrown by one of his own generals, Victoriano Huerta. Huerta was popular with the Catholic Church, the wealthy landowners, and the foreign investors who controlled most of Mexico's natural resources, but he had little support from the Mexican people. One of his first acts was to arrange for the murder of his predecessor, Madero. In the United States, the new president, Woodrow Wilson, was horrified by Huerta's willingness to kill. Wilson gave his support to Huerta's two main opponents, Venustiano Carranza and Francisco Villa.

In April, 1914, Wilson decided to take advantage of a minor conflict at Tampico, Mexico, to have the U.S. Navy seize the port of Veracruz. His goal was to force Huerta's resignation by cutting off his supply of arms from abroad and his source of income, the Veracruz customhouse.

The result was a short battle in which both the Mexicans and the Americans lost many men. Wilson, who had not expected the Mexicans to resist, gladly accepted an offer of mediation from Argentina, Brazil, and Chile. The negotiations did not result in any official agreement, but the Veracruz incident had been enough to topple Huerta's already shaky government. In August, 1914, Villa and Carranza entered Mexico City to set up a new government.

Within a few weeks, however, these revolutionary leaders were fighting each other. Carranza and his general, Álvaro Obregón, led their troops in driving Villa and his forces back to his home state, Chihuahua. On October 19, 1915, the United States, which had formerly favored Villa, recognized Carranza's provisional government as Mexico's legitimate government.

By early 1916, Villa's army had shrunk to a few hundred men hiding from Obregón's troops in the Chihuahua mountains. At 4:15 A.M. on March 9, 1916, Villa and about five hundred of his men attacked the tiny border town of Columbus, New Mexico, and the nearby U.S. Army garrison. Less than an hour later, the Mexicans were retreating toward the border, pursued by a troop of U.S. cavalry. They left behind seventeen American dead—eight of them civilians.

Some historians believe that Villa's goal was to provoke a war between the United States and Mexico so that he could become a hero among the Mexican people for resisting the foreign invaders. If beginning a war was really his aim, he

Hulton Archive

*General John J. Pershing leads U.S. troops across a river in pursuit of Francisco "Pancho" Villa and his followers.*

nearly succeeded. News of the raid raised a storm of anger throughout the United States. President Wilson announced to the press that "an adequate force" would be sent immediately in pursuit of Villa.

The American expedition was made up of about six thousand troops under the command of Brigadier General John J. Pershing. They crossed the border into Mexico in the early hours of March 15, 1916. Meanwhile, Villa's band continued to retreat farther into Mexico. Members of the Seventh and Tenth Cavalry caught up with them near the town of Guerrero. In a series of battles between March 28 and April 1, the Americans either killed or captured most of Villa's men, though Villa himself escaped and fled south toward the town of Parral.

### The Negotiations

President Venustiano Carranza, who had never actually agreed to allow the American troops to enter Mexico, began sending urgent notes to Washington, D.C., requesting that the expeditionary force be withdrawn. It had accomplished its goal, he said, and Mexican troops could now take over the job of guarding the border between the two countries.

As the two governments negotiated over the return of Pershing's troops to the United States, members of the U.S. Thirteenth Cavalry, trying to buy supplies in the town of Parral, were attacked by a mob. In the fight that followed, a large number of Mexicans and two Americans were killed.

As tension between the United States and Mexico increased, Carranza sent Obregón, who had become the minister of war and marine, to meet with General Hugh L. Scott, the U.S. Army chief of staff, in El Paso, Texas. After much argument, the two generals were able to write an agreement for the American troops to be gradually removed from Mexico.

On the day the agreement was signed, however, a band of Mexicans attacked the towns of Glen Springs and Bouguilas in Texas, killing three Americans, including a nine-year-old boy. A new wave of anger and fear swept across the United States, and many Americans began demanding war with Mexico. President Wilson sent out the National Guards of Texas, Arizona, and New Mexico to patrol the border and ordered an additional four thousand army soldiers to be sent to the Southwest.

Meanwhile, the Mexicans also were coming to the end of their patience. On June 16, 1916, Carranza warned the American government that if American soldiers in Mexico made further moves in any direction but north, they would be fired upon. Five days later, a troop of the Tenth Cavalry, led by Captain Charles T. Boyd, tried to enter the town of Carrizel, though Pershing had ordered the troop to avoid "places garrisoned by *carrancista* troops." The Mexican soldiers in the town opened fire. Twelve Americans, including Captain Boyd, were killed, and twenty-three were taken prisoner.

War with Mexico appeared certain, but Wilson was determined to avoid it if possible, and Carranza did not want to see Americans occupying his country. Carranza released the American prisoners and offered to join direct talks to settle the dispute. The negotiations opened at New London, Connecticut, on September 6, 1916, and ended without any real agreement four months later.

## Consequences

Though the New London talks did not produce any official treaty, they were useful in relaxing tensions between the United States and Mexico. Wilson and his advisers decided that it would be best to withdraw the remaining American troops from Mexico. Though Villa was still dangerous, he was now far from the border, and meanwhile the United States was moving closer to war with Germany. On February 5, 1917, the last troops of Pershing's expeditionary force reentered the United States.

*Ronald N. Spector*

# Easter Rebellion Fails to Win Irish Independence

> *Although the Easter Rebellion did not establish Ireland's independence from Great Britain, it helped persuade the Irish people that they must seek freedom from British rule.*

**What:** Civil strife
**When:** April 24-30, 1916
**Where:** Dublin, Ireland
**Who:**
PATRICK PEARSE (1879-1916), a poet and teacher who was commander in chief of the Republican forces in the rebellion
EOIN MACNEILL (1867-1945), a scholar who was chief of staff of the Irish Volunteers
SIR ROGER CASEMENT (1864-1916), an Irish revolutionary who had once been a member of the British Foreign Service
JAMES CONNOLLY (1868-1916), a socialist, leader of the Citizen Army, and commander of the Republican forces in Dublin
THOMAS CLARKE (1857-1916), an organizer of the Irish Republican Brotherhood (IRB)
SEAN MACDERMOTT (1884-1916), a political organizer, journalist, and member of the IRB military council

## Discontent in Ireland

England was first drawn into Irish affairs in 1170, when an Irish chieftain invited an English earl and his men to help defeat other Irish chieftains. The English were so successful that many were soon seeking their fortunes in Ireland, marrying members of the Irish nobility, and adopting Irish laws and customs.

After a time, these English people rebelled against the authority of the monarchs of En-gland. In 1534, King Henry VIII was so annoyed by the Anglo-Irish nobles that he demanded that all lands in Ireland be given up to the Crown, so that they could be parceled out to people who were loyal to England. Henry's daughter, Queen Elizabeth I, carried out this policy.

By breaking away from the Roman Catholic Church and creating the Church of England, with himself as its head, Henry had brought the Protestant Reformation to England. Ireland, however, remained loyal to Catholicism. Eventually the Catholic nobles in Ulster (a province in northeastern Ireland) rebelled against England, and Elizabeth's successor, James I, decided to confiscate the lands of these nobles. The lands were then sold or rented to Protestants.

In the 1640's, and again in 1689, Catholic nobles and their followers revolted against Protestant rule, but they did not succeed. After the revolt of 1689, the English passed laws taking all political rights away from Irish Catholics, who were also denied the right to buy land.

By 1700, Catholics owned less than 12 percent of the land in Ireland and had mostly been replaced by a group of Protestants, called the Ascendancy. For the most part, Irish Catholics were now poor tenant farmers, barely surviving on the estates of Protestant landlords.

At the beginning of the nineteenth century, Ireland was allowed to elect members to the British parliament. When Irish Catholics were given full political rights in 1829, they voted for candidates who called for self-government, or "Home Rule." In 1886, the first Home Rule bill was introduced; yet Home Rule did not become law until 1914, and then it was not enforced because World War I had just begun.

## The Insurrection

Many Irish nationalists did not like the idea of Home Rule anyway, for it would leave Ireland as part of the British Empire. Instead, they wanted Ireland to be an independent republic. Members of a secret nationalist society, the Irish Republican Brotherhood (IRB), saw their chance when the British were busy trying to defeat Germany in World War I. They would lead the Irish Volunteers, a paramilitary group, in an armed uprising to free Ireland.

Meanwhile, to show that Ireland had earned the right to Home Rule, the Irish Volunteers' leaders encouraged them to join the British army. Thousands answered the call, but a minority of thirteen thousand seceded, refusing to serve Great Britain. They were led by a professor of early Irish history, Eoin MacNeill. Though MacNeill did not want to cooperate with the war

effort, he was not a member of the IRB and did not know that the secret society was taking over his organization.

The leading figure among the Irish republicans was Patrick Pearse, a poet and schoolmaster. Others included Tom Clarke, a nineteenth century revolutionary; James Connolly, a trade-union organizer who led a tiny Citizen Army of socialists; Sir Roger Casement, who was already negotiating with the Germans for a shipment of weapons; and Sean MacDermott, a young nationalist from Ulster. They decided that Easter, 1916, would be the date their rebellion would begin.

Many of the Volunteers received confusing orders, so that when the rebellion began in Dublin, only about a thousand Volunteers showed up. Their numbers grew gradually through the week, but the rebels never had more than about eigh-

*Civilians and soldiers engage in a shootout in the streets of Dublin.*

**347**

teen hundred men to fight the twenty-five hundred British troops and more than nine thousand members of the Royal Irish Constabulary (police). Most of the people of Dublin did not support the rebellion.

The IRB had planned to seize the General Post Office building (GPO) in central Dublin, along with other important buildings. The GPO was taken on Monday morning, and Pearse appeared outside to read a proclamation declaring independence and the creation of an Irish republic. Yet most of the people of Dublin were not aware of the rebellion until fighting began around Dublin Castle, the center of British government. Guards there quickly repelled a rebel attack, and British forces in the city began a counterattack.

Over the week of the rebellion, there were several battles, some quite bloody. The British brought artillery in, and the center of Dublin was badly damaged. By Saturday evening, April 29, Pearse agreed to surrender; the word spread quickly to other rebel posts, and by Sunday morning the Easter Rebellion was over.

## Consequences

Within a few days after the surrender, nearly every leader of the IRB had been rounded up and was in jail, soon to be tried for treason. Rank-and-file Volunteers were simply disarmed and sent home, and life in Dublin quickly returned to normal.

After quick trials and fifteen executions, however, many Irish people became more sympathetic toward the rebels. Growing numbers wondered whether the British would ever give them their independence.

Meanwhile, the IRB and the Volunteers reorganized and began to fight political battles, trying to elect Irish republicans to the British parliament. In the elections of 1918, republicans replaced "Home-Rulers" in three-fourths of the Irish districts. Instead of going to London, however, the new Irish members met in Dublin and named themselves the Dail Eireann (Gaelic for "Irish parliament"). With widespread support throughout the country, they declared Ireland an independent republic in January, 1919.

*Thomas C. Schunk*

# British Navy Faces Germany in Battle of Jutland

> *In a North Sea encounter between battle fleets of Great Britain and Germany, the British were successful in keeping the Allies' control of the high seas.*

**What:** War
**When:** May 31, 1916
**Where:** North Sea
**Who:**
SIR JOHN RUSHWORTH JELLICOE (1859-1935), commander of the British Grand Fleet from 1914 to 1916
SIR DAVID BEATTY (1871-1936), commander of the British Battle Cruiser Fleet from 1914 to 1916
REINHARD SCHEER (1863-1928), commander of the German High Seas Fleet from 1916 to 1918
FRANZ VON HIPPER (1863-1932), commander of the German Scouting Forces from 1913 to 1918

## Toward the Battle

Early in 1916, the German Naval Staff decided to begin attacks against the British Grand Fleet in the North Sea. Vice Admiral Reinhard Scheer commanded the German High Seas Fleet of twenty-three dreadnoughts and was aware that the British Grand Fleet, consisting of forty-two dreadnoughts under the command of Vice Admiral Sir John Rushworth Jellicoe, was considerably larger. So Scheer planned to lay mines off the British naval bases and then lure the Grand Fleet out to sea. If German U-boats could sink a good number of British battleships, Scheer thought that he could begin fighting the British on more equal terms.

At the same time, the British became more aggressive. They wanted to attack the zeppelin bases in northern Germany and also to give support to the Russian navy in the Baltic. They shared Germany's goal of luring the enemy fleet out to battle in the North Sea.

Minor battles began in March. German battle cruisers bombarded a few British towns without causing or receiving much damage. The British bombarded selected targets in Germany, also without much success. At the end of May, Scheer took his fleet out to sea, with Vice Admiral Franz von Hipper scouting ahead with battle cruisers. Fourteen U-boats were set in place off British bases. Scheer decided to attack British shipping off the Norwegian coast, hoping that the British would take to sea in large numbers.

His plan started to work when the British intercepted a coded wireless signal which, in their interpretation, meant that the Germans were preparing to advance. On the evening of May 30, the British Grand Fleet moved out to sea from Scapa Flow, Invergordon, and Rosyth. They avoided Scheer's submarines and mines without difficulty, and the Battle Cruiser Fleet, under the command of Vice Admiral Sir David Beatty, went ahead toward the southeast, searching for the Germans.

## Fight at Sea

On May 31, light cruisers of the British and German scouting forces came into contact by chance, and at 2:18 P.M. they guided the heavy ships into action. At 3:48 P.M. the battle cruisers opened fire, and Hipper moved back with his five ships toward the main German force, commanded by Scheer. Beatty pursued him with six cruisers supported by four fast battleships.

A fierce battle followed. Accurate German gunfire and poor British protection led to the blowing up of two of Beatty's cruisers; there were few survivors. Scheer arrived on the scene at

**349**

Library of Congress

*The British navy strikes the German battleship* Oldenburg *during the Battle of Jutland.*

4:48 P.M. and, led by Hipper, chased Beatty toward the north.

Jellicoe hurried southward with twenty-four battleships and joined forces with Beatty; together they turned on the Germans. Between 6:30 and 6:45 P.M., the two fleets exchanged salvos. The Germans blew up a third battle cruiser. The British fired hard on Scheer's battleships, but he skillfully reversed direction, put up a smoke screen, and escaped serious damage; one of Hipper's cruisers was so battered, however, that it had to be sunk later.

Once again, Scheer returned to the attack, but he was forced to pull back at 7:20 P.M. He covered his retreat with an attack by destroyers, from which Jellicoe in turn pulled away. The main bodies of the two fleets were now separated.

When darkness fell, Scheer led his fleet homeward. With accurate gunfire and quick maneuvering, he managed during the night to make his way through the British destroyer flotillas that followed Jellicoe's battleships, and early the next morning his fleet reached home waters.

The battle was over. The British had lost four-

teen ships, including three battle cruisers, totaling 110,000 tons, and more than 6,200 men had
been killed or taken prisoner. The Germans had
lost eleven ships totaling 62,000 tons, including
one battle cruiser and one battleship, and 2,500
men had been killed. Neither side had won a victory, because the battle had not changed their
strength in relation to each other.

After the Battle of Jutland, Jellicoe stopped ordering attacks on the German coast and instead
sat back to await the reappearance of the Germans. In August, Scheer brought out his fleet,
planning to bombard the British coast again.
Once more his submarines lay in ambush. Again
warned by the Germans' careless use of radio,
the British came forward to stop the attack. A
false report from a scouting zeppelin misled
Scheer into turning away from Jellicoe to chase
an isolated British squadron, which turned away
from him. Both sides soon returned to their
bases. German submarines had sunk two British
light cruisers, and the British had damaged one
German dreadnought. Scheer brought his fleet
out a third time in October, but that time the
British Grand Fleet stayed in port.

## Consequences

German attempts to cripple the British Grand
Fleet in 1916, were mostly failures. They did not
succeed in sinking British dreadnoughts with
mines or submarines, yet they lost five of their
own large ships to British mines and submarines.

Geography was important in the British successes. The shallow waters off the German coast
were easy to mine, and German ships had to
move out to sea through a right angle formed by
the German and Danish coasts. The exits from
the British bases were broader, and British
coastal waters were deeper, so that they were
more difficult to mine.

Because Scheer's strategy had not been successful, he stopped attacking the British in the
North Sea. He did make one final try in November, 1918, but the German naval crews could not
forget the Battle of Jutland. They refused to
weigh anchor. The final result of their mutiny
and other similar rebellions took place on November 21, 1918, when Beatty received the surrender of the greater part of the German High
Seas Fleet.

*Samuel K. Eddy*

# Mexican Constitution Establishes Workers' Rights

> *Article 123 of the Constitution of 1917 supported unionization and other workers' rights, so that Mexico became an international leader in labor reform.*

**What:** Labor
**When:** 1917
**Where:** Querétaro, Mexico
**Who:**
FRANCISCO J. MÚGICA (1884-1954), a revolutionary politician
PASTOR ROUAIX (1874-1950), an engineer who entered revolutionary politics and was the main author of Article 123
VENUSTIANO CARRANZA (1859-1920), president of Mexico from 1914 to 1920
LUIS MORONES (1890-1964), a labor organizer who helped build a powerful union in the 1920's
VICENTE LOMBARDO TOLEDANO (1894-1968), a radical labor leader of the 1930's

## Pressure for Change

In the late nineteenth century, railroad construction and foreign investment had brought an economic boom to Mexico. Yet though some Mexicans became very wealthy, those who worked in mines and factories were suffering. Laborers in mines, textile mills, and breweries usually worked a seven-day week for ten or more hours a day. Even with these long hours, pay was so low that the workers could not buy nutritious food and had to live in dilapidated housing.

In 1906, miners in the state of Sonora began to rebel. The Cananea Consolidated Copper Company, which was owned by Americans, paid very low wages and charged high prices at its company stores. To protest the situation, the miners went on strike. During five days of fighting, thirty workers and some of their family members were killed. The strike failed, but labor unrest soon spread to other areas.

In 1911, the dictatorship of Porfirio Díaz was overthrown by Francisco Madero. Madero brought political reforms—elections with greater voter participation—but he did not pay much attention to the demands of workers and peasants. His government was overthrown in 1913 by a military group. By then, workers were active in several revolutionary movements.

In 1916, Venustiano Carranza emerged as the leading politician in Mexico, with support from anarchist labor unions. In October, his government organized elections for a special congress to write a new constitution for Mexico. Only 20 to 30 percent of the eligible voters took part in the elections, and it seems that no labor union or workers' party offered a candidate. Most of the elected delegates came from the middle class, but the years of labor unrest had made them aware of the workers' needs.

## The New Constitution

Francisco J. Múgica, a radical leader, was influential in the writing of the constitution. Having seen the effects of poverty and helped to overthrow Díaz, Múgica was determined to improve the lot of the working class through the new constitution.

The main author of Article 123 was Pastor Rouaix, who was loyal to Carranza and not as radical as Múgica. The son of a construction worker, Rouaix had become an engineer and surveyor. Between 1913 and 1917, he served as governor of Durango and minister of development.

Rouaix and a few others wrote the first draft of Article 123 in a series of meetings in early January, 1917. Several staff members in the Ministry of Development had brought together studies on labor conditions in Mexico and labor laws in other nations, and the Rouaix group used their work in writing the new laws. They chose to give Mexico's national government the main responsibility for improving the life of Mexican workers.

When the group presented its proposal to the constitutional convention, it won a quick victory. Article 123 was discussed during the evening of January 23 and was approved by 10:15 P.M.

Article 123 stated that a workday could not be longer than eight hours. Strikes were declared to be legal means of calling for change, and debt peonage (a system of forced labor to pay off debts) was abolished. Article 123 also set government standards to care for the health and safety of workers. Each state was required to pass minimum-wage laws and other regulations to benefit workers.

Naturally, these changes could not take place immediately. Other national and state laws were needed, along with new government agencies to enforce them. Workers needed time to form unions as well. Yet the history of labor in Mexico had already taken an important new direction.

Carranza signed the constitution into effect on January 31, 1917. Unlike the Bolshevik Revolution in Russia, which began about nine months later, Mexico's new labor code accepted private property and free enterprise. Yet along with Article 27, which restricted rights of property owners, Article 123 seemed certain to change the way business was done in Mexico. In that sense, the Constitution of 1917 was a revolutionary document.

## Consequences

Carranza's government did not enforce Article 123 immediately, but labor unions quickly began to form and grow. The anarchist unions merged into the Regional Confederation of Mexican Workers (CROM), headed by Luis Morones.

Morones, who was a skillful politician, became minister of industry in the cabinet of President Plutarco Elias Calles (1924-1928). In this position, he was able to use Article 123 for the benefit of CROM. Working conditions and wages improved, but not for everyone. The government settled several labor disputes in favor of CROM workers, so that their wages rose; unions outside CROM, however, gained few benefits and often lost their members to CROM. For most Mexican workers, Article 123 was an unfulfilled promise.

In the late 1920's, Morones lost power because of his corruption. CROM was divided into factions, including the Confederation of Mexican Workers (CTM), which was led by Vicente Lombardo Toledano in the 1930's. Working with President Lázaro Cárdenas (1934-1940), Lombardo won many victories for workers. Electrical and textile workers, streetcar operators, and miners went on strike during the mid-1930's and won wage increases. Where the CTM was active, the ten-hour day became mostly a thing of the past.

The most dramatic CTM success came in 1938, when Mexican oil workers around Tampico used strikes to demand better wages and working conditions in the oil fields. The British and American oil companies defied these demands and broke Mexican law. On March 18, 1938, President Cárdenas signed a decree that nationalized the oil properties of these companies. Thus the Mexican government took control of most of the nation's petroleum industry—an action that was called Mexico's "economic declaration of independence."

By 1940, the movement for labor reform seemed to have run out of steam. The CTM was still large and powerful, but it had become deeply involved in politics, and there was less room for groups of workers to take initiative. CTM leaders continued to influence national life, but they often seemed more concerned about political issues than about the needs of ordinary workers. Still, Article 123 remained as a symbol of Mexican workers' rights.

*John A. Britton*

# Birdseye Develops Food Freezing

> *Clarence Birdseye developed a technique for freezing foods quickly and started the modern frozen foods industry.*

**What:** Food science
**When:** 1917
**Where:** Gloucester, Massachusetts
**Who:**
CLARENCE BIRDSEYE (1886-1956), a
    scientist and inventor
DONALD K. TRESSLER (1894-1981), a
    researcher at Cornell University

## Feeding the Family

In 1917, Clarence Birdseye developed a means of quick-freezing meat, fish, vegetables, and fruit without substantially changing their original taste. His system of freezing was called by *Fortune* magazine "one of the most exciting and revolutionary ideas in the history of food." Birdseye went on to refine and perfect his method and to promote the frozen foods industry until it became a commercial success nationwide.

It was during a trip to Labrador, where he worked as a fur trader, that Birdseye was inspired by this idea. Birdeye's new wife and five-week-old baby had accompanied him there. In order to keep his family well fed, he placed barrels of fresh cabbages in salt water and then exposed the vegetables to freezing winds. Successful at preserving vegetables, he went on to freeze a winter's supply of ducks, caribou, and rabbit meat.

In the following years, Birdseye experimented with many freezing techniques. His equipment was crude: an electric fan, ice, and salt water. His earliest experiments were on fish and rabbits, which he froze and packed in old candy boxes. By 1924, he had borrowed money against his life insurance and was lucky enough to find three partners willing to invest in his new General Seafoods Company (later renamed General Foods), located in Gloucester, Massachusetts.

Although it was Birdseye's genius that put the principles of quick-freezing to work, he did not actually invent quick-freezing. The scientific principles involved had been known for some time. As early as 1842, a patent for freezing fish had been issued in England. Nevertheless, the commercial exploitation of the freezing process could not have happened until the end of the 1800's, when mechanical refrigeration was invented. Even then, Birdseye had to overcome major obstacles.

## Finding a Niche

By the 1920's, there still were few mechanical refrigerators in American homes. It would take years before adequate facilities for food freezing and retail distribution would be established across the United States. By the late 1930's, frozen foods had, indeed, found its role in commerce but still could not compete with canned or fresh foods. Birdseye had to work tirelessly to promote the industry, writing and delivering numerous lectures and articles to advance its popularity. His efforts were helped by scientific research conducted at Cornell University by Donald K. Tressler and by C. R. Fellers of what was then Massachusetts State College. Also, during World War II (1939-1945), more Americans began to accept the idea: Rationing, combined with a shortage of canned foods, contributed to the demand for frozen foods. The armed forces made large purchases of these items as well.

General Foods was the first to use a system of extremely rapid freezing of perishable foods in packages. Under the Birdseye system, fresh foods, such as berries or lobster, were packaged snugly in convenient square containers. Then, the packages were pressed between refrigerated metal plates under pressure at 50 degrees below zero. Two types of freezing machines were used. The "double belt" freezer consisted of two metal belts that moved through a 15-meter freezing tunnel, while a special salt solution was sprayed on the surfaces of the belts. This double-belt freezer was used only in permanent installations and was soon replaced by the "multiplate" freezer, which was portable and required only 11.5 square meters of floor space compared to the double belt's 152 square meters.

The multiplate freezer also made it possible to apply quick-freezing to seasonal crops. People were able to transport these freezers easily from one harvesting field to another, where they were used to freeze crops such as peas fresh off the vine. The handy multiplate freezer consisted of an insulated cabinet equipped with refrigerated metal plates. Stacked one above the other, these plates were capable of being opened and closed to receive food products and to compress them with evenly distributed pressure. Each aluminum plate had internal passages through which ammonia flowed and expanded at a temperature of −3.8 degrees Celsius, thus causing the foods to freeze.

A major benefit of the new frozen foods was that their taste and vitamin content were not lost. Ordinarily, when food is frozen slowly, ice crystals form, which slowly rupture food cells, thus altering the taste of the food. With quick-freezing, however, the food looks, tastes, and smells like fresh food. Quick-freezing also cuts down on bacteria.

**Consequences**

During the months between one food harvest and the next, humankind requires trillions of pounds of food to survive. In many parts of the world, an adequate supply of food is available; elsewhere, much food goes to waste and many go hungry. Methods of food preservation such as those developed by Birdseye have done much to help those who cannot obtain proper fresh foods. Preserving perishable foods also means that they will be available in greater quantity and variety all year-round. In all parts of the world, both tropical and arctic delicacies can be eaten in any season of the year.

With the rise in popularity of frozen "fast" foods, nutritionists began to study their effect on the human body. Research has shown that fresh is the most beneficial. In an industrial nation with many people, the distribution of fresh commodities is, however, difficult. It may be many decades before scientists know the long-term effects on generations raised primarily on frozen foods.

*Nan White*

**355**

# Langevin Develops Active Sonar

Active sonar was developed to detect submarines but is also used in navigation, fish location, and ocean mapping.

**What:** Communications; Military capability
**When:** 1917
**Where:** France
**Who:**
Jacques Curie (1855-1941), a French physicist
Pierre Curie (1859-1906), a French physicist
Paul Langevin (1872-1946), a French physicist

## Active Sonar, Submarines, and Piezoelectricity

Sonar, which stands for *so*und *na*vigation and *r*anging, is the American name for a device that the British call "asdic." There are two types of sonar. Active sonar, the more widely used of the two types, detects and locates underwater objects when those objects reflect sound pulses sent out by the sonar. Passive sonar merely listens for sounds made by underwater objects. Passive sonar is used mostly when the loud signals produced by active sonar cannot be used (for example, in submarines).

The invention of active sonar was the result of American, British, and French efforts, although it is often credited to Paul Langevin, who built the first working active sonar system by 1917. Langevin's original reason for developing sonar was to locate icebergs, but the horrors of German submarine warfare in World War I led to the new goal of submarine detection. Both Langevin's short-range system and long-range modern sonar depend on the phenomenon of "piezoelectricity," which was discovered by Pierre and Jacques Curie in 1880. (Piezoelectricity is electricity that is produced by certain materials, such as certain crystals, when they are subjected to

pressure.) Since its invention, active sonar has been improved and its capabilities have been increased. Active sonar systems are used to detect submarines, to navigate safely, to locate schools of fish, and to map the oceans.

## Sonar Theory, Development, and Use

Although active sonar had been developed by 1917, it was not available for military use until World War II. An interesting major use of sonar before that time was measuring the depth of the ocean. That use began when the 1922 German Meteor Oceanographic Expedition was equipped with an active sonar system. The system was to be used to help pay German World War I debts by aiding in the recovery of gold from wrecked vessels. It was not used successfully to recover treasure, but the expedition's use of sonar to determine ocean depth led to the discovery of the Mid-Atlantic Ridge. This development revolutionized underwater geology.

Active sonar operates by sending out sound pulses, often called "pings," that travel through water and are reflected as echoes when they strike large objects. Echoes from these targets are received by the system, amplified, and interpreted. Sound is used instead of light or radar because its absorption by water is much lower. The time that passes between ping transmission and the return of an echo is used to identify the distance of a target from the system by means of a method called "echo ranging." The basis for echo ranging is the normal speed of sound in seawater (5,000 feet per second). The distance of the target from the radar system is calculated by means of a simple equation: range = speed of sound × 0.5 elapsed time. The time is divided in half because it is made up of the time taken to reach the target and the time taken to return.

The ability of active sonar to show detail increases as the energy of transmitted sound pulses

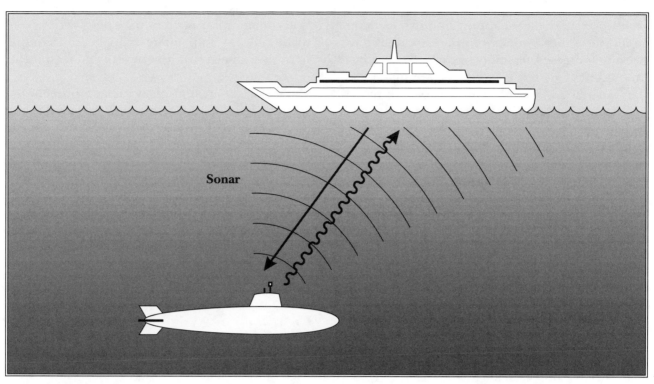

*Active sonar detects and locates underwater objects when those objects reflect sound pulses sent out by the sonar.*

is raised by decreasing the sound wavelength. Figuring out active sonar data is complicated by many factors. These include the roughness of the ocean, which scatters sound and causes the strength of echoes to vary, making it hard to estimate the size and identity of a target; the speed of the sound wave, which changes in accordance with variations in water temperature, pressure, and saltiness; and noise caused by waves, sea animals, and ships, which limits the range of active sonar systems.

A simple active pulse sonar system produces a piezoelectric signal of a given frequency and time duration. Then, the signal is amplified and turned into sound, which enters the water. Any echo that is produced returns to the system to be amplified and used to determine the identity and distance of the target.

Most active sonar systems are mounted near surface vessel keels or on submarine hulls in one of three ways. The first and most popular mounting method permits vertical rotation and scanning of a section of the ocean whose center is the system's location. The second method, which is most often used in depth sounders, directs the beam downward in order to measure ocean depth. The third method, called wide scanning, involves the use of two sonar systems, one mounted on each side of the vessel, in such a way that the two beams that are produced scan the whole ocean at right angles to the direction of the vessel's movement.

Active single-beam sonar operation applies an alternating voltage to a piezoelectric crystal, making it part of an underwater loudspeaker (transducer) that creates a sound beam of a particular frequency. When an echo returns, the system becomes an underwater microphone (receiver) that identifies the target and determines its range. The sound frequency that is used is determined by the sonar's purpose and the fact that the absorption of sound by water increases with frequency. For example, long-range submarine-seeking sonar systems (whose detection range is about ten miles) operate at 3 to 40 kilohertz. In contrast, short-range systems that work at about 500 feet (in mine sweepers, for example) use 150 kilohertz to 2 megahertz.

## Consequences

Modern active sonar has affected military and nonmilitary activities ranging from submarine location to undersea mapping and fish location. In all these uses, two very important goals have been to increase the ability of sonar to identify a target and to increase the effective range of sonar. Much work related to these two goals has involved the development of new piezoelectric materials and the replacement of natural minerals (such as quartz) with synthetic piezoelectric ceramics.

Efforts have also been made to redesign the organization of sonar systems. One very useful development has been changing beam-making transducers from one-beam units to multibeam modules made of many small piezoelectric elements. Systems that incorporate these developments have many advantages, particularly the ability to search simultaneously in many directions. In addition, systems have been redesigned to be able to scan many echo beams simultaneously with electronic scanners that feed into a central receiver.

These changes, along with computer-aided tracking and target classification, have led to the development of greatly improved active sonar systems. Sonar systems have continued to become more powerful and versatile, finding previously unimagined uses. During the late 1990's, for example, improved satellite navigation positioning technology enhanced the value of sonar in environmental studies by making possible dramatic advances in sonar's use as a tool for surveying ocean floors.

*Sanford S. Singer*

# Insecticide Use Increases in American South

*Entomologists discovered a technique for poisoning the cotton boll weevil, reinforcing the widespread use of insecticides and discouraging further support of cultural and biological controls.*

**What:** Agriculture
**When:** 1917
**Where:** Louisiana, Arkansas, and Mississippi
**Who:**
BERT RAYMOND COAD (born 1890), an American entomologist
LELAND OSSIAN HOWARD (1857-1950), an American entomologist
CHARLES VALENTINE RILEY (1843-1895), an American naturalist and entomologist

## The Need for Insect Control

Insecticides had become established in the United States by the nineteenth century, when exotic pest insects arrived from Europe on ships. As early as 1868, an unknown farmer had discovered that Paris green (a brightly colored dye often used to paint window shutters) killed the Colorado potato beetle. The active toxic ingredient in Paris green was arsenic, a poison that was also employed in later insecticides such as London purple and lead arsenate. In the 1870's, Paris green was found to be effective against other pests as well, and it soon became a standard insecticide for the American farmer.

There was a real need for insect control in the nation's agricultural industry. In 1870, for example, journalist Horace Greeley estimated that the average annual loss to farmers from insect damage exceeded $100 million. Assisted by the Division of Entomology of the American Department of Agriculture (USDA), farmers found themselves with three basic strategies for fighting pest insects: insecticides, biological controls, and cultural controls. By the 1920's, however, insecticides had emerged as the principal means of insect control. The reasons for this can be grasped by reviewing the bureau's experience with three pests: the cottony-cushion scale, the gypsy moth, and the cotton boll weevil.

## Three Major Pests

The cottony-cushion scale had been accidentally imported from Australia or New Zealand in the 1870's. Arsenicals had limited effect on the pest, so there was little to prevent the rapid spread of the scale through the orange and lemon groves of California in the 1880's. Charles Valentine Riley, first chief of the USDA Division of Entomology, sent his assistant, Albert Koebele, to Australia to search for natural predators. Koebele returned in 1889 with a small beetle that preyed upon the scale. Known as the Australian ladybird, or vedalia beetle, the new predator became so effective that the scale was brought under control in the first season after its release. The results of this experiment aroused great enthusiasm among farmers and entomologists, many of whom saw biological control as the solution to the war on insects.

The gypsy moth had been introduced into the United States in 1869 by Leopold Trouvelot, a French-born astronomer who was interested in the breeding of silkworms. Some of the insects escaped from Trouvelot's laboratory and gradually became established near his home in Medford, Massachusetts. Twenty years after the accidental release, their population exploded. Writing in 1930, the chief of the USDA Division of Entomology, Leland Ossian Howard, described the infestation of caterpillars that invaded the town in 1889: "The numbers were so great that in the still, summer nights the sound of their feeding could plainly be heard, while the pattering of their excremental pellets on the ground sounded like rain."

The caterpillars created a nightmare for Medford, defoliating trees, covering sidewalks and fences, and invading food and bedding inside houses. They were found to be resistant to

Paris green and able to consume nearly ten times the amount of arsenic required to kill caterpillars of other species. Increasing the proportion of arsenic merely burned the foliage. An extensive search for natural predators made little impact.

Relief came in 1892, when the chemist F. C. Moulton found that lead arsenate could kill the caterpillar without causing as much injury to foliage as that produced by Paris green. Lead arsenate proved to be effective on the moth and on other insects; in the early 1900's, it became the most popular insecticide in the country, and it remained so until it was replaced by dichlorodiphenyl-trichloroethane (DDT) in the 1940's.

The USDA was alerted to the boll weevil problem in 1894, when it received word from Corpus Christi, Texas, that a peculiar weevil had destroyed much of the "top crop" of cotton (a late harvest that can be obtained whenever the first frost arrives late). Local farmers found that ordinary poisons had no effect on the pest. Howard immediately dispatched entomologist C. H. Tyler Townsend to investigate the infestation.

Townsend reported extensive crop damage and recommended cultural control measures, such as burning or flooding the stalks after the main harvest to eliminate the weevil's food source prior to hibernation, and a fifty-mile-wide noncotton zone along the Texas international border to prevent further migration of the insects from Mexico. Strong opposition from farmers forced state legislators to decide against a noncotton zone on the Mexican border. Other cultural control methods recommended by the bureau also were rejected by farmers because of the loss of income they represented. Although it was known that the feeding habits of the weevils enabled them to avoid poisons that remained on the surface tissues of the cotton plant, Texas farmers used an estimated twenty-five boxcarloads of Paris green during a three-month period of 1904 in futile attempts to destroy the boll weevils.

A breakthrough finally occurred in 1914, when entomologist Bert Raymond Coad's experimental work showed that the weevils could be poisoned by taking advantage of their habit of drinking water from the plant's surface. Over the next three years, he experimented with this idea on cotton plantations in Louisiana, Arkansas, and Mississippi. In the process, Coad found that calcium arsenate was more poisonous to the insect than other arsenicals were. He recommended that cotton plants be dusted with calcium arsenate at night, when the plants were especially moist from the dew.

**Consequences**

During the 1920's, farmers became increasingly reliant upon insecticides. No other method, it seemed, would stop insects as effectively as chemicals. The manufacture of insecticides developed into a large industry, which provided further encouragement and support to farmers who were already inclined to dust and spray. Publicly funded entomologists were under pressure to base their recommendations on control methods they knew would produce immediate results, and chemical insecticides had become the most efficient weapon in the war on insects.

Public health concerns about chemical residues on food were expressed shortly after Paris green first gained popularity as an insecticide. Tolerance levels for arsenic and lead residues, however, were not set by the government until the 1920's and 1930's. When such levels were finally set by the USDA, they were based more on what the department believed could be achieved by the proper washing of foods than on safe exposure levels. With few exceptions, government officials were looking for evidence of acute poisoning rather than chronic health effects from long-term exposure to spray residues. The effects of ingesting these chemicals over the course of a lifetime, as well as the effects of the chemicals on the environment, did not become an important issue for public debate until after the publication of Rachel Carson's *Silent Spring* in 1962.

*Robert Lovely*

# Congress Passes Espionage and Sedition Acts

> *In the atmosphere of enthusiastic patriotism that swept across the United States during World War I, Congress passed two laws that were used to stifle any criticism of the government's policies.*

**What:** National politics; Civil rights and liberties

**When:** 1917-1918

**Where:** Washington, D.C.

**Who:**

WOODROW WILSON (1856-1924), president of the United States from 1913 to 1921

GEORGE CREEL (1876-1953), chairman of the Committee on Public Information from 1917 to 1919

## War Propaganda

On the evening of April 6, 1917, United States president Woodrow Wilson delivered his war message to Congress. The nation, he said, was beginning a crusade to "make the world safe for democracy." Unfortunately for minority groups such as the socialists, pacifists, German Americans, and leaders of the Industrial Workers of the World—all of whom opposed U.S. involvement in the war—the president said nothing about protecting democracy at home.

Two related problems faced the U.S. government. First, American citizens had to be persuaded to support a war effort that did not involve an enemy's direct attack on them and that had been accepted slowly and unwillingly. Second, there was the need to guard the country against spies and other possible enemies.

To solve the first problem, Wilson, on April 13, 1917, established the Committee on Public Information (CPI), under the leadership of George Creel. The committee's job was to convince uncertain Americans that the war was a righteous one and to educate all Americans about the aims of the war. It was the first time in American history that the government carefully

planned a large-scale campaign of propaganda to "sell" a war to its citizens.

As the war proceeded, Creel hired approximately 150,000 artists, writers, lecturers, actors, and scholars to make the war popular. Colorful committee posters urged citizens to join the army or navy, to subscribe to Liberty Loans, to knit socks for soldiers, and to be aware of the danger of spies and saboteurs. Writers produced hundreds of "true" stories to show how cruel and barbaric the Germans were, and teams of speakers toured the country, giving anti-German talks. Motion-picture audiences were excited by the heroism portrayed in *Pershing's Crusaders* (1918), and they learned to hate the enemy as they watched *The Prussian Cur* (1918) and *The Kaiser, the Beast of Berlin* (1918).

Creel was a Progressive who had gained a reputation as a crusading journalist and had been one of Wilson's earliest supporters. Because of his reputation as a reformer, the press cheered his appointment as chairman of CPI. Surely a man such as Creel would not impose government censorship on the press. And he did not. He did, however, call for voluntary press censorship, and usually the editors of newspapers and magazines cooperated.

## The New Laws

In his war message, President Wilson had promised that if there was disloyalty, it would be corrected with "a firm hand of stern repression." The Espionage Act, passed two months later, put teeth to the promise: It gave the government authority to limit freedoms of speech and press.

Title I, Section 3, of the Espionage Act made it a crime to make false reports that could aid the enemy, to encourage rebellion among the armed forces, and to hinder military recruiting or the

**361**

draft. These provisions of the law were used during the war to stifle disagreement with and criticism of the government. Those prosecuted under Title I included socialists Victor Berger and Eugene V. Debs, as well as "Big Bill" Haywood, a leader of the Industrial Workers of the World. Under Title XII of the act, socialist and pacifist newspapers were denied use of the mails. In October, 1917, another law required foreign-language newspapers to turn in translations of all war-related stories and editorials before the papers could be distributed to local readers.

Somehow most Americans came to believe that the Espionage Act, passed by Congress on June 15, 1917, gave the CPI power to enforce censorship. It did not, but Creel never tried to make the truth clear. Because of the misconception, it was easy for the CPI to get people to cooperate with its goals. Yet the legal power to impose censorship actually belonged to the Post Office Department and the Justice Department.

The Espionage Act was strengthened in May, 1918, by a companion law, the Sedition Act. This law set penalties of up to ten thousand dollars and twenty years' imprisonment for the person who might write, speak, or publish material abusing the government, showing contempt for the Constitution, urging others to resist the government, supporting the enemy, or hindering production of war supplies.

The Sedition Act did not make it necessary to prove that the person's words had actually af-

*A scene from the 1918 film* The Kaiser, the Beast of Berlin. *Anti-German features such as this built American public support for the laws against espionage and sedition.*

fected anyone or caused harm to the government. The postmaster general was given authority to deny use of the mails to anyone who, in his opinion, had used them to disobey the act.

## Consequences

The Espionage and Sedition Acts, together with the propaganda produced by CPI, did encourage American commitment to participation in World War I. It became quite risky to speak out against the war.

One of the worst effects of the censorship laws and the propaganda was the birth of many volunteer organizations energized by "superpatriot-ism" and determined to root out all enemies of the United States. The most influential of these groups were the National Security League and the National Protective Association; less powerful were groups with such colorful names as the Boy Spies of America, the Sedition Slammers, and the Terrible Threateners. They helped produce a wave of war hysteria that took loyalty to extremes. The teaching of German was banned in some schools, and eating German foods was considered disloyal. Many people were harmed physically and mentally by the zealots of these organizations.

*Thomas J. Edward Walker*

# United States Enters World War I

*Two months after Germany announced a policy of unrestricted submarine warfare, the United States left behind its position of neutrality and entered World War I.*

**What:** War
**When:** April 6, 1917
**Where:** Washington, D.C.
**Who:**
WOODROW WILSON (1856-1924), president of the United States from 1913 to 1921
THEOBALD VON BETHMANN-HOLLWEG (1856-1921), chancellor of the German Empire from 1907 to 1917

## Staying Neutral

For two and a half years following the outbreak of World War I in August, 1914, the United States had practiced a policy of neutrality. The United States had loaned money and supplies to the Entente Allies—Great Britain and France—but also had tried to keep from antagonizing the Central Powers—Imperial Germany, Austria-Hungary, and the Ottoman Empire.

Neutral American ships did carry arms and munitions to the Allies. Because its war effort was being threatened, Germany tried to restrict this trade by sinking American ships, mainly through submarine attacks. In 1914, 1915, and 1916, Germany announced increases in the kinds of neutral ships that would be considered fair game for submarine attacks, but each time it backed off from its threats when the United States threatened, in turn, to break off diplomatic relations. During these years, U.S. president Woodrow Wilson tried several times to help negotiate an end to the European conflict. His sincere attempts failed.

At a meeting of the German Crown Council at Pless on January 9, 1917, German leaders debated whether they should step up submarine attacks. Chancellor Theobald von Bethmann-Hollweg was in favor of avoiding that risk, but he was overruled by the military leaders. The Germans soon announced publicly that as of February 1, their submarine warfare would be unrestricted: Any ship, armed or not, that tried to bring supplies to the British or French would be attacked. Two days later, Wilson broke off U.S. diplomatic relations with Germany.

## Declaring War

Two months later, on April 2, 1917, President Wilson stood before the two houses of Congress and asked for a declaration of war against Germany. "It is a fearful thing to lead this great peaceful people into war," he said. "But the right is more precious than peace." On April 6, Congress agreed to his request, and the United States was officially at war.

Wilson's reason for asking for the declaration of war two months after Germany's new policy was announced has been debated by many historians. Some have suggested that American bankers and manufacturers of weapons argued that if the Allies lost the war they would not repay their debts; so it was that these wealthy, powerful people forced Wilson to declare war for economic reasons.

Another argument has been that clever British propaganda tricked Americans into seeing the Allies as pure and the Germans as completely evil. Perhaps the American public was so eager to enter the war that Wilson felt compelled to please them. According to some historians, Wilson believed that the national security of the United States would be endangered if Germany were to win the war, since the Germans were determined to fight wars and make conquests all over the world.

Most historians emphasize two explanations: First, the Germans' submarine warfare forced the United States into the war, and second, Wilson sincerely believed that by joining the war, the

*U.S. soldiers leave for France.*

United States could make sure that the peace settlement was just and fair. This last explanation is especially important. Wilson firmly believed that he could bring about an idealistic peace in Europe only by helping the Allies defeat the Central Powers. The United States would bring a quick end to this "war to end all wars," and then Wilson would finally be able to help create "peace without victory."

### Consequences

The United States' entrance into the war was important in bringing about an Allied victory. The British and French forces were exhausted and had lost hundreds of thousands of men. Once the American troops were trained and ready to start fighting, they brought fresh strength to the Allied armies and were able to push back the Germans, who ended the war by signing an armistice on November 11, 1918.

Wilson's sincere, thoughtful desire to help create a lasting peace was unfulfilled. Though he did have influence in the writing of the Treaty of Versailles, the treaty failed to put an end to European hatreds. Even the United States failed to join the League of Nations, which Wilson had worked tirelessly to establish.

*Burton Kaufman*

# War Industries Board Is Organized

> *The War Industries Board was organized to coordinate and control the industrial resources of the United States in its fight against Germany in World War I.*

**What:** Economics
**When:** July 8, 1917
**Where:** Washington, D.C.
**Who:**
F. A. Scott (1873-1949), chairman of the War Industries Board in 1917
Daniel Willard (1861-1942), chairman of the War Industries Board from 1917 to 1918
Bernard Mannes Baruch (1870-1965), chairman of the War Industries Board from 1918 to 1919
Woodrow Wilson (1856-1924), president of the United States from 1913 to 1921

## Organizing for War

In 1915, the United States Congress agreed to the creation of the Committee on Industrial Preparedness, which was to study the supply needs of the army and the navy. This committee's main goal was to make a list of factories that would, in case of war, be able to manufacture weapons and ammunition.

As the United States moved toward involvement in World War I, the federal government extended its power over the nation's industries. By the National Defense Act of June, 1916, Congress authorized the president to place orders for war materials with any supplier and to take over plants and factories for war use when he considered it to be in the nation's best interests.

The Military Appropriations Act passed two months later set up the Council of National Defense, which included the secretaries of war, the Navy, the Interior, Agriculture, Commerce, and Labor. It also created an advisory commission composed of civilian representatives from all the major branches of the country's economy. The two committees shared one main purpose: to plan for the most efficient use of the country's resources in case of war.

The Advisory Commission, which served as the executive committee of the Council of National Defense, did most of the work and made complicated plans to meet wartime needs. Each of the seven members of the commission took charge of a particular segment of the economy—transportation, engineering and education, munitions and manufacturing, medicine and surgery, raw materials, supplies, and labor. Bernard Baruch, a Wall Street investor and the commission member in charge of raw materials, devised an elaborate plan for organizing representatives of various businesses into "committees of the industries" to work with the council in coordinating the country's resources.

The Council of National Defense had been formed to plan ways to make use of U.S. industries during a war, not actually to direct and coordinate that mobilization. Also, it was organized quite loosely; several of its best members served only on a part-time basis. For these reasons, the council was not well prepared to begin directing the mobilization of industry after the United States entered the war in 1917.

To coordinate the purchases of the Departments of War and the Navy, the Advisory Commission set up what was first called the Munitions Standard Board and then the General Munitions Board. Members of the commission served on this board along with representatives from the military bureaus in charge of purchasing. There had been no really clear definition of the board's authority and responsibilities, and under the pressure of handling war orders it nearly broke down.

Within a month after the board was established, its leaders found out that its responsibilities actually overlapped those of many other

**366**

committees set up by the Advisory Committee. Furthermore, the military bureaus in charge of purchasing wanted to keep their decision-making rights, and they tended to go their own way and ignore the board.

## The Board Is Formed

Realizing that a central coordinating agency was needed, the Council of National Defense on July 8, 1917, replaced the General Munitions Board with the War Industries Board. It consisted of five civilians and one representative each from the army and the navy.

Yet there were still problems. The War Industries Board was given quite a bit of responsibility for directing the war industry, yet it had little power to make sure that its decisions were carried out. As a result, the government's attempts to organize U.S. industries to help the war effort made little progress. The board's first chairman, F. A. Scott, had to resign because he was unable to take the stress of the wartime crisis. Daniel Willard, who succeeded him, soon resigned as well, complaining that the board lacked authority.

Finally, in the spring of 1918, President Woodrow Wilson took matters into his own hands. He reorganized the board and named Bernard Baruch as its chairman. In effect, Wilson passed on to Baruch the authority Congress had given him in the National Defense and Military Appropriations Acts of 1916. Baruch also gained certain controls over the military.

With this authority, Baruch was able to set priorities, requisition supplies, conserve resources, take over plants for government use, and make purchases for the United States and the Allies. The only important control he did not exercise directly was that of fixing prices, which was left to a separate committee within the board.

## Consequences

Though some members of Congress became critical of how much power the War Industries Board was given, the board was quite effective in coordinating the nation's industrial and military effort. The pattern of organization created by the board through trial and error was used by the Allies to regulate the use of industry during World War II.

On December 31, 1918, President Wilson directed that the board should be dissolved, and that process was finally finished on July 22, 1919. Other agencies that had been created during the war to help in national mobilization—such as the United States Railroad Administration, which controlled the nation's railroads—were also gradually dissolved after the war had ended.

*Burton Kaufman*

# Hooker Telescope Begins Operation on Mount Wilson

George Ellery Hale oversaw the installation of the 254-centimeter Hooker telescope on Mount Wilson, the world's largest until 1948.

**What:** Astronomy

**When:** November, 1917

**Where:** Mount Wilson, near Pasadena, California

**Who:**

GEORGE ELLERY HALE (1868-1938), an American astronomer who oversaw the building of the Hooker telescope

JOHN DAGGETT HOOKER (1838-1911), an American businessman who donated money for the Mount Wilson telescope

GEORGE WILLIS RITCHEY (1876-1956), an American optician who began the grinding of the mirror for the Hooker telescope

WALTER SYDNEY ADAMS (1876-1956), an American astronomer who contributed to the design of the Hooker telescope

## The World's First Telescopes

Since the Italian scientist Galileo first used a telescope to view celestial objects in 1609, astronomers have worked to make bigger and better telescopes. Galileo's first telescope was a refracting telescope; that is, it worked by using lenses to bend, or refract, the light. In the seventeenth century, the English physicist and mathematician Sir Isaac Newton and the French inventor N. Cassegrain worked to perfect reflecting telescopes, which use mirrors to reflect the image being observed.

There is a limit to how big a refracting telescope can be, because a lens can be supported only around its edges while it is in use. A lens that is too large will thus bend under its own weight, distorting the images observed through it. The largest refracting telescope is the 102-centimeter telescope at Yerkes Observatory in Wisconsin. Reflecting telescopes can be made much larger than this, however, since a mirror can be supported across its whole lower surface. Astronomers want big telescopes because the larger a telescope's aperture (the width of the mirror or lens receiving the image), the more light is gathered by the telescope. The more light-gathering capacity an astronomical telescope has, the greater the detail that can be seen with it.

## The World's Largest Telescope

In 1906, George Ellery Hale was overseeing the construction of a 153-centimeter reflecting telescope at the Mount Wilson Observatory in California. With characteristic enthusiasm and energy, he described to John Daggett Hooker how a larger telescope would perform even greater wonders. Hooker, a wealthy businessman who had already contributed money to the Mount Wilson Observatory, offered $45,000 to the observatory to have a 213-centimeter mirror cast. He decided later to fund a 254-centimeter mirror instead. He wanted the telescope to be the biggest in the world, and he thought that a 254-centimeter mirror would be difficult to surpass.

The Saint-Gobain company in France was the only company willing to cast such a large mirror. Workers had to melt the glass for the mirror in three separate batches, which were all poured into the mold at the same time. The meeting of the three streams of molten glass formed large clouds of air bubbles in the glass, and the glass also crystallized partially during its cooling. Nevertheless, the Saint-Gobain company shipped the disk to Pasadena, where it arrived on December 7, 1908. Both Hale and Hooker were disap-

*Astronomer Francis G. Pease looks through the Hooker telescope.*

The Observatories of the Carnegie Institution of Washington

A special road was built for the dangerous task of carrying the huge mirror and its massive mounting and machinery up the mountain. It arrived on July 1, 1917, and was assembled by November. On the evening of November 1, Hale pointed the telescope at Jupiter. Despite the 90,000 kilogram weight of the telescope, it moved smoothly. Unfortunately, the image of Jupiter was distorted disastrously. Rather than one clear image, there were six or seven overlapping images. The horrified astronomers could only hope that the mirror had been distorted by the sun and would return to normal once it had had a chance to cool. The astronomers went home to wait and sleep, but at around 2:30 A.M. Hale returned to the observatory, where he was joined by Adams. Jupiter had set, so Hale moved the telescope to the bright star Vega. One can only imagine how delighted he must have been to see at last a clear, sharp image. This was the beginning of the long and useful career of the Hooker telescope.

## Consequences

The Hooker telescope began immediately to prove its worth. Photographs of the moon and of nebulas revealed previously unseen details of great fineness. Rather than make many unrelated observations, however, Hale decided to focus on one of the great controversies of the time: whether spiral nebulas were parts of the Milky Way or whether they were huge, distant, independent systems. The Hooker telescope photographed the Andromeda nebula, and the American astronomer Edwin Powell Hubble used the photographs to settle the debate. The photographs were so detailed that he was able to identify a star of a type known as "Cepheid variables." These stars vary in light output, and the period

pointed with the quality of the mirror blank, and George Willis Ritchey—who was to grind the mirror—believed that the glass would never hold the necessary shape. Roughly 51,000 square centimeters of the mirror's surface had to be polished and shaped accurately to within millionths of a centimeter. Hooker refused to support the enterprise any further, and the remainder of the expenses were paid by a gift from the American industrialist and philanthropist Andrew Carnegie. Walter Sydney Adams tested the disk and found that it would work. In 1910, Ritchey began this painstaking job, which was completed by W. L. Kinney in early 1916.

**369**

of variation is related to the absolute brightness of the star; the absolute brightness can be compared to the brightness as seen from Earth, and this comparison can be used to calculate the star's distance. Hubble calculated the distance of this star, and thus of the entire Andromeda nebula: They are, indeed, remote, separate galaxies that are not part of the Milky Way. After this 1924 discovery, Hubble and others used the Hooker telescope to discover the speeds at which galaxies are receding from one another. They were able to use the speeds to estimate the age of the universe.

This work and other work being done with the Hooker telescope led to results that could not be investigated further without a larger telescope. Thus, the success of this instrument led to Hale's plans for the 508-centimeter Hale Telescope, which was eventually built on Mount Palomar and completed in 1948.

*Mary Hrovat*

# Balfour Declaration Supports Jewish Homeland in Palestine

*The British government, in a letter sent to Lionel Walter Rothschild, announced that it supported setting up a Jewish homeland in Palestine.*

**What:** Political independence

**When:** November 2, 1917

**Where:** London

**Who:**

ARTHUR BALFOUR (1848-1930), British foreign secretary from 1916 to 1919

CHAIM WEIZMANN (1874-1952), a scientist and Zionist leader

LIONEL WALTER ROTHSCHILD (1868-1937), president of the British Zionist Federation

## Problem of Jewish Loyalty

About a century after the Romans conquered Israel in 63 C.E. (common era), the Jews revolted against the rule of Rome but could not shake it off. As punishment, most Jews who survived were forced to leave their land. They became exiles with no political rights. During the Middle Ages and the Renaissance, some Jews made their way back to Palestine, for they never lost the religious connection to their ancient home.

After the American and French revolutions, the new democracies began giving citizenship to Jews for the first time. Now the Jews faced a new problem. How could they be loyal to France, for example, while holding onto their belief that they would one day return to Palestine?

Seeing that prejudice and persecution continued even though Jews had gained legal rights, some Jewish leaders argued that Jews needed to live in their own land. The first World Zionist Congress was held in 1897, and about twenty-five thousand Jews moved to Palestine between 1902 and 1903.

During World War I, there were Jews on both sides of the conflict. Some Jews of Palestine wanted to join the Turkish army to fight czarist Russia, a country that had persecuted many Jews. Others, however, had been persecuted by the Ottoman Turks and wanted to support the British side. It was not clear which side would help the Jews gain independence in Palestine.

For its part, the British government wanted to win the war and keep control over the Holy Land. To do this, it was important to get the support of the people in areas controlled by the Ottomans, and so the British made promises to both the Arabs and the Jews.

Also, there were some in the British government who truly wanted to help the Jews establish a homeland. Arthur Balfour was one of these. Because of his Christian upbringing, Balfour believed that the Jews had a special claim on the Holy Land. Prime Minister David Lloyd George agreed with him.

A Zionist leader, Chaim Weizmann, worked hard to make British officials see that it would be to their advantage to support a Jewish homeland. A Russian Jew who had moved to England in 1904, Weizmann had done a great service to the British by inventing a synthetic material that could be used in explosives. He became friends with many prominent British leaders, and he helped persuade them of the Zionist cause.

## The Declaration

The British government decided that it would be wise to support the Zionists, and government representatives began negotiating with Zionist leaders over what the "Balfour Declaration" would say. The Zionists wanted the government to say that Palestine should be the Jewish homeland. Yet the government, wanting to keep friendship with the Arabs of Palestine, chose to state its support for a Jewish homeland *in* Palestine.

The Balfour Declaration was a letter, dated November 2, 1917, from Foreign Secretary Arthur Balfour to Lionel Walter Rothschild, president of the British Zionist Federation:

> His Majesty's Government views with favour the establishment in Palestine of a national home for the Jewish people, and will use their best endeavours to facilitate the achievement of this object, it being clearly understood that nothing shall be done which may prejudice the civil and religious rights of existing non-Jewish communities in Palestine, or the rights and political status enjoyed by Jews in any other country.

This simple message brought much joy in Jewish communities around the world. The Jews had no influence in the newly formed Soviet Union, but in Great Britain, the United States, and France, most Jews threw their support behind the Allied war effort. Most German Jews did not commit themselves to the Allied cause; in Palestine, the vast majority of Jews gave their allegiance to the Allies. At the end of the war, the Balfour Declaration became part of the document that granted control over Palestine to Great Britain.

Zionists believed the declaration meant that unlimited numbers of Jews would be allowed to move to Palestine. Because the Arabs feared that the Jews were taking over their land, however, the British issued the Churchill White Paper in 1922. This document limited Jewish settlement to the area of Palestine west of the Jordan.

Still, violence erupted between Jews and Arabs in Palestine. By 1939, the situation was so bad that the British government changed its position on the Zionist question. The White Paper of May, 1939, stated that within ten years Great Britain would create a State of Palestine that would not exclude either Arabs or Jews. More important, the White Paper announced that no more land would be sold to Jews, and that Jewish immigration to Palestine would be ended in five years.

Tragically, the White Paper came just as the Nazis were beginning the Holocaust in Germany

and Eastern Europe. With the new restrictions on immigration to Palestine, many Jews had nowhere to go to escape the persecution of Adolf Hitler, and millions of them were killed.

## Consequences

Like the Jews, the Arabs of Palestine lacked political rights and had often been persecuted. At first, many Arabs were willing to support the goal of Jewish independence, for they believed that Arabs and Jews could work together toward common goals. For many of them, however, the Balfour Declaration meant that the British would no longer support Arab independence.

*Arthur Balfour.*

After the Balfour Declaration encouraged more Jews to move into Palestine, hostility between Arabs and Jews increased. The two independence movements began to fight each other.

On the Jewish side, however, the Balfour Declaration gave Zionism the official recognition and support of a world power. Now the Zionists hoped to convince the other nations of the world that a Jewish homeland was both right and possible.

Before World War I, there had been about eighty-five thousand Jews in Palestine. Many died from the war, however, or the famine that came with it, so that the Jewish population fell to fifty-five thousand. The Balfour Declaration brought many Jewish immigrants—a flood that peaked in 1935, when more than sixty thousand Jews moved to Palestine.

*Renee Marlin-Bennett*

# Bolsheviks Seize Power During October Revolution

*Following many serious defeats for Russia in World War I, a Marxist group known as the Bolsheviks toppled the parliamentary provisional government that had been in place for less than a year.*

**What:** Political reform; Coups

**When:** November 6-7 (Old Calendar: October 24-25), 1917

**Where:** Petrograd (Saint Petersburg), Russia

**Who:**

NICHOLAS II (1868-1918), czar of Russia from 1894 to 1917

ALEKSANDR FEODOROVICH KERENSKI (1881-1970), Social Democrat premier of the provisional government from July to November, 1917

VLADIMIR ILICH LENIN (1870-1924), a leader of the Bolsheviks and later chairman of the Soviet of People's Commissars

LEON TROTSKY (1879-1940), a leading Bolshevik and later commissar for foreign affairs

JOSEPH STALIN (1879-1953), a leading Bolshevik and later commissar for national minorities

LAVR GEORGIEVICH KORNILOV (1870-1918), commander of Russian troops in Petrograd

## Time of Turmoil

In March, 1917, some workers and soldiers in Petrograd (as Saint Petersburg had been renamed during World War I) joined forces and revolted against the Russian government. The new government they established became known as the Provisional Government, which took control of Russia away from the last Romanov czar, Nicholas II. It was a difficult time. Russia's involvement in World War I had brought nothing but defeat for its armies, and the Russian people had been impoverished by the war effort.

The Provisional Government was created through negotiations between the previously elected Duma (parliament), which was a moderate group, and the recently established Petrograd Soviet (council) of Workers' and Soldiers' Deputies, which leaned more toward socialism. The Petrograd Soviet saw itself as the guardian of the masses of working people, who would in time change the "bourgeois" revolution into a socialist one. Its leaders decided not to participate in the Provisional Government. Because of this split, central power in Russia was divided between the Provisional Government and the Petrograd Soviet, but it was the latter that had the most power.

The lack of a strong central authority made it impossible for anything to be done about the terrible conditions facing Russia. The two most serious problems were the peasants' demands for land and Russia's continuing role in World War I. Though the peasants insisted that the great estates should be taken over and divided among them without delay, the Provisional Government said that such a decision could be made legally only by a constituent assembly, and elections for the assembly would be held later that year.

Furthermore, the Provisional Government made plans to continue Russia's cooperation with the Allies in fighting Germany. One of the government's main goals in pursuing the war was to annex certain territories: the Turkish Straits between the Mediterranean and the Black Sea, including the Bosporus and the Dardanelles. The Petrograd Soviet was displeased, for its members had approved continuing the war but not fighting for the annexations.

In May, however, the Provisional Government was reorganized to include several members of the Petrograd Soviet as ministers. Aleksandr Kerenski, a socialist who had already been a member of the Provisional Government, became minister of war under this "First Coalition." The coalition was an attempt to bring together the liberals and the socialists in order to keep the Bolsheviks from gaining power.

## The Bolsheviks Rise Up

The Bolsheviks were calling for an immediate end to the war, immediate seizure of land by the peasants, control of industry by committees of workers, and a transfer of power from the Provisional Government to the soviets, which were growing throughout Russia. This set of demands was being well received by the peasants and working people of Russia, who did not believe that their concerns mattered to the leaders of the Provisional Government.

The Bolsheviks' main leader was Vladimir Ilich Lenin, who had been a spokesman for Marxist revolutionaries in Russia since 1895. After a time of exile in Switzerland, he had returned to Russia with the help of the German High Command, for the Germans believed that he would help to topple the Provisional Government and pull Russia out of the war.

Along with Leon Trotsky and Joseph Stalin, Lenin began plotting to take over the Petrograd Soviet and overthrow the Provisional Government. Their first opportunity came in July, 1917. Thinking that a great war victory would lessen the Russian people's discontent, Kerenski decided in June to attack the Austrian and German armies. The Russian forces were soundly defeated by July 7, and between July 16 and 18, soldiers, sailors, and workers staged an uprising in Petrograd.

The Bolsheviks were not yet prepared to take power; though they did make an attempt, it was

*Bolsheviks form a barricade on the streets of Petrograd in 1917.*

easily foiled by the government. Lenin fled into hiding in Finland; Trotsky and other Bolsheviks were arrested.

The government now formed a "Second Coalition," in which Kerenski served as premier. Its next challenge came from the Right, with a September uprising led by General Lavr Georgievich Kornilov, recently appointed commander of Russian troops in Petrograd. In response, the government released many Bolshevik radicals, including Trotsky, to win away support from Kornilov. This tactic was successful: Kornilov's troops refused to obey him, and the coup soon failed.

The Bolshevik Party had been steadily growing stronger. By August, its membership had reached a quarter of a million—ten times its size in March of the same year. In mid-September, the Bolsheviks won a majority in both the Petrograd and Moscow soviets. Early in October, Trotsky was elected chairman of the Petrograd Soviet, and he soon became head of the Military Revolutionary Committee as well. In this way he had control over the Bolshevik militia and the Red Guard, and he also managed to gain power over other sectors of the military.

Lenin came out of hiding and, on October 23, convinced the Bolshevik Party's Central Committee that the moment had come to seize power. On the night of November 6, Bolshevik troops occupied key points in Petrograd; the next morning, the Bolsheviks announced the fall of the Provisional Government. The Winter Palace, Kerenski's headquarters, held out until evening, when it was captured after a short battle. Kerenski fled the country.

## Consequences

The Bolsheviks faced the immediate task of solidifying their power. It was not an easy process. Outside the capital, there was often strong resistance. Moscow, for example, came under Lenin's control only after a week of hard fighting. The Bolsheviks' power would not be complete until the end of the Great Civil War in 1921.

Having laid down a decree handing land over to the peasants, Lenin created a new cabinet, the Council of People's Commissars, in which Stalin and Trotsky held important posts. The scheduled elections to the Constituent Assembly did take place on November 25, but after allowing the assembly one meeting in January, 1918, Lenin dissolved it. He also ended all ties between the church and the state and decreed that the land assigned to the peasants would be nationalized—it would become the property of the state.

In March, 1918, Lenin concluded a formal peace with Germany at Brest-Litovsk, made Moscow the Russian capital, and adopted the name "Communist Party" for his movement.

# Influenza Epidemic Kills Millions Worldwide

> *Beginning in the spring of 1918 and continuing through 1919, a new strain of deadly influenza circled the globe, leaving a trail of death and chaos behind it. At least 30 million people died throughout the world.*

**What:** Health; Medicine; Disasters
**When:** 1918
**Where:** Worldwide
**Who:**
WILLIAM HENRY WELCH (1850-1934), medical researcher, pathologist, and physician
WILLIAM CRAWFORD GORGAS (1854-1920), U.S. surgeon general

## A New Flu

Influenza, or "flu," is an illness caused by a virus. Many people become ill each year with influenza because the virus mutates, or changes, from year to year. People who have built up immunity to one strain of the flu are still susceptible to other forms of the disease. Fortunately, influenza, although widespread, is not usually a serious illness, and people generally recover without serious side effects. In general, only the very young and the very old are at risk for serious complications.

In 1918, however, a new strain of influenza surfaced, quickly spreading throughout the world and leaving a trail of death behind it. Although symptoms associated with influenza are usually not worse than those associated with a bad cold, the 1918 virus caused bleeding from the nose and from the lungs. People exposed to the disease were often dead within three days.

## The First and Second Waves

Researchers are still unable to pinpoint the exact location of the origin of the 1918 influenza. Some argue that although the disease was commonly called the "Spanish Influenza," or even the "Spanish Lady," it is likely that the 1918 flu began in the American Midwest.

Regardless of its origin, cases began to appear in the United States in the spring of 1918. Because local doctors were not required to report cases of flu at that time, it is difficult to trace the first wave of what would become the pandemic, a worldwide epidemic. However, military and prison records do reveal what must have been occurring in the country at large. First, in addition to the expected deaths among the very young and very old, a surprising number of young, healthy adults were dying. In addition, many of the deaths were caused by the pneumonia that often followed an attack of the flu. Nevertheless, the death rate from this first wave was not terribly high.

Of special concern, however, was the high rate of infection among soldiers and sailors. In 1918, Europe was engulfed in the last days of World War I, and the United States was preparing large numbers of men to join the fight. In many cases, these men became sick on their way to Europe. They spread the disease to people in the places where they landed. By April, 1918, both British and German troops fell ill. By May, French troops also began to suffer from the illness. About four months after the flu first erupted in the United States, the disease had spread worldwide.

As the disease spread, it mutated and became ever more dangerous. For example, British troops landed in Sierra Leone with about two hundred of their members mildly ill with the flu. The local people, however, contracted the disease and grew violently ill. When the local workers passed the disease back to other British sailors, these sailors contracted the more deadly strain of the disease and died in alarming numbers.

The second wave of the flu virus hit the United States in Boston in September, 1918, exploding in the military camp of Fort Devens. When word of the disaster reached Surgeon General William Crawford Gorgas, known as the man who defeated yellow fever in Cuba, he quickly assembled a team of researchers led by William Henry Welch, a doctor well known for his work in public health, to investigate the outbreak. The team was appalled at what they found at Fort Devens. Indeed, some even doubted that this was flu at all, thinking that it might be an outbreak of some new hemorrhagic fever, a disease that would cause terrible bleeding in the lungs and respiratory system.

The epidemic rapidly spread from the coastal areas across the country. Some cities, such as Philadelphia, were particularly hard hit. The disease first appeared in Philadelphia on September 29, 1918. By November 2, barely four weeks later, 12,162 people had died, mainly young people between the ages of twenty-five and thirty-

five. Similar scenes developed across the country. There were shortages of nurses, doctors, hospital beds, coffins, morgues, and gravediggers.

Meanwhile, the second wave of the pandemic had reached the rest of the world. Although the disease was terrible in the United States, other countries suffered far more greatly. Russia, Latin America, Tahiti, Africa, and India were all hard hit. In India alone, it is estimated that between 17 million and 20 million people died. Colonial Africa also suffered extremely high death tolls, in large part because of the many European troops passing through its ports. On the South Pacific islands of Tahiti and Samoa, between 10 percent and 20 percent of the entire population died in about three weeks.

## Consequences

The worldwide influenza epidemic of 1918 finally ended in the spring of 1919, leaving at least 30 million people dead in its wake. Across the world, longevity statistics dipped sharply for this year. Birthrates plummeted because of the large numbers of young women who had died. The influenza epidemic was all the more catastrophic because it followed hard on the devastation wrought by World War I.

It is difficult to determine why this particular strain of flu virus wreaked such havoc with the human body. Indeed, researchers have been unable to determine conclusively the identity of the particular strain of flu that caused the pandemic, although research conducted in the 1990's on frozen lung tissues taken from victims points toward a swine influenza. Questions that continue to plague scientists include why the virus affected young, healthy adults; whether such a virus could return in the future; and whether any means exists to forecast or prevent such a worldwide catastrophe. Although many of the secrets of the 1918 pandemic lie buried with the dead, ongoing research continues to unravel the mysteries.

*Diane Andrews Henningfeld*

*American Red Cross*

*Medical professionals check on a soldier suffering from influenza in a New York Army hospital.*

**378**

# Soviets and Anti-Soviets Clash in Great Russian Civil War

> *The new Soviet government succeeded in turning back a series of uprisings and ensured the triumph of the Communist experiment in Russia.*

**What:** Civil war
**When:** 1918-1921
**Where:** Soviet Union
**Who:**
VLADIMIR ILICH LENIN (1870-1924), Bolshevik leader and head of Communist Russia from 1918 to 1924
LAVR GEORGIEVICH KORNILOV (1870-1918), ANTON IVANOVICH DENIKIN (1872-1947), and PËTR NIKOLAEVICH WRANGEL (1878-1928), commanders of White forces in southern Russia
ALEKSANDR VASILIEVICH KOLCHAK (1874-1920), white commander in Siberia and white "supreme ruler" from 1918 to 1920
NIKOLAI VASILIEVICH CHAIKOVSKI (1850-1926), socialist revolutionary party leader of anti-Soviet movement in northern Russia
NIKOLAI NIKOLAEVICH YUDENICH (1862-1933), white commander in northwest Russia
LEON TROTSKY (1879-1940), Red Army commander and commissar of war from 1918 to 1924
NESTOR IVANOVICH MAKHNO (1889-1934), a Ukrainian anarchist

## The Whites Organize

After Vladimir Ilich Lenin and his well-organized Bolshevik followers seized power from Russia's Provisional Government in November, 1917 (October according to the Julian calendar in use in Russia at the time), the new Soviet gov-ernment met with several kinds of opposition. Many wealthy Russians, realizing that the Bol-sheviks (or Reds) had an ambitious plan of redis-tributing land and wealth, fled the country for Paris, New York, and other Western cities. There they continued to oppose the Bolshevik regime from a safe distance.

By far the most serious threat to the Bolshevik Revolution, however, was a series of armed upris-ings at the edges of Soviet Russia between 1918 and 1921. These revolts are traditionally grouped together as the Great Russian Civil War.

Throughout Russia, most opponents to the new regime were known as Whites. In southern Russia, Whites raised a Volunteer Army and also had the help of Cossack forces. Led in turn by Generals Lavr Kornilov, Anton Denikin, and Pëtr Wrangel between 1918 and 1920, the Volunteer Army tried to fight the Soviets and the Germans at the same time. Their aim was the restoration of a parliamentary form of government with tradi-tional political parties. This southern movement was weakened, however, by a lack of careful coor-dination of its daily military operations.

Another important center of White activity was Siberia. Here fighting broke out in 1918 be-tween Soviet troops and the famous Czechoslo-vak Legion, which had helped czarist forces in at-tacking Austria-Hungary during World War I. The legion was on its way through Siberia, plan-ning to cross into North America and from there travel to the Western Front in Europe. After the clashes with the Soviets, however, the legion—which included thirty-five thousand troops—decided not to leave Russia. It became dedicated to protecting the Siberian east, creating a safe zone for anti-Soviet movements. The Siberian White movement came to be led by Admiral Aleksandr Kolchak.

## RUSSIAN CIVIL WAR, 1918-1921

The Whites' crusade was also joined by forces from Russia's minority nationalities, especially in the Ukraine. In opposing the Soviets, the fiercely nationalistic Ukrainians hoped to gain political independence. Their key leader was the peasant Nestor Ivanovich Makhno, an anarchist.

### Victories and Defeats

In Siberia, Admiral Kolchak began an offensive against certain sites near the Volga River. Kolchak had a string of military successes through March, 1919, but the Red Army was able to keep his forces back during April. Northern Whites based in Archangel tried to link up with Kolchak's troops but failed.

In the north, Chaikovski's and Yudenich's forces won some impressive victories at first. So did Denikin, in the south; most important was his capture of Kiev in September, 1919. By November, however, the Red Army moved quickly to defeat the Volunteer Army and to drive what remained of it into the Crimea.

With the end of World War I, there was no longer any need to maintain the Triple Entente between Russia, Great Britain, and France, so the Allies withdrew their forces from Russia in 1919. By the winter of 1919-1920, major White offensives had ceased. In 1920, the Soviets were able to crush the Ukrainian independence movement.

That same year, the Soviets chased the Volunteer Army out of Crimea; its remaining troops had to be evacuated through Istanbul to the Kingdom of Serbs, Croats, and Slovenes (as Yugoslavia was then called). This defeat ended all southern resistance. In Siberia, the tired and demoralized Czechoslovak Legion, which had been helping the eastern White forces, quit the field and handed Kolchak over to the Soviets, who promptly executed the admiral in February, 1920.

Northern anti-Soviet movements, led by Nikolai V. Chaikovski, relied too heavily on British troops brought in as part of the Allied intervention. So did the White forces in Estonia, under the control of General Nikolai N. Yudenich.

Great Britain, France, the United States, and Japan had intervened in Russia in 1918. Landing its forces at the southeastern port city of Vladivostok, Japan had the goal of conquering territory for itself. That goal was not shared by the other Allied nations. British, French, and American troops were sent to Vladivostok as well and to the Gulf of Archangel (later renamed Dvina Gulf) in northern Russia, but they had no direct combat role. Instead, they supplied arms and advice to the White forces.

By the beginning of 1921, all serious revolt against the Bolshevik regime had been wiped out. The Marxist leaders of the Soviet Union had solidified their power and could begin to impose a Communist structure on the entire country.

## Consequences

The Soviets had had the advantage of defending the central part of the country, while the Whites worked from the edges. Transportation and communication systems were not adequate to allow White leaders to stay in touch and work together. In any case, the White movement was actually a set of very different movements, from parliamentary liberalism to a reactionary right wing, and the Ukrainian nationalists had never really been a part. The Whites had never had political unity.

Throughout the Great Civil War, the Bolsheviks had had a psychological advantage. The peasants and working classes associated the Whites with the hated czarist regime and saw them as the guardians of aristocratic privilege. By contrast, the Reds appeared to give oppressed peoples—Jews, peasants, or workers—the opportunity to rise through the ranks and take part in building a new society.

In the next decade the Reds lost whatever favor they had gained among the peasants, especially once the First Five-Year Plan was introduced. Yet the Communist Party was able to keep tight control over the country until the last two decades of the twentieth century, when economic problems, independence movements, and other factors combined to force a basic reshaping of the Soviet Union's structure.

*Edward A. Zivich*

# Shapley Proves Solar System Is Near Edge of Galaxy

*From his studies of star clusters, Harlow Shapley determined the size of the Milky Way galaxy and Earth's location within it.*

**What:** Astronomy
**When:** January 8, 1918
**Where:** Mount Wilson Observatory, California
**Who:**
HARLOW SHAPLEY (1885-1972), an American astronomer, humanitarian, and civil libertarian
JACOBUS CORNELIS KAPTEYN (1851-1922), a Dutch astronomer
ADRIAAN VAN MAANEN (1884-1946), a Dutch-American astronomer
HEBER DOUST CURTIS (1872-1942), an American astronomer
HENRIETTA SWAN LEAVITT (1868-1921), an American astronomer

## The Grindstone Galaxy

After giving the matter long and careful thought, the ancient Greek philosopher Aristotle concluded that Earth lay in the center of the universe, which ended at the sphere of stars that lay just beyond the planets. He considered the possibility that Earth might move around the sun, but he could find no evidence to support this idea. Some nineteen hundred years later, the Polish astronomer Nicolaus Copernicus established the sun as the center of the universe. Since it seemed that the sun was largely here for their benefit, human beings could still consider themselves specially located. At about this same time, Christian scholars began to suggest that God, being infinite, must have a domain far more extensive than Earth and the heavens that surround it. The English astronomer Thomas Diggs supposed that the stars were distributed throughout an infinite universe and that myriad planets orbited these stars and were inhabited by the creations of God.

Jacobus Cornelis Kapteyn attempted to form a model of this universe by doing "star counts." In its simplest form, the star-counting method requires counting the number of stars seen in a patch of the sky. After this process has been repeated for many patches across the sky, one can construct a model of the galaxy. Kapteyn determined the distances to nearby stars using the principle of parallax. Perhaps the most familiar example of parallax is seen while driving: Objects near the road appear to move rapidly, while the farther from the road an object is, the more slowly it seems to move. In a similar fashion, as Earth moves around the sun, nearby stars appear to move as seen from Earth. Measuring how much the stars seem to move allows one to find the distance to nearby stars. The parallax of even the nearby stars is quite small and can be detected only with cameras and telescopes. This explains why Aristotle was unable to detect stellar parallax even though he searched carefully for it.

Kapteyn's model of the galaxy was shaped like a grindstone, was ragged around the edges, and placed the sun near the center. Overall, it was 30,000 light-years across and 6,000 light-years thick. (A light-year is the distance light travels in a vacuum in one year; it takes slightly more than eight minutes for light to make the trip from the sun to Earth.)

## A Bigger Model

Harlow Shapley took a different approach. Scattered throughout the sky are ball-shaped groups of stars called globular clusters, which contain tens to hundreds of thousands of stars. To determine how far away the clusters are, Shapley used a new method developed by Henrietta Swan Leavitt using variable stars. Variable stars grow bright and dim, and some do so in a very predictable cycle. In 1924, Leavitt discovered that, for Cepheid variables (so called because the first one discovered was in the constellation Cepheus), the brighter the star, the longer is its cycle. Shapley measured the cycles of the Cepheids he found in globular clusters and used Leavitt's results to calculate how bright they should be. Comparing this value with how bright they looked, Shapley could then calculate how far away they (and the clusters they were in) must be.

The globular clusters did not fit into the grindstone galaxy model. Instead, the clusters would float above and below the grindstone and would occupy a roughly spherical volume ten times the size of Kapteyn's grindstone. The center of the sphere would be well off to one side of the grindstone. Shapley found that the sphere was 300,000 light-years across, with its center 50,000 light-years from the sun, in the direction of Sagittarius. Later, Shapley would write in his book *The View from a Distant Star* (1963), "It was a shocking thought—this sudden realization that the center of our universe was not where we stood but far off in space, that our heliocentric picture of the universe must be replaced by a strange sort of eccentric universe."

## Consequences

The full impact of Shapley's findings was not felt for years; it took that long to separate truth from error. The relevant questions were brought into focus by the "great debate" held between Shapley and another astronomer, Heber Doust Curtis, in 1920. One important issue was that of the nebulas. *Nebula* is Latin for cloud, a meaning retained in the English word "nebulous." In 1920, astronomers were mostly concerned with two kinds of nebulas: those that consisted of irregular patches of glowing gas near hot stars and those that appeared as glowing patches with a somewhat spiral shape. Curtis, while championing Kapteyn's view of the galaxy, supposed that spiral nebulas were galaxies like the Milky Way but were so far away that individual stars could not be discerned.

Astronomers eventually discovered the presence of novas in a few of these spiral nebulas. A nova is a star that explodes with tremendous energy; for a brief time, a nova can shine with the brightness of billions of stars. The presence of novas in the spiral nebulas enabled astronomers to estimate the distance to the nebulas. Curtis's theory—that the nebulas must be galaxies themselves—was supported by the evidence.

Another important factor in resolving the debate was the presence of interstellar dust. A handful of dust particles scattered throughout a volume of cubic kilometers is not enough to change it from a good vacuum. Yet thousands of light-years of such dust in the vacuum of space is enough to dim, or even to block, starlight. The chief effect dust was to have on Shapley's measurements was to explain why he did not see globular clusters close to the plane of the Milky Way, which is dusty. The effect of dust was far greater on Kapteyn's results. The sun appeared to be near the center of the galaxy only because Kapteyn could see into the dust about the same distance in all directions. Also, since his view was limited by dust, Kapteyn's "universe" was only a fraction of the real Milky Way galaxy.

The final important factor in the resolution of the issue arose when Shapley discovered an error in the measurement of the distances to various stars. He had originally supposed that the Milky Way galaxy was an "island universe" containing all the stars and nebula that exist, surrounded by an infinite dark, empty space. However, modern scientists believe that the universe consists of a multitude of galaxies like the Milky Way.

*Charles W. Rogers*

# Allies Dominate in Meuse-Argonne Offensive

> *Hundreds of thousands of American troops poured into France in mid-1918 to join the British and French armies in driving back the Germans.*

**What:** War
**When:** September 26-November 11, 1918
**Where:** Northeastern France
**Who:**
WOODROW WILSON (1856-1924), president of the United States from 1913 to 1921
NEWTON DIEHL BAKER (1871-1937), U.S. secretary of war from 1916 to 1921
JOHN J. PERSHING (1860-1948), commander in chief of the American Expeditionary Force from 1917 to 1924
FERDINAND FOCH (1851-1929), supreme commander of the Allied Armies from 1914 to 1918
SIR DOUGLAS HAIG (1861-1928), commander in chief of the Expeditionary Forces in France and Flanders from 1915 to 1919

## Getting Ready to Fight

When President Woodrow Wilson went before Congress on April 2, 1917, to ask for a declaration of war on the German Empire, neither he nor his advisers were clear on what kind of involvement the United States would have in World War I. At that time the U.S. Army numbered about 128,000 officers and men—a number smaller than the French had lost in the Battle of Verdun in 1916. The government began at once to raise and train a larger army to serve in France. Yet some American leaders thought at first that U.S. help for the Allies would consist mainly of money, equipment, and naval support, for the United States was the world's greatest industrial power and had the world's third largest navy.

Political and military leaders of the Allied nations, however, soon made it clear to American leaders that what the Allies needed most were human resources. They asked to be allowed to recruit Americans directly into their armies or, if that plan was not acceptable, to bring whole American battalion or regiments into British and French divisions.

Secretary of War Newton D. Baker and American military leaders rejected these proposals. They said that Americans must serve under their own flag and their own officers. American leaders had little confidence in the abilities of the French and British generals who, in three years of war, had lost record numbers of soldiers without achieving more than a standoff. The Americans were also aware that by sending a separate American army, the United States would be in a better position to bargain at any future peace conference.

Although it would take many months to raise and train an army of the size the Allies said would be needed, the United States decided to send one American division to France immediately, to "show the flag" and encourage the Allies. To command this American division, Secretary Baker chose Major General John J. Pershing. By the time he arrived in France in mid-June, 1917, Pershing had been named commander of all the American forces to be sent to the Western Front.

Because of a lack of shipping and crowded ports, the American buildup in France went slowly before the spring of 1918. In March, 1918, there were only five American divisions in France. By that time, Russia had signed its own peace agreement with Germany at Brest-Litovsk, and hundreds of thousands of German troops could be transferred from the Eastern to the Western Front. A combined Austrian and German attack against Italy in the fall of 1917 had crippled the Italian army.

Gathering troops from Russia and Italy, the German High Command formed them into a gigantic force of 192 divisions, which they planned to send quickly to face the British and French in order to win the war before the American army could be formed. After their long, hard fighting throughout 1916 and 1917, the Allied armies were exhausted, and it was not difficult for the Germans to push them back in France. Late in May, the Germans broke through the Chemin des Dames sector of the French line and advanced west toward Paris. Three days later, they stood on the banks of the Marne River, less than fifty miles from the French capital.

During that spring, Allied leaders had renewed their requests for American troops to join their armies individually or in regiments. With the backing of Wilson and Baker, Pershing resisted these requests but did "loan" American divisions already in France to the Allies. The American Expeditionary Force was now building up rapidly and soon would be ready to join the Allies as an independent army.

## The Offensive

Beginning on May 31, the American troops in France started to fight. Parts of the Second Division stationed themselves in the vicinity of the Marne River, near a small hunting preserve called Belleau Wood. Their goal was to stop the Germans from advancing farther. For two days the Germans tried to force them out, but the Americans beat the Germans back and, on June 6, counterattacked. Fierce fighting continued in Belleau Wood and a nearby village for almost three weeks, but it was soon clear that the German drive toward Paris had been stopped.

By early August, there were almost thirty American divisions in France. On August 10, nineteen of these divisions were formed into the First U.S. Army under Pershing's command. The exhausted Germans now faced a powerful new Allied army, completely fresh and growing stronger every day.

General Ferdinand Foch, Supreme Commander of the Allied Armies, and Sir Douglas Haig, Commander of the British Armies, planned a new offensive: a giant "pincers" attack on both flanks of the German line. The British would form the left pincer, while the Americans, supported by the French in the center, would form the right pincer.

The First U.S. Army proceeded with its plans to attack the St. Mihiel salient, a triangle-shaped bulge in the German line, with the town of St. Mihiel at its apex. On September 12, 1918, the Americans attacked the German troops on both sides of St. Mihiel, catching them in a trap. The Americans captured fifteen thousand prisoners and lost seven thousand of their own soldiers.

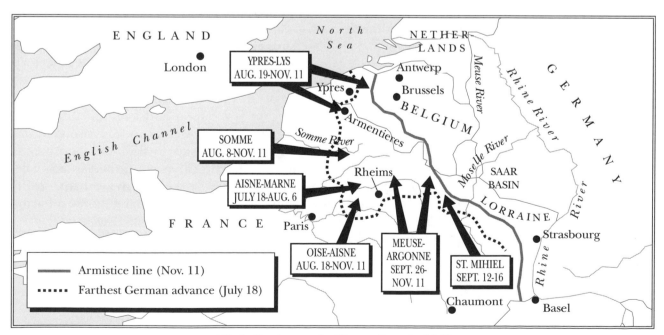

*Allied offensives on the Western Front in 1918.*

National Archives

*Members of the U.S. Thirty-fifth Coast Artillery load a mobile railroad gun on the Argonne front.*

The Americans' job now was to begin a much larger attack in the area bounded by the Meuse River on the east and the Argonne Forest on the west, about forty miles northwest of St. Mihiel. With the help of surprise, the Americans hoped to overwhelm the Germans and penetrate the main German defensive line on the first day of the attack.

The American attack began well but was soon slowed down by the rugged terrain. Also, the leaders believed that all three corps should advance at the same pace; though Corps I and III made good progress at first, they were required to stop after a time and wait for Corps V to catch up. By that time, the Germans had recovered from their surprise and were quickly sending in reinforcements. By the end of September, the attack had ground to a halt. A second attack on October 14 made little further progress.

## Consequences

Though the Americans' Meuse-Argonne offensive did not at first result in a clear victory, the combined weight of Allied attacks forced Germany to ask for an armistice. Meanwhile, the Americans regrouped and began a third drive forward on November 1. Without resisting much, the Germans fell back to positions beyond the Meuse River. The Allies continued to advance, Kaiser William II of Germany went into exile, and on November 11, Germany signed an armistice. The "war to end all wars" was over.

*Ronald N. Spector*

# Harding Is Elected U.S. President

*Though World War I was won under Woodrow Wilson's Democratic administration, the strains of wartime politics led to Republican victories in the elections of 1918 and 1920.*

**What:** National politics
**When:** November 5, 1918
**Where:** United States
**Who:**

WOODROW WILSON (1856-1924), president of the United States from 1913 to 1921

WARREN GAMALIEL HARDING (1865-1923), senator from Ohio from 1915 to 1921 and president of the United States from 1921 to 1923

CALVIN COOLIDGE (1872-1933), governor of Massachusetts from 1919 to 1920 and vice president of the United States from 1921 to 1923

JAMES M. COX (1870-1957), governor of Ohio from 1913 to 1915 and 1917 to 1921, and Democratic presidential candidate in 1920

FRANKLIN DELANO ROOSEVELT (1882-1945), assistant secretary of the Navy from 1913 to 1920 and Democratic vice-presidential candidate in 1920

HENRY CABOT LODGE (1850-1924), senator from Massachusetts from 1893 to 1924 and chairman of the 1920 Republican National Convention

## 1918 Elections

With traditional Democratic supporters such as farm, labor, and business groups divided over various national issues, the Republicans were in a good position to gain control of the House of Representatives and to add a few Senate seats in the elections of 1918. World War I was not yet formally ended, but the Allies' victory was assured, and President Woodrow Wilson was looking forward to the peace conference at Versailles.

Though many Republicans had supported his foreign policies, on October 25, 1918, Wilson ap-pealed to the voters to keep the Congress under the Democrats' control. Peacemaking, he said, was no time for a divided leadership.

For Wilson, the election was a troubling upset. On November 5, the Republicans won majorities in both houses of Congress: 49 to 47 in the Senate and 237 to 193 in the House. The Republican Party now had a strong base from which to prepare for the 1920 presidential election.

Between 1918 and 1920, Senator Henry Cabot Lodge of Massachusetts, the new chairman of the foreign relations committee, had the difficult task of building party unity among Republicans in Congress. Senate Republicans were badly divided over the Treaty of Versailles; a minority of isola-

*Warren G. Harding.*

**387**

tionists opposed it in any form, while the majority favored ratifying it with certain amendments. If the isolationists made good on their threat to reject the treaty, they might cause the party's defeat in 1920. As Lodge knew, Wilson owed his victory in 1912 to a division among the Republicans.

Then there was the League of Nations issue. If Congress agreed to make the United States part of the league, one result would be increased prestige for the Democrats. The Republicans insisted that American participation in the League of Nations must come from both major political parties, but Wilson refused to accept their amendments to either the Treaty of Versailles or the League of Nations agreement. In the end, Lodge managed to lead the Republicans in defeating both without dividing the party.

## 1920 Conventions and Election

When the Republicans gathered in Chicago for their 1920 convention, they were confident of success. Pennsylvania party boss Boies Penrose boasted, "Any good Republican can be nominated for president and can defeat any Democrat." The party professionals did not want a strong-willed candidate; they had in mind someone who was loyal to the party and did not have many strong personal opinions. That sort of person could be more easily controlled by party leaders.

In the balloting, neither of the two leading candidates, Army chief of staff Leonard Wood and governor Frank O. Lowden of Illinois, could win the nomination. A "dark horse" was needed to break the standoff. A number of the old-time party leaders met until 2:00 A.M. and agreed to back Senator Warren G. Harding of Ohio.

Harding had made many friends and no enemies during his brief career in the Senate. Tall and handsome, with silver hair and a suntan, Harding even looked like a president. Best of all, he had the kind of controllable personality the party leaders were looking for. The next day, the convention nominated him, with Massachusetts governor Calvin Coolidge as his running mate.

At the Democratic convention in San Francisco, delegates were sobered by the knowledge that President Wilson was still partially paralyzed after a stroke. After the Senate failed to ratify the Treaty of Versailles, Wilson called on the Ameri-

can people to show their support for the League of Nations by supporting him and the Democratic Party in the 1920 election. He was hoping to be nominated again by the party.

Yet that nomination never came. Through four grueling days of balloting, three contenders battled one another at the convention: Treasury Secretary William G. McAdoo, Attorney General Mitchell Palmer, and Governor James M. Cox of Ohio. On the forty-fourth ballot, Cox became the nominee, with assistant secretary of war Franklin D. Roosevelt as his running mate.

The election campaign took place in an atmosphere of upheaval. Since the Armistice, a wave of strikes, bombings and attempted bombings, race riots, and lynchings had frayed the country's nerves. Harding had captured the country's mood in May, 1920: "America's present need," he said, "is not heroics, but healing."

With his country united behind him, Harding had only to sit back and wait for the most smashing victory up to that time in American presidential elections. On election night, radio station KDKA made the first commercial radio broadcast in the United States from its facilities in East Pittsburgh, Pennsylvania. A small scattering of people, straining to hear through the static in their earphones, heard a voice reporting the election returns. With a 60 percent majority, Harding had carried every state outside the South—and even made a dent there by winning in Tennessee.

## Consequences

After the nation's mobilization in World War I, the American people wanted desperately to return to normal, everyday life. They were ready to leave behind Wilson's idealistic efforts to make the United States a world crusader for peace and justice. Many Americans wanted their leaders to pay less attention to foreign policy and focus on issues and needs inside the country. Progressivism was losing popularity.

The 1918 and 1920 elections reflected this national mood, and in the decade that became known as the "Roaring Twenties" the Republicans established a policy that has been called "isolationism"—drawing back from commitments and involvements overseas.

*Donald Holley*

# Habsburg Monarchy Ends

*Just after Germany signed an armistice with the Allies, Emperor Charles I Habsburg of Austria abdicated, bringing to an end Europe's oldest ruling monarchy.*

**What:** Political reform
**When:** November 13, 1918
**Where:** Austria, Hungary, and Bohemia
**Who:**
FRANCIS JOSEPH I (1830-1916), emperor of Austria from 1848 to 1916 and king of Hungary from 1867 to 1916
FRANCIS FERDINAND (1863-1914), archduke of Austria and heir apparent to the thrones of Austria-Hungary
CHARLES I (1887-1922), emperor of Austria and king of Hungary from 1916 to 1918
TOMÁŠ GARRIGUE MASARYK (1850-1937), professor who was Czechoslovakia's first president, from 1918 to 1935

## A Shaky Monarchy

The Habsburg dynasty had a long history of empire in Europe. By 1815, the monarchy had lost its former control over the Holy Roman Empire, Spain, and Belgium, but it still ruled over Alpine Austria, Bohemia, Hungary, and parts of Italy, Poland, and the Balkans. By the beginning of the twentieth century, its territories had shrunk further, but the Habsburg land area, located in central and southeastern Europe, was about the size of Texas. The fifty-two million people in these lands included a dozen nationalities and many different cultures.

Several factors had kept these diverse lands together until the twentieth century: the aristocrats' loyalty to the dynasty; the army; the Catholic Church; the city of Vienna as a cultural and administrative center; and the German language. Yet throughout the nineteenth century, another set of factors had worked to weaken the monarchy: liberal political thinking, industrialization, and nationalistic feelings among the various ethnic groups in the realm. A number of minority groups rose up in 1848 in an attempted liberal revolution; though the attempt failed, it foreshadowed the changes that would come seventy years later.

In the 1860's and 1870's, the Habsburgs lost the friendship of Russia and had to let go of their lands in Germany and Italy. The Compromise of 1867 transformed the former Austrian Empire (which had been formally created in 1804) into the Dual Monarchy of Austria-Hungary. Though the German Austrians and the Magyars were minorities among the other nationalities, these two groups had the real power within the state.

From 1870 to 1914, Austria-Hungary went through many internal difficulties. The problem of nationalities was the most serious. The Czechs and South Slavs demanded independence; though they received some limited control over their own affairs, they remained dissatisfied. The Magyars refused to grant greater rights to the Slovaks and Croats under their control. Many German Austrians and Magyars became very nationalistic themselves and began to resent the Habsburgs; among them was the young Adolf Hitler, who was living in Vienna just before World War I. Though the Habsburgs granted universal voting rights in 1907, they then began ruling by decree so that no new reforms could be made.

In the early twentieth century, Austria-Hungary was fighting a losing battle to keep its position as a world power. The Habsburg monarchy became more and more dependent on the might of the new German Empire (which had been unified in 1871). In 1908, Austria-Hungary's annexation of the mostly Serbian area of Bosnia-Herzegovina infuriated the leaders of the new kingdom of Serbia.

## World War Disaster

The problem of nationalism in Austria-Hungary caused a crisis that led to the outbreak

**389**

of World War I. On June 28, 1914, Archduke Francis Ferdinand, heir apparent to the Habsburg throne, was assassinated in the Bosnian capital, Sarajevo, by a Bosnia-Serbian nationalist named Gavrilo Princip. Exactly a month later, Austria declared war on Serbia; Austria's ally Germany went to war with Russia and France, and World War I began.

The majority of the peoples of Austria-Hungary at first rallied behind their old emperor, Francis Joseph I. By 1915, however, the strains of war began to widen the cracks in the Dual Monarchy's structure. The government began to resemble a military dictatorship, and many groups within the realm became hostile. In 1915, radical Czechs and Croats formed national committees abroad. In October, 1916, Austrian prime minister Count Karl von Stürgkh was assassinated by a young socialist.

On November 21, 1916, Emperor Francis Joseph I died after a reign of sixty-eight years. He had been a hardworking ruler, but he had not known how to respond to changing times. He was succeeded by his grandnephew Charles I, who had good intentions but lacked experience.

Charles reassembled the Reichsrat (parliament) and declared an amnesty, but these moves were not enough to win him many friends. He began trying to make peace with the Allies, yet he was shackled by his dependence on Germany. Most serious of all was his unwillingness to grant complete autonomy to the national groups within the realm.

In 1917, the Russian revolutions and the United States' entry into the war encouraged more calls for reform within Austria-Hungary, and Germany was on its way to a defeat. Economic hardships in the winter of 1917-1918 resulted in rioting, desertions from the army, and strikes. When the German offensive in France began in the spring of 1918, the Allies decided to speed victory by encouraging the breakup of the Habsburg monarchy. In May, 1918, the exiled Czech leader Tomáš Garrigue Masaryk, the influential British journalist H. Wickham Steed, and President Woodrow Wilson of the United States all urged an end to the monarchy.

By October, 1918, the Czechs, Poles, and South Slavs had all seceded from the monarchy. The Austrian and Hungarian parliaments drew up plans for their own independent republics. At noon on November 11, 1918, Charles I penciled his resignation, without formally abdicating his powers, as Emperor of Austria. Two days later, he resigned the kingship of Hungary.

## Consequences

Charles I had not been driven from office; his resignation simply reflected the realities of the situation. The forces of nationalism and liberalism had fractured the empire, and with the Central Powers' defeat in World War I the Habsburgs were left without strength to rebuild their realm.

With the fall of the Habsburgs, the national and ethnic groups of Eastern Europe were left free to create new republics, but these small new states were quite vulnerable to the ambitions of larger, more established nations. Within two decades they were being threatened by Nazi Germany, and with the defeat of the Nazis in World War II most of them came under the domination of the Soviet Union. It would be more than fifty years before they could throw off the repressive control of the Soviet Communists.

# Aston Builds First Mass Spectrograph

*Francis William Aston invented the first mass spectrograph to measure the mass of atoms and discovered that the atomic mass of elements is the result of the combination of isotopes.*

**What:** Physics; Photography
**When:** 1919
**Where:** Cambridge, England
**Who:**
FRANCIS WILLIAM ASTON (1877-1945), an English physicist who was awarded the 1922 Nobel Prize in Chemistry
SIR JOSEPH JOHN THOMSON (1856-1940), an English physicist
WILLIAM PROUT (1785-1850), an English biochemist
ERNEST RUTHERFORD (1871-1937), an English physicist

## Same Element, Different Weights

Isotopes are different forms of a chemical element that act similarly in chemical or physical reactions. Isotopes differ from "normal" atoms of an element in two ways: They possess different atomic weights and different radioactive transformations. In 1803, John Dalton proposed a new atomic theory of chemistry that claimed that chemical elements in a compound combine by weight in whole number proportions to one another. By 1815, William Prout had taken Dalton's hypothesis one step further and claimed that the atomic weights of elements were integral (the integers are the positive and negative whole numbers and zero) multiples of the hydrogen atom. For example, if the weight of hydrogen was 1, then the weight of carbon was 12, and that of oxygen 16. Over the next decade, several carefully controlled experiments were conducted to determine the atomic weights of a number of elements. Unfortunately, the results of these experiments did not support Prout's hypothesis. For example, the atomic weight of chlorine was found to be 35.5. It took a theory of isotopes, developed in the early part of the twentieth century, to verify Prout's original theory.

After his discovery of the electron, Sir Joseph John Thomson, the leading physicist at the Cavendish Laboratory in Cambridge, England, devoted much of his remaining research years to determining the nature of "positive electricity." (Since electrons are negatively charged, most electricity is negative.) While developing an instrument sensitive enough to analyze the positive electron, Thomson invited Francis William Aston to work with him at the Cavendish Laboratory. Recommended by J. H. Poynting, who had taught Aston physics at Mason College, Aston began a lifelong association at Cavendish, and Trinity College became his home.

When electrons are stripped from an atom, the atom becomes positively charged. Through the use of magnetic and electrical fields, it is possible to channel the resulting positive rays into parabolic tracks. By examining photographic plates of these tracks, Thomson was able to identify the atoms of different elements. Aston's first contribution at Cavendish was to improve the instrument used to photograph the parabolic tracks. He developed a more efficient pump to create the required vacuum and devised a camera that would provide sharper photographs. By 1912, the improved apparatus had provided proof that the individual molecules of a substance have the same mass. While working on the element neon, however, Thomson obtained two parabolas, one with a mass of 20 and the other with a mass of 22, which seemed to contradict the previous findings that molecules of any substance have the same mass. Aston was given the task of resolving this mystery.

## Treating Particles Like Light

In 1919, Aston began to build a device called a "mass spectrograph." The idea was to treat ionized or positive atoms like light. He reasoned that, because light can be dispersed into a rainbowlike spectrum and analyzed by means of its different colors, the same procedure could be used with atoms of an element such as neon. By creating a device that used magnetic fields to focus the stream of particles emitted by neon, he was able to create a mass spectrum and record it on a photographic plate. The heavier mass of neon (the first neon isotope) was collected on one part of a spectrum and the lighter neon (the second neon isotope) showed up on another. This mass spectrograph was a magnificent apparatus: The masses could be analyzed without reference to the velocity of the particles, which was a problem with the parabola method devised by Thomson. Neon possessed two isotopes: one with a mass of 20 and the other with a mass of 22, in a ratio of 10:1. When combined, this gave the atomic weight 20.20, which was the accepted weight of neon.

Aston's accomplishment in developing the mass spectrograph was recognized immediately by the scientific community. His was a simple device that was capable of accomplishing a large amount of research quickly. The field of isotope research, which had been opened up by Aston's research, ultimately played an important part in other areas of physics.

## Consequences

The years following 1919 were highly charged with excitement, since month after month new isotopes were announced. Chlorine had two; bromine had isotopes of 79 and 81, which gave an almost exact atomic weight of 80; krypton had six isotopes; and xenon had even more. In addition to the discovery of nonradioactive isotopes, the "whole-number rule" for chemistry was verified: Protons were the basic building blocks for different atoms, and they occurred exclusively in whole numbers.

Aston's original mass spectrograph had an accuracy of 1 in 1,000. In 1927, he built an even more accurate instrument, which was ten times more accurate. The new apparatus was sensitive enough to measure Albert Einstein's law of mass energy conversion during a nuclear reaction. Between 1927 and 1935, Aston reviewed all the elements that he had worked on earlier and published updated results. He also began to build a still more accurate instrument, which proved to be of great value to nuclear chemistry.

The discovery of isotopes opened the way to further research in nuclear physics and completed the speculations begun by Prout during the previous century. Although radioactivity was discovered separately, isotopes played a central role in the field of nuclear physics and chain reactions.

*Victor W. Chen*

# Frisch Learns That Bees Communicate

---

*Karl von Frisch discovered that honeybees returning to the hive use a so-called round dance to communicate to their comrades that food is nearby.*

---

**What:** Biology
**When:** Spring, 1919
**Where:** Munich, Germany
**Who:**
KARL VON FRISCH (1886-1982), an
   Austrian physiologist and student of
   animal behavior who was a cowinner
   of the 1973 Nobel Prize in Physiology
   or Medicine

### Bee Watching

Karl von Frisch can be credited, in part, for several lines of experimentation, such as the study of color perception in bees and hearing in fish. It was his study of communication in bees, however, that brought him world fame and, in 1973, a Nobel Prize, which he shared with Konrad Lorenz and Nikolaas Tinbergen.

Many observers, including the ancient Greek philosopher Aristotle, have noticed that when honey or sugar water is placed near a hive, it may be many hours before a wandering bee discovers the food. Yet once the food is discovered, hordes of bees soon descend upon the new find. Obviously, in some way, the forager bee communicates information about the presence of food to other members of the hive. A few naturalists noticed the dancing movements of bees and speculated about what their meaning might be, but it remained for Frisch to perform the many years of exacting experiments that were needed to substantiate that dancing in bees is actually a form of communication.

Frisch's autobiography recounts the experiment that led to the most far-reaching observation of his life. At the time, he was at the Munich Zoological Institute, studying bees in a queen-breeding cage, which has glass sides so that all the bees can be seen easily. Frisch put out a dish of sugar water to feed foraging bees from the little glass-sided hive. He marked the bees that fed on the sugar water with small dots of red paint. He then removed the dish of sugar water, and the bees came less and less frequently. Finally, he once again put the sugar water out and allowed a bee to feed. He watched the behavior of the bee once it returned to the hive. As Frisch describes in his autobiography, "I could scarcely believe my eyes. She performed a round dance on the honeycomb which greatly excited the marked foragers around her and caused them to fly back to the feeding place." When Frisch and his family moved to Brunnwinkl, Austria, he continued his studies of the round dance as a form of communication in honeybees. The results of these early studies were published in 1920.

### Learning to Dance

When the dancing bee performs the round dance, it moves in a tight circle to the right and then to the left, describing between one and two circles in each direction, and repeating the turning movements for half a minute or longer. The sweeter the food source, the more vigorous the dancing becomes. Typically, a group of bees surround the dancing bee and extend their antennas over the body of the dancing bee. This behavior allows the new recruits to detect odors adhering to the dancer's body. These odors enable the recruits to find the particular species of flower that is producing nectar, at distances of up to about fifty meters. During pauses in the dance, the dancer regurgitates nectar from her honey stomach and feeds the bees around her. This nectar carries the scent of the flower that was visited.

Frisch also demonstrated that bees have color vision and can learn to seek out a given color that

**393**

they have associated with food. He found that bees cannot distinguish red from black and can see ultraviolet as a distinct color. The patterns of color on flowers thus appear different to bees from the way they appear to humans. Individual bees use color vision to locate flowers they have already visited, but there is no indication that they can communicate colors to other bees.

It was another twenty years before Frisch discovered the workings of the more incredible "wagging dance" that was often performed by bees returning from a distance with loads of pollen. Upon closer inspection of bees fed four hundred meters north of their hives at Brunnwinkl in June of 1945, he discovered that this more elaborate dance communicates both direction and distance and is used for finds at more than fifty meters from the hive. The dancing bee moves in a figure-eight pattern, its movements followed by the outstretched antennae of forager bees. The greater the distance to the food, the slower the tail wagging performed by the dancing bee. If the tail wagging is in an upward direction, then the food is toward the sun; and if the tail wagging part is downward, the food is away from the sun. The biological clock of the bee gradually corrects the dance for changes in the apparent position of the sun.

## Consequences

Frisch's publication of his 1919 observations on the round dance as a form of communication in bees did not result in either immediate fame or controversy, although eventually both would come. It was the steady stream of scientific papers on animal physiology and behavior, especially on fish and bees, that eventually established Frisch as the most widely known German biologist. By World War II, his reputation allowed him to continue his work for a time, even though his mother was partially of Jewish descent. Later, forced into isolation at Brunnwinkl, Frisch continued his research during the war years. It was there in 1943 that Frisch discovered the importance of the wagging dance.

Frisch's studies on the round dance honed the experimental techniques that he later applied to the wagging dance. The work with the round dance generally furthered Frisch's reputation, and it played a role in the development of the slowly emerging field of animal behavior.

*John T. Burns*

# Red Scare Threatens Personal Liberties

> *In the year following World War I, many Americans feared that Communists were soon to take over the United States; the result was a flood of raids, arrests, and attacks on suspected "Reds."*

**What:** National politics; Civil rights and liberties
**When:** 1919-1920
**Where:** United States
**Who:**
Alexander Mitchell Palmer (1872-1936), attorney general of the United States from 1919 to 1921
J. Edgar Hoover (1895-1972), director of the General Intelligence Division, Department of Justice from 1924 to 1972
Woodrow Wilson (1856-1924), president of the United States from 1913 to 1921

## A Growing Panic

When Woodrow Wilson took the steps that led the United States into World War I in 1917, he said that fighting a war abroad would require cutting back freedoms at home; he feared the "spirit of ruthless brutality" that would enter American life as a result. Patriotic propaganda poured out of the government-funded Committee on Public Information, and those who were suspected of sympathizing with the German enemy or opposing the U.S. government in other ways were likely to be censored when they tried to make their opinions known.

After the Armistice in November, 1918, the wartime crusade against Germany did not slow down; instead, it simply changed its direction. It became a crusade against anything "un-American"—which in 1919 meant "radical" or "red" (Communist). Postmaster General Albert Burleson, who during the war had been given power to censor mail when he considered it to be hindering the war effort, now censored mail when it was used to communicate radical ideas. Conscientious objectors—people who refused to participate in the war because of their moral or religious beliefs—had been imprisoned during the war, and now they remained in jail, for the Wilson administration refused to grant them amnesty.

The American Civil Liberties Union, which had been formed during the war to defend the constitutional rights of Americans, now found even more to do. There were new patriotic societies that preached "one hundred percent Americanism" and tried to use schools and fraternal orders to spread their propaganda. In Congress, an investigation into German propaganda activities quickly turned into an anti-Bolshevik investigation.

Several events in 1919 led many Americans to believe that the "Reds" (Bolsheviks or Communists) had made a conspiracy to destroy the United States. First came a strike by police in Boston, which the newspapers promptly labeled "Bolshevik." Next there was a widespread strike in the steel industry, and the U.S. Steel Corporation accused the strike leaders of associating with Bolshevik groups. Then there was a strike in the coal industry. Meanwhile, crude homemade bombs had been sent in the mail to several public officials, or had been thrown at them. Throughout the year, there were street riots between whites and African Americans, and between war veterans and socialists.

## The Crest of the Wave

By the fall of 1919, the American people were clamoring for some kind of government action in response to the "conspiracy." In August, Attorney General A. Mitchell Palmer had established the General Intelligence Division in his department to investigate and take action against radicals. He appointed J. Edgar Hoover as its director, and Hoover began to make a huge index of radical organizations, publications, and leaders.

**395**

On November 7, Hoover's agents raided the headquarters and branch offices of a labor society known as the Union of Russian Workers. Throughout the country, state and local officials carried out smaller raids on suspected radicals. Meanwhile, Congress began considering several bills requiring that foreign radicals be deported from the United States. One senator even proposed that native-born Americans who were radical be expelled to a special prison colony on Guam.

Action was taken after the Labor Department proclaimed that membership in the Union of Russian Workers was a deportable offense. On December 21, 249 deportees set sail from New York aboard the old army transport ship *Buford*, nicknamed "the Soviet Ark." Most of the deportees were members of the Union of Russian Workers and the others were labeled as anarchists, public charges, criminals, and misfits. Two weeks later came the last and greatest raids, as the wave of the Red Scare crested and finally broke.

Late in the afternoon of Friday, January 2, 1920, Justice Department agents began arresting thousands of persons in major cities throughout the nation. They poured into private homes, clubs, pool halls, and coffee shops, seizing citizens and foreigners, Communists and noncommunists; they tore apart meeting halls and damaged and destroyed other property. Agents jailed their victims, prevented them from contacting family members or lawyers, and interrogated them.

Prisoners who could prove their American citizenship were released, though they were often placed under the custody of state officials who hoped to bring them to court. Aliens were released a few days later, unless they were members of the Communist Party or the Communist Labor Party. The Department of Justice hoped to deport these individuals.

In two days, nearly five thousand people had been arrested, and possibly another thousand were seized in follow-up actions during the next two weeks. The arrests did not conform to the process of law, and many of those who were arrested were treated very badly. These raids, which became known as the Palmer Raids, were the climax of the Red Scare.

**Consequences**

After January, 1920, conscientious objectors still found it hard to obtain amnesty, and restrictions on immigration grew tighter under the 1924 National Origins Act. Throughout the 1920's, the U.S. Army held antiradical training seminars. School textbooks were censored, and those who wished to take certain civilian jobs were often required to sign oaths of their loyalty to the U.S. government.

On the other hand, raids and deportations decreased quickly after the winter of 1920. Many people were outraged at the violation of constitutional rights that had occurred. In June, 1920, Federal Judge George W. Anderson, in the case of *Colyer v. Skeffington*, declared that the Justice Department's methods were brutal and unjust and that its raids were "sordid and disgraceful." Attorney General Palmer tried to win the presidential nomination at the Democratic National Convention in 1920, but failed.

*Burl L. Noggle*

# Bjerknes Discovers Atmospheric Weather Fronts

> *Vilhelm Bjerknes's model of the atmosphere emphasized the idea of "fronts," those boundaries along which masses of warm and cold air clash and converge to produce the weather.*

**What:** Earth science
**When:** 1919-1921
**Where:** Bergen, Norway
**Who:**
VILHELM BJERKNES (1862-1951), a
    Norwegian meteorologist
JACOB BJERKNES (1897-1975), a
    Norwegian meteorologist
CARL-GUSTAV ARVID ROSSBY (1898-1957),
    a Swedish American meteorologist
TOR BERGERON (1891-1977), a Swedish
    meteorologist

## Gathering Information

Vilhelm Bjerknes, a Norwegian geophysicist and meteorologist, knew in the early 1900's that accurate forecasting of weather required much information. He also knew that the weather in one place was part of a huge mass that covered the world. While teaching at Stockholm University from 1895 to 1907, he proposed that movements in the atmosphere are stimulated by heat from the sun. At the same time, these movements radiate heat as air masses rub up against one another, causing friction.

Bjerknes was motivated by the need for improved weather prediction for commercial fishing and agriculture. In part, the urgent need for better domestic food production arose from restrictions of imports and communications as a result of World War I (1914-1918). He persuaded the Norwegian government to help set up strategically located observing stations. In addition to the stations, he founded a school at Bergen that attracted meteorologists from all over the world, including his son Jacob Bjerknes; Carl-Gustav Arvid Rossby, a Swedish American

meteorologist; and Tor Bergeron, a Swedish meteorologist.

Weather types and changes, along with moving masses of air and their interaction, have been studied and noted for centuries. In the nineteenth century, Luke Howard, an English physicist, had written of northerly and southerly winds blowing alongside each other, with the colder wedging in under the warmer and the warmer gliding up over the colder and causing extensive and continued rains. In 1852, evidence had been found of a polar wind advancing under a warm, nearly saturated tropical wind and pushing it upward producing cumulus clouds.

## Explaining the Weather

Vilhelm Bjerknes was a pioneer in the development of a mathematical theory of fronts and their effects. In addition, along with his son, he was the first to study extratropical cyclones and use them to forecast the weather. Extratropical cyclones are cyclones that may cross an ocean in ten days, lose most of their intensity, and then develop again into large and vigorous storms. In the years following World War I, Norwegian meteorologists had a fairly good understanding of the action in the big storms sweeping across the Atlantic. From this knowledge, Vilhelm Bjerknes theorized that the main idea in storm development is a clashing of two air masses, one warm, the other cold, along a well-defined boundary, or front.

Aside from the idea of storm fronts, his view of cyclone development produced another important idea. At the beginning of the life cycle of a storm, there is an undisturbed state in which cold and warm air masses flow side by side, separated by a front. Each air mass flows along its side of the front until some of the warmer air begins to invade the cooler air, leading to a wave distur-

**397**

bance. This disturbance spreads and grows, creating low-pressure areas at the tip of the wave. Air motions try to spiral into these areas, and both fronts begin to advance. The cold air generally moves faster, catching up with and moving under the lighter warm air. As the storm grows deeper, the cold front becomes more pronounced. The whole process—from the time the polar air meets the northward-flowing warm air to the point at which the area of low pressure is filled completely—is known as the "life cycle of a frontal system." This description is based on the wave theory, which was originally developed by Vilhelm Bjerknes in 1921.

In 1919, when this work began, upper atmosphere studies were limited by the lack of knowledge of such things as radar images, lasers, computers, and satellites. Vilhelm Bjerknes showed that the atmosphere is composed of distinct masses of air meeting at various places to produce different meteorological effects. He published the study *On the Dynamics of the Circular Vortex with Applications to the Atmosphere and Atmospheric Vortex and Wave Motion* in 1921.

## Consequences

The weather forecasting stations established by Vilhelm Bjerknes in the 1920's were a monumental accomplishment, considering the limited amount of information and the lack of high-speed, worldwide communications. All the computations were done without the assistance of a computer to analyze and model the data. Bjerknes also did not have the advantage of satellites and high-flying aircraft; however, his work was well developed and reasoned.

In the 1980's, the National Meteorological Center (NMC) used national, international, public, and private resources to predict the weather. Throughout the Northern Hemisphere, approximately two thousand stations regularly transmit weather data to national or regional collecting centers four times a day, seven days a week. On an average day, the NMC also picks up about thirty-two hundred ship reports on weather throughout the seven seas, about one thousand reports from commercial aircraft in flight, and at least two hundred reports from scheduled reconnaissance flights by military aircraft. A model of the atmosphere can be created entirely by entering weather data from around the world into a computer. Great progress has been made as a result of new technology and a universe of data, but it was pioneers such as Vilhelm Bjerknes who paved the way.

*Earl G. Hoover*

# Mises Develops Frequency Theory of Probability

> *Richard von Mises clarified the fundamental concept of probability and established probability theory as a field on a par with geometry and theoretical mechanics.*

**What:** Mathematics
**When:** 1919-1933
**Where:** Germany and Austria
**Who:**
Pafnuti L. Tchebychev (1821-1894), a
    Russian mathematician
Richard von Mises (1883-1953), an
    Austrian mathematician
Andrei Nikolaevich Kolmogorov (1903-
    1987), a Russian mathematician

## Early Theories of Probability

In mathematics, words and phrases such as "probable" and "chances are" are used to indicate an event that is likely, but not certain, to occur. The philosopher Immanuel Kant (1724-1804) thought that, on a scale that ran from truth to error, probable events would be those that were closer to the truth than to error. What does it mean, however, to say that the number 3 will appear about one-sixth of the time when a die is thrown or that there is a 30 percent chance that it will rain tomorrow?

One of the factors that led to the development of probability theory was the need to determine insurance rates. During the fourteenth century, the cost of insuring goods against loss was 12 to 15 percent of the value of those goods if they were shipped by sea and 6 to 8 percent if they were shipped by land. These rates were set by experience, and this intuitive form of probability theory was sufficient for that time.

Another motivating factor in the development of probability theory was the desire to understand games of chance. For a person attempting to make a living as a professional gambler, a precise knowledge of odds was (and is) imperative.

In the seventeenth century, gamblers turned to mathematicians for this knowledge. Assuming that games were fair, mathematicians defined the probability that an event would occur as the number of ways the particular event could occur divided by the total number of events. Therefore, the probability of throwing a 3 with a fair die is one-sixth, since a die has six sides.

## The Frequency Definition

The theory of probability based on equally probable events was developed further in the nineteenth century, but it was too restrictive to meet the growing demands that were made of it. In 1867, Tchebychev showed that probabilities could be estimated on the basis of a large number of experiments. His result is sometimes referred to as the "law of averages." By the beginning of the twentieth century, many researchers were searching for an approach based on a large number of trials. Such an approach appeared in 1919 with the publication of Mises's book *Probability, Statistics, and Truth*.

First, Mises emphasized that the set of outcomes of the experiments had to be carefully delineated. (He called this set a "collective," but it is now called a "sample space.") Next, he assumed that the experiments were repeatable. In a set of experiments, the ratio of the number of outcomes of a specified event to the total number of times the experiment has been conducted is called the "frequency" for that set of experiments. Thus, if an experiment is performed $n$ times and the number of favorable outcomes is $m$, the frequency is $m/n$. As the number of experiments is increased, the ratio will stabilize. It is an empirical (provable) fact that the value around which this stabilization takes place is the probability of the event.

**399**

*Richard von Mises.*

taken by Kolmogorov, who described it in 1933 in his book *The Foundations of Probability.*

## Consequences

Many common occurrences are too complicated to be described in any way except in terms of probability. In the Mises approach, probabilities can be determined on the basis of collecting large amounts of data. The approach is so natural, so widely accepted, and so much in keeping with modern philosophical views that many people are unaware of the struggles that were involved in setting up and implementing the theory. Probability theory touches every aspect of modern life. Insurance rates (medical, car, property, life, and so forth) are set up on the basis of this theory, and probability theory has permeated all aspects of data collection in the natural, biological, and social sciences. Medical research and drug testing also apply its methods. It is still used in gambling, and it can be useful even when games are biased. With the advent of quantum theory, it has also become fundamental in the study of atomic and nuclear processes. Probability theory itself is still undergoing rapid development, and its range of applications is continually increasing. The insights of Mises have affected modern life in innumerable ways.

*Ronald B. Guenther*

After the concept of probability had been clarified, the final step was to give a minimal set of logically consistent rules that allowed one to treat problems rigorously. The final step was

# Students Call for Reform in China's May Fourth Movement

*After the Paris Peace Conference decided to give China's Shandong (Shantung) Province to Japan, college students in Beijing became angry and began demanding international justice and reforms within their nation.*

**What:** Political reform; Social reform
**When:** May 4, 1919
**Where:** Beijing
**Who:**

CHEN DUXIU (CH'EN TU-HSIU; 1879-1942), dean of the school of letters at National Beijing University and founder of the magazine *New Youth*

HU SHI (HU SHIH; 1891-1962), a professor at National Beijing University

LI DAZHAO (LI TA-CHAO; 1889-1927), head of the library at National Beijing University

LUO JIALUN (LO CHIA-LUN; 1897-1969), a student at National Beijing University, editor of *New Tide* magazine, and author of the May Fourth Manifesto

CAI YUANPEI (TS'AI YUAN-P'EI; 1867-1940), chancellor of National Beijing University

## China Is Humiliated

In 1911, a revolution in China overthrew the Qing (Ch'ing; Manchu) Dynasty and established the Republic of China. Sun Yat-sen, who had founded the Guomindang (Kuomintang; Nationalist Party), became president, but he soon resigned because he had no real power. The weak governments that followed were dominated by warlords, and there were no democratic reforms. With civil wars continuing, China had no strength to stand up to aggressive nations.

In 1914, as World War I was beginning, Japan conquered China's Shandong Province, which had been dominated by Germany since 1898. In 1915, Japan forced China to accept the Twenty-one Demands, which gave the Japanese many special privileges in China—including permanent rights in Shandong. Secret agreements with Great Britain, France, and Italy also promised Japan that after the war it would continue to control Shandong.

Meanwhile, the war benefited China in one important way. As fewer products were available from the West, many factories sprang up in China. Now there were working-class people in the cities, along with middle-class managers and business leaders.

At the same time, new schools were formed. In 1915, the Ministry of Education listed 120,000 government schools of all sorts, and there were also private schools, such as those run by Western missionaries.

Many newly educated Chinese people became excited to learn of the Fourteen Points proclaimed by American president Woodrow Wilson. Wilson called for self-determination for oppressed people, and the Chinese believed that this freedom would come to them. When World War I ended, thousands of people celebrated in the streets of Chinese cities, hoping that a new day of justice was dawning for China.

At the Paris Peace Conference in 1919, China tried to win back Shandong and to abolish the unequal treaties that had made it a sort of joint colony of the powerful Western nations and Japan. Yet Wilson agreed with British and French leaders that Germany's rights in Shandong should be given to Japan.

## The Protest Movement

Within China, intellectuals had already been looking for ways to reform their country. Many of these people had studied in Europe and the

United States, and they admired the democracies and prosperity of the West. After returning from studies in France, Chen Duxiu (Ch'en Tu-hsiu) had founded a magazine called *New Youth* that called for political and social reforms. Hu Shi, a professor who wanted the Chinese to begin writing in their everyday language instead of the classical form, contributed articles to *New Youth.*

In 1916, Cai Yuanpei, who had been part of the Guomindang during the 1911 revolution, became chancellor of National Beijing University. He began reforming the school to allow opinions to be expressed freely, and the best teachers and brightest students began gathering there.

In 1918, another new magazine, *New Tide,* was founded by National Beijing students who wanted to encourage scientific thinking, reformed language, and freedom of speech. *New Youth* and *New Tide* became voices for the New Culture Movement (also known as the Intellectual Revolution) in China.

When news of the Allies' decision to give Shandong to Japan reached China, students at the National Beijing University organized a protest demonstration. Five thousand students from thirteen universities and colleges gathered at the Gate of Heavenly Peace (Tiananmen) on May 4. They demanded that China not sign the Treaty of Versailles and that the Chinese officials who had signed secret treaties with Japan be punished.

In the May Fourth Manifesto, the protesters asked fellow Chinese to be ready to die for their nation and never to agree to give up their territory. Petitions of protest were presented to the American and British diplomatic missions. When the demonstrators marched past the house of a pro-Japanese official, some of them broke into it and set it afire. Police arrived and arrested ten students.

News of the arrests prompted all students in Beijing to join in a strike, which quickly spread throughout other major cities of China. Many factory workers joined in the strike, and shop owners closed their doors and began refusing to sell products made in Japan. Almost all newspapers supported the new patriotic movement as well.

Faced with demands that China not sign the Treaty of Versailles, government leaders finally told the Chinese delegates in Paris to do as they liked. Chinese students in Western Europe then surrounded the hotel where the Chinese delegates stayed, and they did not allow the delegates to travel to Versailles for the signing ceremony.

In late May and June, student leaders gathered in Shanghai and formed the All China Student Union. Sun Yat-sen and other Guomindang leaders tried to persuade them to support the Guomindang program of "nationalism, democracy, and livelihood."

In Beijing, the warlord government tried to stop the movement by imprisoning 1,150 students. Since there was no jail space, part of the National Beijing campus was turned into a prison. A new wave of strikes and boycotts, however, forced the government to give in. As the students marched triumphantly out of jail, the government apologized, pro-Japanese officials were fired, and the cabinet resigned.

**Consequences**

The May Fourth Movement had many long-term effects. It was the first time that students became a force in Chinese national politics. They encouraged new feelings of nationalism among the Chinese people—and a determination to stop giving in to the control of other nations. Later generations of students followed this example in demanding social reforms, greater democracy, and economic reform.

Sun Yat-sen recruited many students for the Guomindang, and they helped make it a disciplined political party. Other Chinese students, however, were so disappointed with the West and with President Wilson that they turned to Marxism. Chen Duxiu and Li Dazhao formed a Marxist study club in Beijing in 1919, and similar groups were formed in other cities. In 1921, Chen, Li, Mao Zedong (who had been hired by Li as a library assistant), and others founded the Chinese Communist Party.

Meanwhile, the United States Senate recognized the unfairness of giving Shandong to Japan, and this was one of the reasons the Treaty of Versailles was rejected. At the Washington Conference of 1921-1922, President Warren Harding and the British leaders persuaded Japan to return Shandong to Chinese control.

*Jean-Robert Leguey-Feilleux*

# League of Nations Is Created

*Leaders of the world's Great Powers formed the League of Nations after World War I in the hope that such a terrible war would never be fought again.*

**What:** International relations
**When:** June 28, 1919
**Where:** Versailles, near Paris
**Who:**
WOODROW WILSON (1856-1924), president of the United States from 1913 to 1921
EDWARD MANDELL HOUSE (1858-1938), an American diplomat
DAVID LLOYD GEORGE (1863-1945), prime minister of Great Britain from 1916 to 1922
JAN CHRISTIAN SMUTS (1870-1950), prime minister of the Union of South Africa from 1919 to 1924 and from 1939 to 1948
LORD ROBERT CECIL (1864-1958), assistant foreign secretary of Great Britain from 1918 to 1919
LÉON-VICTOR-AUGUSTE BOURGEOIS (1851-1925), a French statesman who served as chairman of the first meeting of the League of Nations
SIR JAMES ERIC DRUMMOND (1876-1951), secretary-general of the League of Nations from 1919 to 1933

## A League for Peace

In the hundred years before 1914, the people of Europe grew accustomed to peace. Wars continued, but they were usually confined to fairly small areas. People came to assume that peace was simply a characteristic of modern society.

Yet the European nations assumed in dealing with one another that there was no higher authority than the individual nation-state. Over the years, they tangled themselves in a web of secret alliances and rivalries that eventually led to the outbreak of World War I. The idea that peace would come naturally in a modern society was proved false.

As the war dragged on through four bloody years, many statesmen and leaders in Europe and the United States decided that some kind of world forum should be set up to give a structure to relations among nations. In January, 1918, United States president Woodrow Wilson addressed Congress and proposed his Fourteen Points as the basis for world peace. The Fourteen Points can be summarized as follows:

> open covenants of peace; freedom of navigation on the seas; removal of all economic barriers and establishment of equal trade relations among all nations; reduction of each nation's armaments; fair judgment of colonial claims; friendly settling of questions regarding Russia; evacuation and restoration of Belgium; restoration of France's invaded territory and return of Alsace-Lorraine; readjustment of Italy's frontiers; autonomy of Austro-Hungarian peoples; guarantees of free determination to the Balkan states; sovereignty of Turkey, but independent development of nationalities under Turkish rule and free passage through the Dardanelles Strait; creation of an independent Polish state with access to the sea; formation of a permanent association of nations to preserve peace.

The last point, the establishment of an organization of nations, was especially important to both President Wilson and British prime minister David Lloyd George. By November, most of the nations involved in the war had committed themselves—some more enthusiastically than others—to support the League of Nations.

The Peace Conference at Versailles opened on January 10, 1919. At Wilson's insistence, a plenary session of January 25, 1919, decided that the Covenant of the League of Nations should be made part of the peace treaty with Germany. A

Hulton Archive

*The Council of the League of Nations meets for its first session in November, 1920.*

committee was set up to make a plan for the league's constitution and structure. The committee included Wilson as chairman; "Colonel" Edward Mandell House, representing the United States; Jan Christian Smuts of South Africa, representing the British Empire; Lord Robert Cecil, representing Great Britain; and Léon-Victor-Auguste Bourgeois, representing France.

By April, the committee had drafted a constitution. It made use of plans submitted by Wilson, House, and others, and it especially incorporated the ideas from a pamphlet by Smuts titled *The League of Nations: A Practical Suggestion.* Smuts clearly defined the three primary purposes of the league: to safeguard the peace, to organize and regulate international business of all kinds, and to assist states needing help.

On April 28, President Wilson presented the Covenant of the League of Nations to a plenary session of the peace conference. It became part of the Treaty of Versailles, which was signed on June 28, 1919. When that treaty officially became

effective on January 10, 1920, the League of Nations also began to function.

## Covenant and Structure

The Covenant of the League of Nations, which became the first twenty-six articles of the Treaty of Versailles, defined conditions for membership, the league's structure, how a Permanent Court of International Justice was to be created, methods for settling international disputes, the mandate system, and the league's role in matters such as disarmament, treaties, and various humanitarian activities (for example, attempts to prevent disease, stop illegal traffic in drugs, and improve labor conditions).

The charter (original) members of the league included the nations that signed the Versailles treaty, along with "invited" states that accepted the covenant as it stood within two months of the league's official inauguration. Other nations could become members by a two-thirds vote of the Assembly. The Assembly, along with the

**404**

Council and permanent Secretariat, made up the basic structure of the league. All member states were represented in the Assembly.

The Council was composed of one delegate each from five permanent and four nonpermanent seats. The Great Powers were to fill the five permanent seats; by 1933, the Assembly had increased the number of nonpermanent seats to ten. The Council had broad authority, but it concentrated on such matters as disarmament and the peaceful settling of disputes.

The permanent Secretariat was the league's bureaucracy: the secretary-general and his staff. Sir James Eric Drummond, a British diplomat, was named the first secretary-general; later secretaries-general were appointed by the Council and approved by the Assembly.

The Permanent Court of International Justice, or World Court, was established in 1921 by Article XIV of the Covenant. The International Labor Organization, provided for by the Treaty of Versailles, also came into existence around this time.

## Consequences

The league, located permanently in Geneva, Switzerland, was weakened from the beginning by the nonparticipation of the United States and the absence of the Soviet Union. The U.S. government rejected the league because the U.S. Congress was dominated by Republicans who favored isolationism (avoiding involvement in the affairs of other countries). The Soviet Union was excluded from the league until 1934 because of its open goal of promoting worldwide revolution—a goal that clearly contradicted the aims of the league's covenant.

The most important work of the league after 1920 was the effort to create a program of disarmament and the peaceful settling of unresolved disputes among the Great Powers. The league completely failed in these tasks. Because it was unable to force Fascist Italy to stop attacking Ethiopia in the 1930's, the league practically came to a standstill in 1936. After the Ethiopian crisis, Italy withdrew from the league, joining Japan and Nazi Germany, which had withdrawn earlier.

The league did have several little-known successes—especially its supervising of the financial rebuilding of Austria between 1922 and 1926. Yet these successes were overshadowed by its inability to preserve the peace settlement. In the end, the League of Nations failed to prevent the outbreak of a second world war, which proved even more disastrous than the first had been.

*Edward P. Keleher*

**405**

# Treaty of Versailles Is Concluded

*The Treaty of Versailles included the Covenant of the League of Nations and specified that Germany must give up certain territories and colonies as well as a part of its military.*

**What:** International relations; War
**When:** June 28, 1919
**Where:** Versailles, near Paris
**Who:**
GEORGES CLEMENCEAU (1841-1929), premier of France from 1906 to 1909 and from 1917 to 1920
DAVID LLOYD GEORGE (1863-1945), prime minister of Great Britain from 1916 to 1922
WOODROW WILSON (1856-1924), president of the United States from 1913 to 1921
VITTORIO EMANUELE ORLANDO (1860-1952), prime minister of Italy from 1917 to 1919
COUNT ULRICH VON BROCKDORFF-RANTZAU (1869-1928), foreign minister of Germany in 1919

## Shaping the Treaty

The Allied leaders who gathered in Versailles, France, to write a treaty to end World War I faced several great challenges. Somehow the treaty would need not only to impose peace terms on Germany but also to take account of major political changes in Europe. The principal changes had to do with the collapse of the Russian, Austrian, German, and Ottoman Empires, along with their ruling dynasties.

In Russia, the Romanov czarist government had been overthrown in the Russian Revolution of 1917; only a few months later, the Bolsheviks had expelled the Provisional Government and seized control. As World War I ended, Austria-Hungary had begun to divide up into small independent states. Under the terms of peace, Germany was sure to lose territory to France and to the new Polish state. The loss of Ottoman control over the empire's former European lands had

helped set off the war in the first place; now the Turks' former non-Turkish subjects in the Near East had also risen up against them.

In place of these empires, Eastern Europe was divided up into a set of new states stretching from Finland in the north to Yugoslavia in the south. In all these countries, the spirit of nationalism was burning brightly. Given their position between Germany and Russia, the Eastern European nations could easily come to be dominated by either. The Allies wanted badly to prevent that from happening.

The Peace Conference of Paris opened on January 18, 1919. After about two months, the four top-ranking leaders at the conference had begun meeting separately to write the part of the treaty that concerned Germany. These leaders were Georges Clemenceau, premier of France and president of the Peace Conference; David Lloyd George, prime minister of Great Britain; Woodrow Wilson, president of the United States; and Vittorio Emanuele Orlando, prime minister of Italy.

During the three months in which these men worked on the treaty, they found it hard to agree on many issues. Each one brought his own nation's particular concerns. Clemenceau wanted to take away from Germany the Rhineland and the Saar—lands that bordered France; he also wanted to make sure that Germany did not annex Austria. Lloyd George did not wish to see Germany weakened too much by France's demands; instead, he called for the traditional balance of power in Europe to be restored, but with guarantees that Germany would not attack France again. Orlando seemed most interested in trying to obtain the Adriatic territories that Great Britain and France had promised as Italy's reward for entering the war on the Allied side.

Wilson had a number of concerns for a fair settlement of conflicts and independence for

Eastern European national groups; these he had listed in his famous Fourteen Points. He was prepared, however, to give up some of his goals in order to gain what he saw as the most important one: the creation of a League of Nations. He believed that once the league was established, it could eventually iron out whatever problems remained after the peace treaty was signed.

## Signing the Treaty

The Council of Four, as these Allied leaders were called, managed to settle their differences and finish their work on the German treaty by the end of April. They then instructed the provisional government of Germany to send representatives to Paris to receive the treaty.

Germany at first insisted that it be allowed to bargain on the treaty's terms, but the Allies stood firm in their insistence that the treaty was to be simply "received." On May 7, they formally presented the document to the German delegation, led by Foreign Minister Count Ulrich von Brockdorff-Rantzau. The Germans were given three weeks in which to write and submit their comments. The Germans did have some strong objections—they argued, for example, that the German-speaking peoples of Austria and Czechoslovakia should be allowed to unite with Germany—but the Allies made only a few changes in response.

The Germans reluctantly accepted the final treaty on June 22. On June 28, 1919, the fifth anniversary of the assassination of Archduke Francis Ferdinand (the event that had sparked the war), the treaty was signed in the Hall of Mirrors of Versailles. Allied representatives then turned their attention to writing treaties for Austria, Bulgaria, Hungary, and Turkey; the Peace Conference of Paris came to a close on January 21, 1920, after all this work had been done.

*Diplomats and delegates assemble in the Hall of Mirrors at Versailles to watch the signing of the treaty.*

**407**

The Treaty of Versailles included the Covenant of the League of Nations as its first twenty-six articles. Germany was required to return Alsace-Lorraine (taken in 1871) to France, while other small territories were to be given to Belgium. The Rhineland was to have no German military presence, and a joint Allied force would protect it for fifteen years. The Saar was dealt with in the same way, except that at the end of the fifteen years, the Saar's people were to be allowed to vote on whether they wanted to join France or Germany.

To its east, Germany had to recognize a new, independent Polish state and to give it some Prussian territory so that it would have an outlet to the sea. The German city of Danzig now became a free state, and Germany was forbidden to annex Austria.

Beyond Europe, Germany was forced to give up all its colonies in Africa and Oceania. Most of these lands now came under the protection of the League of Nations and were assigned to the oversight of France, Great Britain, Japan, and other countries.

Germany was required to reduce the size of its military forces and armaments; the German navy, for example, could no longer include submarines, and Germany was not allowed to have any air force at all. Germany was to acknowledge its responsibility for starting the war and for all the damage it caused. The Allies wanted Germany to pay for those damages and later set the amount at five billion dollars, to be paid by May 1, 1921.

## Consequences

The Treaty of Versailles was harsh in its treatment of Germany; it reflected the bitterness the Allies felt after a long war that had taken millions of lives. It soon became apparent that since Germany had lost many of its mineral-rich territories, it was much too poor to pay the reparations that had been demanded.

Other parts of the treaty were also difficult to enforce. Great Britain and France were left with the main responsibility, since Russia was not involved in it and the U.S. Senate refused to ratify it. With some support from Italy, France and Great Britain did manage to uphold the Treaty of Versailles until 1935, when they began to give in to the demands of Nazi Germany in the hope of preventing another major war.

*Edward P. Keleher*

# Weimar Constitution Guides New German Republic

> *After World War I and the German Empire came to an end together, German leaders approved a democratic constitution that gave broad powers to the central government.*

**What:** Political reform
**When:** August 11, 1919
**Where:** Weimar, Germany
**Who:**

MAX OF BADEN (1867-1929), prince and chancellor of the German Empire from October to November, 1918

FRIEDRICH EBERT (1871-1925), German chancellor from 1918 to 1919, and president of the Weimar Republic in 1919

PHILIPP SCHEIDEMANN (1865-1939), prime minister of the Weimar Republic in 1919

HUGO PREUSS (1860-1925), a German professor of law, secretary of the interior from 1918 to 1919, and chief author of the Weimar Constitution

## The Republic Is Launched

On November 9, 1918, two days before the signing of the armistice ending World War I, Germany was proclaimed a republic. Exhausted by four years of warfare, the German Empire that had been created in 1871 by Prince Otto Eduard Leopold von Bismarck Schönhausen came to an end. Prince Max of Baden, who had just begun serving as chancellor, put pressure on Emperor William II to resign. Soon William fled to Holland, where he formally renounced his throne on November 28.

Prince Max turned over the government to the Majority Socialists, led by Friedrich Ebert and Philipp Scheidemann. Ebert, the new chancellor, set up a "directory" of six men, three Majority Socialists and three Independent Socialists, to run the country until a constituent assembly could be elected.

Elections for the National Assembly took place on January 19, 1919. More than 80 percent of eligible voters cast their ballots, and three-quarters of the total vote went to the political parties committed to forming a German republic. Clearly, most of the German people wanted a republic.

The National Assembly met in Weimar on February 6, 1919, to ratify the Treaty of Versailles and then to draft a constitution. Hugo Preuss, a professor of constitutional law who had become secretary of the interior in November, 1918, brought a constitution he had written to the assembly on February 24, 1919. After three days of debate in the assembly, the draft was sent to a constitutional committee. Through about forty sessions, from March to July, the committee examined each of the constitution's articles. The new German constitution became official on August 11, 1919.

## The Constitution

The constitution provided Germany with a liberal, democratic government. Many of its ideas were borrowed from the constitutions of Western Europe and the United States, yet it also made use of parts of the old constitution of the German Empire. For example, the word *Reich,* or "empire," was kept as part of the new republic's name, while the individual states were called *Länder.*

Under the new constitution, a *Reichspräsident* would be elected by the people to hold office for seven years with the possibility of reelection. The Majority Socialists opposed a strong presidency, for they feared that an ambitious person in that position might turn himself into a king or emperor, as Napoleon III had done in France. At their insistence, the powers of the president were weakened: All the president's orders would have to be countersigned by the chancellor.

The German president, who stood somewhere between the strong American president and the weak French president, appointed the chancellor. The president also had the authority to dissolve the Reichstag and call for new elections. If he did not want to enact a law passed by the Reichstag, he could call for a referendum in which the people of Germany would vote on the matter.

A legislature of two houses was set up: a Reichstag, representing the people, and a Reichsrat, representing the states. Reichstag members were to be elected for a four-year term by secret ballots, with voting rights being universal; in this way, the Reichstag was similar to the U.S. House of Representatives. As the expression of the will of the people, the Reichstag had the power to create and enact laws. The nation's cabinet was made responsible to the Reichstag.

In the Reichsrat each state was to have at least one member, but no state could control more than two-fifths of the membership. The Reichsrat could also write new laws, and it had veto power over bills passed by the Reichstag. The Reichstag could, however, override the veto by a two-thirds vote.

The most controversial part of the new constitution was Article 48. Under this article, the basic rights guaranteed by the constitution could be suspended by the president in an emergency that threatened public safety and order. Because Germany was in a chaotic state after the war, this provision was considered necessary to prevent the new republic from being overthrown by radicals.

## Consequences

Although the Weimar Republic did give Germany a parliamentary democracy after World War I, the freedom to vote for their leaders did not put an end to the German people's frustration. Losing the war and suffering the further loss of numerous territories and colonies led to a widespread feeling of national humiliation. The people's desire for a restoration of Germany's greatness became fuel for the Nazi Party in the 1930's.

In spite of the changes called for by the Majority Socialists, the Weimar Constitution gave great powers to the nation's president. Most significant was Article 48, which allowed the president alone to decide when the nation was in an emergency that called for basic civil rights to be suspended. Adolf Hitler later took advantage of this article to make himself dictator.

*Harry E. Wade*

# Einstein's Theory of Gravitation Is Confirmed

> *Measurements made during a total solar eclipse showed that the sun's gravitational field bends the path of starlight.*

**What:** Physics; Earth science
**When:** November 6, 1919
**Where:** London, England
**Who:**
ALBERT EINSTEIN (1879-1955), a German American physicist
SIR ARTHUR STANLEY EDDINGTON (1882-1944), an English astronomer and physicist
SIR FRANK DYSON (1868-1939), the English Astronomer Royal
WILLEM DE SITTER (1872-1934), a Dutch astronomer

## Bending Light

Sir Isaac Newton's law of gravitation, published in 1687, states that every mass (every object composed of matter) attracts every other mass through gravitation. Gravitation is the tendency of every particle of matter in the universe to be attracted to every other particle. The more massive the bodies, the stronger the gravitational force between them, and the farther apart the bodies, the weaker the force between them.

Although this law was extremely successful, it seemed impossible that one body could exert a force on another body some distance away. To avoid this problem, scientists began to speak of a mass modifying the space around it by establishing a gravitational field. Another mass interacted with the field at its location but not directly with the first mass. How mass could establish a field was not explained.

Albert Einstein took gravitational theory a giant step forward. Taking a hint from the well-known result that all objects—no matter how massive—accelerate at the same rate when acted upon by gravity, Einstein deduced that gravita-tion and acceleration are equivalent in some fashion. He thought about how a beam of light might look to someone moving upward very quickly in an elevator. If the light beam were actually parallel with the floor as it entered the elevator, it would appear to bend downward slightly as the elevator accelerated. If acceleration and gravitation are equivalent, then gravity must also deflect light rays downward. Since light has no mass, as one normally thinks of it, this result was completely unexpected.

Einstein was perplexed about why a light beam took a curved path when traversing a gravitational field in an otherwise empty space. After all, the path taken by a light ray in an empty space is the definition of a straight line. Yet when is a straight line also a curved line? The answer is clear when the line is drawn on a curved surface such as a globe. For example, one may travel in a straight line along the earth's equator and eventually return to the starting point without ever turning around.

## Curving Space-Time

In 1915 and 1916, Einstein announced his general theory of relativity. This theory interpreted gravitation not as a force but as the result of curved space-time. Einstein's idea of space-time imagines the universe as being one unified "continuum" made up of four dimensions. These dimensions are length, width, and height—all of which are defined as "space"—and time. Within that space-time continuum, physical events occur and can be precisely located or mapped. A moving mass would produce a ripple in the curvature of space-time that would expand at the speed of light. By contrast, a weak gravitational field corresponds to almost no curvature of space-time, meaning that space-time is nearly flat.

**411**

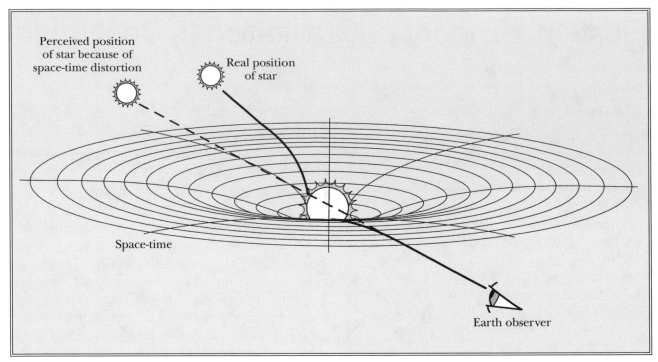

*Perceived position of star because of space-time distortion*

*Real position of star*

*Space-time*

*Earth observer*

*The sun's gravity causes space-time to curve, which bends the star's light and makes it appear to be located where it is not.*

Einstein suggested three effects that could be measured to see if his theory was accurate: the gravitational "redshift" of light, the advancement of the "perihelion" of the planet Mercury (the part of the planet's orbit that takes it closest to the sun), and the deflection of starlight by the sun. Einstein calculated that a ray of starlight just grazing the sun should be deflected by only about 1.75 seconds of arc. Stars cannot normally be seen when the sun is out, so Einstein suggested the measurement be made during a total solar eclipse.

Sir Frank Dyson, the British Astronomer Royal, sent out two expeditions to photograph the eclipse of May 29, 1919. Charles Rundle Davidson led one expedition to Sobral in northern Brazil, while Sir Arthur Stanley Eddington, an English astronomer and physicist, headed the other expedition to Príncipe Island in the Gulf of Guinea. Eddington's expedition took sixteen photographs of the eclipse. Comparing one of them with another photograph of the same star field taken six months earlier when the sun was not present, Eddington was delighted to find the star images shifted by the same amount that Einstein had predicted.

On November 6, 1919, Dyson reported on the eclipse expeditions to a joint meeting of the Fellows of the Royal Society and the Fellows of the Royal Astronomical Society held at the Burlington House in London. Sir Joseph John Thomson, credited with the discovery of the electron and then president of the Royal Society, called Einstein's theory "one of the greatest achievements in the history of human thought."

**Consequences**

After the confirmation of Einstein's general theory of relativity, the public was eager to learn more about him and his theory. Within one year, more than one hundred books on the subject had been published. Leopold Infeld, who cowrote a book with Einstein on relativity, suggested that the intensity of the public reaction was a result of the timing—World War I had just ended. "People were weary of hatred, of killing. . . . Here was something which captured the imagination . . . the mystery of the sun's eclipse and of the penetrating power of the human mind." The general theory of relativity was a great achievement in which all of humankind could take pride.

Einstein's theory of gravitation will continue to be tested under any circumstance that can be devised, for that is the nature of science. The general theory of relativity has passed the three tests (gravitational redshift, perihelion advance of Mercury, and bending of starlight) Einstein suggested as well as many more tests using radar, radio telescopes, pulsars, and quasars.

Einstein has shown that space is not simply an empty place. Space and time are not independent but must be considered together; furthermore, they are curved by mass. Perhaps most exciting is the picture of the universe that the theory predicts. Ironically, when Einstein's theory led to the conclusion that the universe was expanding, he rejected it at first, until 1929, when the American astronomer Edwin Powell Hubble offered experimental evidence to show that the universe is, in fact, expanding. Although the properties of the universe as a whole are not yet known, there is every reason to suppose that they will be consistent with the general theory of relativity.

*Charles W. Rogers*

# Slipher Measures Redshifts of Spiral Nebulas

*Vesto Melvin Slipher pioneered the spectroscopy of galaxies, discovering that most galaxies are receding at high velocities.*

**What:** Astronomy
**When:** Early 1920's
**Where:** Flagstaff, Arizona
**Who:**
VESTO MELVIN SLIPHER (1875-1969), an American astronomer
PERCIVAL LOWELL (1855-1916), an American astronomer
EDWIN POWELL HUBBLE (1889-1953), an American astronomer
MILTON L. HUMASON (1891-1972), an American astronomer
ALBERT EINSTEIN (1879-1955), a German American physicist

## What Does a Galaxy Include?

The general view held by astronomers during the first two decades of the twentieth century was that the universe consisted entirely of the Milky Way galaxy—in which the earth is located—which was a system of stars that was estimated to be about 10,000 light-years in diameter. (A light-year is the distance that light can travel in one year, about 9.5 trillion kilometers.) All objects that were visible in the heavens through the largest telescopes were thought to be part of the Milky Way galaxy; beyond, a trackless, undifferentiated void extended infinitely far.

By the early part of the twentieth century, the most mysterious parts of the galaxy were "nebulas," misty patches of light in the sky that were thought to be made up primarily of gases. Photographs of nebulas showed that some had irregular shapes but that many had a spiral structure.

In 1901, Vesto Melvin Slipher was hired as an observer at Lowell Observatory in Flagstaff, Arizona, and assigned to a project on spiral nebulas.

Slipher was to analyze the spectra (the unique patterns of radiation) of the brighter nebulas and look for "Doppler shifts" in their light, which would reveal motions taking place within them. Percival Lowell expected the results to support the theory that spiral nebulas were rotating, contracting clouds of gas that would eventually form a planetary system or a cluster of stars. Slipher, however, did not agree; he believed that the spiral nebulas were probably systems of stars outside the Milky Way galaxy.

By late 1912, using the 61-centimeter telescope at the Lowell Observatory, Slipher had found that the spectrum of the Andromeda nebula, one of the largest and brightest of the spirals, was similar to that of the sun and those of other stars. This was the first clue that a spiral nebula was a system of stars rather than a collection of gases. Without knowing the distance to the Andromeda nebula, however, Slipher was not able to demonstrate conclusively that the nebula was outside the Milky Way galaxy. Like many other astronomers who learned of his work, Slipher interpreted the failure to detect individual stars in the spirals as an indication of their remoteness.

## Nebula Doppler Shifts

In the spectra of the Andromeda nebula, Slipher noticed that there was a Doppler shift of all the lines toward the blue end of the spectrum. This phenomenon, called a "blueshift," indicates that the nebula is moving toward the observer on Earth. If the nebula were moving away, there would be a corresponding "redshift" (a movement toward the red end of the spectrum). The size of the shift in the position of the dark lines provides a direct measure of the speed with which the object is moving toward or away from the observer. Slipher's data indicated that the Andromeda nebula was approaching at a speed of 300 kilometers per second, faster than any astronomical object that had been measured at the time.

By 1914, Slipher had analyzed the spectra of twelve additional spirals. Only one of these exhibited a blueshift—the rest all exhibited a redshift. If the spirals were part of the Milky Way galaxy, roughly half would have been approaching and half would have been receding. Moreover, the speeds of the spiral nebulas (ranging up to an astounding 1,100 kilometers per second) seemed to be too great for them to be gravitationally bound to the Milky Way galaxy.

By 1923, Slipher had measured Doppler shifts in forty-one different nebulas; thirty-six had redshifts, and the remaining five had blueshifts. Meanwhile, other astronomers had added four more nebulas to the list; these all had redshifts. It was discovered in 1925 that the Milky Way galaxy as a whole was rotating and that the sun, because of the rotation, was moving at a speed of 250 kilometers per second in a direction generally toward the Andromeda nebula. Thus, the 300-kilometer-per-second approach of the Andromedanebula is composed of the 250-kilometer-per-second motion of the sun toward the spiral and the 50-

kilometer-per-second approach of the spiral. When this information was taken into account, forty-three of the forty-five nebular Doppler shifts were found to be redshifts. The two exceptions were the largest—and therefore probably the nearest—nebulas.

**Consequences**

Following the construction of the giant 254-centimeter telescope on Mount Wilson, California, Milton J. Humason and Edwin Powell Hubble were able to study the spiral nebulas in much greater detail than had been possible for Slipher. In fact, with the new telescope, individual stars could be discerned in some of the larger spirals, particularly in the Andromeda nebula. By this time, all doubt had been removed regarding the spiral nebulas: They were galaxies, vast systems of stars similar to the Milky Way. With individual stars available for study, the distance to the galaxy could be estimated. The Andromeda galaxy lay at a distance of 750,000 light-years, far outside the Milky Way. Other spirals were even farther away. Hubble combined the redshift data inherited from Slipher with the distance measurements and arrived at a relationship between distance and recessional velocity: The greater the recessional velocity, the greater the distance.

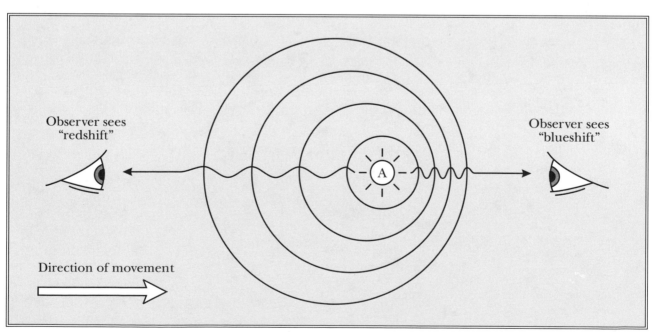

*In the phenomenon called the "Doppler shift," lightwaves appear bluer when moving toward the observer and redder when moving away from the observer. Here, "A" is the source of light, the Andromeda nebula.*

**415**

To astronomers of the time, the cause of the recession was now clear. If the galaxies were rushing away from one another, the universe must be expanding. This idea had its roots in the general theory of relativity published by Albert Einstein in 1916. Since that time, Einstein and other astronomers and physicists had examined the consequences of the theory and had found that the universe should generally be in a state of either expansion or contraction.

Slipher's redshift measurements and Hubble's correlation with distance demonstrated that the universe is expanding, an observation of central importance to cosmology. Since the universe expands as time progresses, it must have been the case that in earlier epochs, the galaxies were very close together and, at one time, entirely coalesced. Astronomers therefore reached the startling conclusion that the universe had been born in a gigantic explosion that threw out the material from which the galaxies eventually were formed. The high-speed rush of galaxies away from one another is direct evidence of that genesis, which is now known as the "big bang."

*Anthony J. Nicastro*

# GREAT EVENTS

# 1900-2001

# CATEGORY INDEX

## LIST OF CATEGORIES

## AGRICULTURE

Congress Reduces Federal Farm Subsidies and Price Supports, **7**-2760

Gericke Reveals Significance of Hydroponics, **2**-569

Insecticide Use Increases in American South, **1**-359

Ivanov Develops Artificial Insemination, **1**-44

Morel Multiplies Plants in Vitro, Revolutionizing Agriculture, **3**-1114

Müller Develops Potent Insecticide DDT, **2**-824

## ANTHROPOLOGY

Anthropologists Find Earliest Evidence of Modern Humans, **6**-2220

Benedict Publishes *Patterns of Culture*, **2**-698

Boas Lays Foundations of Cultural Anthropology, **1**-216

Boule Reconstructs Neanderthal Man Skeleton, **1**-183

Dart Finds Fossil Linking Apes and Humans, **2**-484

Humans and Chimpanzees Are Found to Be Genetically Linked, **5**-2129

Johanson Discovers "Lucy," Three-Million-Year-Old Hominid Skeleton, **5**-1808

Last Common Ancestor of Humans and Neanderthals Found in Spain, **7**-2841

Leakeys Find 1.75-Million-Year-Old Hominid Fossil, **4**-1328

Mead Publishes *Coming of Age in Samoa*, **2**-556

117,000-Year-Old Human Footprints Are Found Near South African Lagoon, **7**-2860

Pottery Suggests Early Contact Between Asia and South America, **3**-1230

Simons Identifies Thirty-Million-Year-Old Primate Skull, **4**-1553

Weidenreich Reconstructs Face of Peking Man, **2**-786

Zdansky Discovers Peking Man, **2**-476

**IV**

## COMMUNICATIONS

## COMPUTER SCIENCE

XIII

## LABOR

**XXII**

## POLITICS